The Japan Handbook

Regional Handbooks of Economic Development
Prospects onto the 21st Century

The China Handbook
The India Handbook
The Japan Handbook

Forthcoming

The Southeast Asia Handbook

The Japan Handbook

Edited by

Patrick Heenan

Advisers

William M. Tsutsui
University of Kansas

Julie Gilson
University of Birmingham

Aurelia George Mulgan
University of New South Wales

FITZROY DEARBORN PUBLISHERS

LONDON • CHICAGO

FITZROY DEARBORN PUBLISHERS
70 East Walton Street
Chicago, Illinois 60611
USA
or
310 Regent Street
London W1R 5AJ
England

British Library Cataloguing in Publication Data
The Japan Handbook
1. Japan – History – Heisei period, 1989–
I. Heenan, Patrick
952′.049

ISBN 1–57958–055–6

Library of Congress Cataloging in Publication Data is available.

First published in the USA and UK 1998
Typeset by Florencetype Ltd, Stoodleigh, Devon
Printed by the Bath Press, UK

Contents

Society and Culture

International Relations

Appendices

Editor's Note

The Japan Handbook is the third volume in the series *Regional Handbooks of Economic Development: Prospects onto the 21st Century*. Its main purpose is to provide an overview of the country at the end of a century in which it has undergone huge changes. Inevitably, Japan's prospects onto the 21st century will be colored by its economic and social development up to this point, and by the varying interpretations of that development which weigh upon the minds of decision-makers, opinion-formers and citizens, both in Japan and in the countries with which its fate is bound up. Just as inevitably, there are more changes to come, and some of the chapters in this book address their likely effects.

Of course, no single book can hope to provide a complete account of such a complex and ever-changing country. Our goal here has been to treat a defined range of complex problems and data, concisely, accessibly, and evenhandedly. The book is aimed at students, scholars and the general educated public alike, and is largely restricted to events and circumstances directly affecting Japan's development over the past 50 years or so, and on into the early years of the coming century. We hope that, whether readers want to engage with the whole book or consult it on just a few specialized topics, they will find useful information and thought-provoking analysis in all 19 chapters, as well as in the appendices which supplement them.

I would like to thank all those who have helped in the making of this book. Our contributors have been patient and cooperative throughout the process of commissioning, writing, and editing; most have also contributed to the appendices. Our three academic advisers – Professor William M. Tsutsui of the University of Kansas in the United States, Dr Julie Gilson of the University of Birmingham in Britain, and Dr Aurelia George Mulgan of the University of New South Wales in Australia – all did sterling work, recommending most of the contributors and helping to shape the overall structure of the book. Professor R. L. Sims of the School of Oriental and African Studies at the University of London and Dr Ann Waswo of the Nissan Institute of Japanese Studies at the University of Oxford also gave some invaluable encouragement and advice. Roda Morrison has proved to be an ideal supplier of editorial support at every stage, and Monique Lamontagne's help was indispensable.

Linguistic Conventions

Names of Japanese individuals appear in the customary order, with family name before personal name: hence, among other results, the absence of commas from their names in the reading lists, the Bibliography, and the Index.

A modified version of the Hepburn system is used to transliterate Japanese words. The difference between short and long vowels is not indicated for "a", "o", or "u"; and the sound usually represented as "n" is given as "m" when it comes before "b" or "p". Words are given capital initials according to English conventions (even though the Japanese writing system does not have capital letters).

As for English usage: Japan's national legislature (*Kokkai*) is referred to as "Parliament", not "the Diet" (as explained in Chapter 2); the Occupation is referred to as "Allied", not "American", since non-American personnel also took part in it, and its legal basis was in multilateral agreements; the present Constitution is referred to by the year when it came into effect (1947), not the year when it was promulgated (1946); and "East Asia" is used in its broader sense, encompassing not only Japan, China, Taiwan and the two Koreas (together called "Northeast Asia" here), but also the countries of Southeast Asia.

Abbreviations

The following abbreviations are used throughout this book:

EU	European Union
GATT	General Agreement on Tariffs and Trade
GDP / GNP	Gross Domestic Product / Gross National Product
IMF	International Monetary Fund
NHK	Nippon Hoso Kyokai (Japan's public service broadcasting company)
OECD	Organization for Economic Cooperation and Development
SCAP	Supreme Commander for the Allied Powers (referring both to the holder of this post and to the Occupation authorities in general)
UK	United Kingdom (when used as an adjective)
UN	United Nations
US	United States (when used as an adjective)

In other cases, the meaning of each abbreviation or acronym is stated at its first appearance within each chapter.

The Regions, Prefectures, and Major Cities of Japan

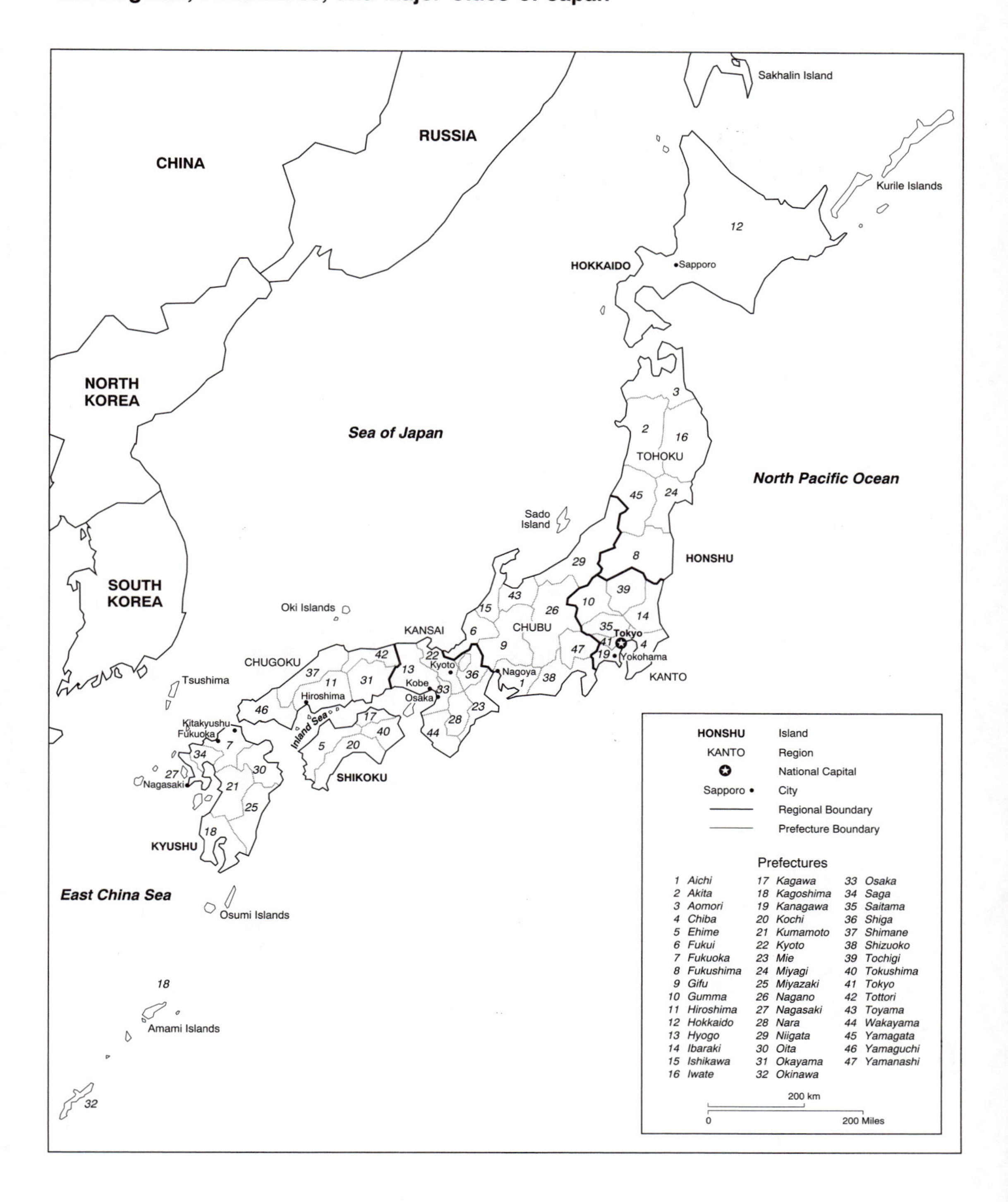

History and Context

Chapter One

The Japanese Economy Since 1945

Dennis B. Smith

In 1955, even after the first spurt of postwar economic recovery, Japan's GNP was a pitiful one fifteenth that of the United States, while income per capita, at US$220 a year, was only 35th in the world (see Uchino). By 1990, however, Japan had the world's second largest economy, while the Japanese enjoyed one of the world's largest per-capita incomes, exceeded only by the likes of Switzerland and the oil-rich sheikhdoms of the Middle East. Rates of growth in the three decades after the Pacific War were awesome. Between 1946, when the economy was barely functioning, and 1976, it grew 55-fold. Between 1951 and 1960, national income more than doubled, and then it trebled again between 1960 and 1973. The structure of the economy also changed at astronomic speed. In the immediate postwar period Japan was still a predominantly rural society, partly because of migration from city to countryside in the final stages of the Pacific War. In 1950, 50% of school leavers went into agriculture, while in 1965 only 5% did so, and by 1992 less than 3% of the workforce were engaged in full-time agriculture (see Smith).

After 1973, the economy performed much less gloriously. It was hit by soaring oil prices from that year onwards, and, along with structural difficulties with the economy, they contributed to pinning growth to 5% a year or less. Certain of the industries which had led high-speed growth declined relatively as they hit competition from the newly industrialized countries of East Asia. The expansion of other industries, notably automobiles, and rising exports caused serious friction with international competitors, particularly the United States. However, during the 1980s Japan performed better than any other mature industrial economy.

Postwar Japan's economic achievement was brought virtually to its knees by the consequences of policies and actions in the late 1980s, which defy rational explanation. When the bizarre "bubble economy" burst in 1990, the economy was thrown into its longest recession since 1945, and there is little indication that Japan will scramble out of this recession in the near future.

The Foundations of the Modern Economy, 1868–1945

Of course, Japan's rise to economic greatness did not begin in 1945. After the Meiji Restoration in 1868, when the last of the Tokugawa shoguns (hereditary military dictators) was overthrown, the new leadership encouraged industrialization to create the essential base for national power thought necessary to defend Japan in the dangerous world of the late 19th century. In the two and a half centuries of Tokugawa rule (1600–1868), the economy had become prosperous and sophisticated, and governments in the Meiji period (1868–1912) were able to milk this prosperity to finance the importation of foreign industrial technology and expertise. At first, the government was the prime mover in industrialization, but by 1880 state finances could no longer sustain the cost. From then on, private enterprise and money were responsible for the growth of industry, albeit with continued formal and informal government and bureaucratic interference.

Official encouragement after the Meiji Restoration sponsored the emergence of the great multisectoral conglomerates which were to become such a distinctive feature of modern Japan's industrial structure (see Chapter 10). These conglomerates, the *zaibatsu*, immediately became the powerhouses of prewar Japan's economic transformation. The four original *zaibatsu*, Mitsui, Mitsubishi, Sumitomo and Yasuda, steadily moved into manufacturing, including textiles, metals and shipbuilding; into communications and transportation; and into financial services, such as banking and insurance. Factories, textile mills, shipyards, railroads and banks sprouted throughout much of the country. However, the speed of Japan's early industrialization should not be exaggerated. The economy entered a stage of sustained modern economic growth only at the time of the Russo-Japanese War (1904–05), by which time 15.5% of the workforce was employed in some form of manufacturing. As usual in newly industrializing economies, textile-making was the largest single industry in the manufacturing sector. The growth of more advanced manufacturing was slow, and well into the 20th century Japan continued to depend on Europe and the United States for most specialist and sophisticated machinery and manufactured goods.

World War I was a powerful stimulant to Japanese industrial development. The war deprived Asia, including Japan, of European manufactured goods and Japanese manufacturers seized the opportunity, stepping into the commercial vacuum to supply manufactured goods to those Asian markets deserted by the European combatants. Between 1915 and 1920, manufacturing in Japan grew at an average of 9.3% a year, with metals increasing at 10.7% and machinery and tool manufacture growing at 28.1% (see Nakamura 1983). After the wartime boom ended, in 1920, the economy hit problems, and growth, which had been 62.5% between 1910 and 1920, managed only 33.4% in the following decade. Beneath this gloomy surface, however, the 1920s proved to be a key decade for Japan's economic, and especially technological, development. There was a radical switch in use of energy, as electricity supplanted steam power as the principal source, not only changing production methods in older-established industries but also stimulating new industries, such as chemicals and electrical machinery. High-technology manufacturing methods spawned new industries, including aircraft manufacture, which honed skills in the use of advanced metal alloys, while the 1920s saw the beginnings of radio and electrical appliance manufacture (see Nakamura 1983), and even the beginnings of Japan's future prowess in information technology (see Fransman 1990).

In this period of rapid industrialization, Japan's economy began to develop the dual structure which has given it considerable flexibility ever since. The myriad small companies in this dual structure served as subcontractors to the larger manufacturers, and would provide a cushion for them when the economy experienced a downturn. Then, the larger companies could vary their production and costs by squeezing their subcontractors (see Francks). Also in the 1910s and 1920s, some of Japan's large companies developed personnel management methods, such as giving guarantees of "lifetime employment" to key skilled workers, which were to be revived after the Pacific War. Suitably altered and refined, they became vital ingredients in postwar economic growth.

For Japan World War II began in September 1931, when the Imperial Army began the forcible takeover of Manchuria in Northeast China. That action began during, and was partially provoked by, the world depression, but, after it had initially hit some of Japan's luxury industries, especially silk production, not only was Japan less severely affected by the depression than the United States, but the 1930s and early 1940s formed a period of continued economic, especially manufacturing, growth and change. In 1931, the Ministry of Finance (MoF) began a proto-Keynesian policy of deficit financing of the economy, which led to steady growth. Between 1931 and 1938 the economy grew by 5% a year, higher than any other major economy (see Minami). From the outbreak of war with China, in July 1937, manufacturing industry was further stimulated by demand for equipment and the wherewithal. Between 1937 and 1944, output of steel increased by 46%, of non-ferrous metals by 70%, and of machinery by 252%. In 1930 only 27% of industrial workers were employed in heavy industry; by 1942 the figure was 68%. By 1940 textiles, which had been at the center of early industrialization, constituted only 17% of manufacturing output, while heavy industry, much of it in sophisticated areas such as aircraft manufacture, automobiles, optics, electronics and chemicals, contributed 59%. During the 1930s the powerful military establishment added important new players to Japan's economy by encouraging the emergence of "new *zaibatsu*" to develop new industries, such as truck and automobile manufacture. The most notable new *zaibatsu* from this period were Toyota and Nissan.

Before and during World War II Japanese industry, and its technological base, advanced significantly, but it still could not compete with that of the United States. During the war the industrial and technological gap between Japan and the West, and especially with the United States, widened further, and Japan was steadily ground down. The human and material cost was horrific. Perhaps 8 million Japanese were killed or wounded, and 2.5 million buildings had been destroyed or damaged, 709,906 in Tokyo alone. The war cost Japan the value of all of the substantial economic development which had taken place since 1935. All assets put into military production and activity had been lost, while 25% of "peaceful assets" had been destroyed. Japan's colonial empire, and all the resources invested in it, were also lost.

The economic picture was not uniformly black. Following Japan's surrender industrial production virtually ground to a halt, but bomb damage was only partially responsible. The four "home" islands of Japan had been spared a land campaign, which would have brought even greater devastation to its industrial and other facilities, while US bombing tactics, which used incendiary rather than demolition bombs, had spared much of Japan's industrial and transportation infrastructure. For instance, in August 1945 there were still eight blast furnaces in operation; by September 1946 there were only three. Between August and November 1945 coal production plummeted from 1.67 million tons to 550,000 tons. The collapse of industrial production was due as much to organizational chaos, lack of raw materials and rampant inflation, some of it induced by governments in 1945, as it was to the direct devastation of war. Finally, the toll of war on Japan's managers and skilled workers had been relatively small, and much of prewar Japan's skilled human capital had thus been preserved.

The Prostrate Economy, 1945–50

Unconditional surrender brought with it foreign occupation, in theory by all the victorious Allies but, for most practical purposes, by the United States alone. The occupiers offered no real hope of a Japanese economic revival. The basic instructions issued to General Douglas MacArthur and his administration – both described by the abbreviation "SCAP" (for Supreme Commander for the Allied Powers) – focused on the political objective of ensuring that Japan did not again become a threat to its neighbors. In part, this was to be accomplished by avoiding "undue measures" to revive Japan's economy. Japan's neighbors in East Asia were to be helped up to Japan's level of economic development by transfers of Japanese industrial plant. Further, the occupiers believed that the *zaibatsu* had been closely associated with the internal extremism and external aggression of the 1930s and early 1940s, and were determined to try to break them up, an act which could only have a disruptive effect on the already semi-moribund economy.

In the event, the occupiers found the burden of supplying relief to an impoverished Japan unbearable, while they feared that the economic plight of the population might push many more Japanese towards Communism. By May 1947 reparations had died a quiet death. *Zaibatsu* dissolution was begun by SCAP in September 1946 and continued into 1947, by which time the occupiers had lost enthusiasm for it. SCAP did dispossess the old families and groupings which had owned the *zaibatsu*, and excluded the existing managers from any role in the companies. However, only 18 out of 325 firms designated as members of *zaibatsu* were, to use the jargon of the day, "deconcentrated" (broken away from the parent company), and virtually all of the prewar *zaibatsu* survived to become members of the post-1952 *keiretsu*, although some would change their names to disguise their parenthood. The younger managers who replaced those excluded from office by SCAP tended to be more innovative and effective, and in the 1950s they were responsible for shaping distinctive Japanese management methods. The essential divorce of ownership from management which SCAP's partial attack on the *zaibatsu* brought about enabled the new managers to devise strategies aimed at long-term development rather than short-term profit, and to devise management methods much more sensitive to the interests of the mass of employees than might otherwise have been the case.

Many of the reforms which SCAP imposed, in the course of implementing their basic policies of demilitarizing, punishing and democratizing Japan, had profound, if unforeseen, effects on the postwar economy. Land reform, carried through to further democratization, led to a substantial rise in agricultural productivity, which greatly assisted general economic recovery in the 1950s and early 1960s (see Dore). Education reform was also implemented by SCAP as part of the democratization strategy. Prewar Japan's education system had been highly elitist, with only a small proportion of the population progressing beyond elementary school. For essentially political reasons, the Occupation imposed a new educational structure, modeled on that of the United States, which provided for much easier entry into higher levels of education (see Chapter 14). This new education system was to be the basis upon which Japan developed a very highly educated workforce, capable of using the increasingly advanced technology which was imported from abroad in the 1950s and 1960s, and developed by Japanese companies themselves in the 1970s and 1980s. Finally, the occupiers refined methods of controlling key parts of the economy, such as interest rates, foreign exchange controls, bank regulation, and labor union rights, which Japanese governments after the Occupation used to "guide" significant parts of the economy.

Serious economic recovery was impeded by chronic inflation, caused by the general dislocation of the economy and by the failure of successive Japanese governments to balance their budgets (see Tsuru). By 1948 Occupation officials, anxious to stabilize Japan's currency as part of building Japan into a bulwark against Communism, took resolute action, backed by a new, economically orthodox Minister of Finance, Ikeda Hayato (later to become Prime Minister). In December 1948, Joseph Dodge, a Detroit banker who had curbed inflation in the western zones of Germany, was sent to Japan to repeat the trick. The "Dodge Line" which the government was required to follow, from March 1949, was fierce. With inflation still running at 165%, it had to balance its budget; cut off excess money supply by suspending loans from the Reconstruction Finance Bank, which it had set up only two years earlier; dismiss government employees; and end government subsidies to industry. Dodge also saw to the pegging of the yen at ¥360 to the US dollar, an exchange rate which survived until 1971, providing long-term currency stability which was very beneficial to Japan's economic development. It was a dose of severe medicine for the Japanese economy, but Dodge's policies helped lay the basis of future economic growth. In 1950, however, the economy was still at a profoundly low ebb.

The Recovering Economy, 1950–60

Japan's postwar economy received its first decisive stimulus with what Yoshida Shigeru, then Prime Minister, called "a gift from the gods" in June 1950. Communist North Korea invaded the South, sucking the United States into a large-scale conflict in Korea until an armistice brought the conflict to an end in July 1953. When the Korean War broke out, the deflationary effect of the Dodge Line had pinched off much of the modest industrial growth which had taken place in Japan in the late 1940s. As the war intensified, the United States used Japan as a vital source of supplies for its forces in Korea, and as a vast repair depot for equipment and vehicles damaged in the conflict. These "special procurements" pumped US$930 million into Japan between 1950 and 1953, stimulating basic industries such as iron and steel, while profits from repairing war-damaged US military vehicles resuscitated virtually moribund Japanese industries, including automotive engineering. Between 1951 and 1956, US military spending paid for more than 25% of Japan's imports (see Hein). Without the foreign exchange earned the Japanese could not have begun to buy the latest industrial technology from abroad, and without this foreign technology there could have been no high-speed growth in the 1950s and 1960s (see Denison and Chung).

From the late 1940s the Japanese government sought to "rationalize" basic industries, making them more efficient by updating their technologies, restructuring them, and achieving economies of scale. Between 1951 and 1954, the government focused the rationalization policy on iron and steel, and on coalmining. Around ¥120 billion was invested in iron and steel. This first rationalization drive culminated in the opening of the Kawasaki Steel Company's integrated rolling strip mill in Chiba Prefecture, the world's most advanced steel-making facility. Responding first to an explosion in the need for high-quality steel, to meet US requirements during the Korean War, and then to the growth of the shipbuilding, automobile and domestic appliance industries, two government rationalization plans poured ¥620 billion into Japan's steel industry. Using the latest imported steel-making technology, more integrated steel mills were built around the coasts, while combinations of supporting industries and technologies were sited in the most geographically advantageous positions, developing into vast and efficient industrial parks. By the end of the 1950s, only a decade after it had emerged prostrate from World War II, Japan's steel industry was second only to that of the United States (see Kosai).

However, plans to develop the coal industry ended in failure: the geological problems which cursed Japan's coal industry were compounded by serious industrial strife. In 1952, a major strike by coalminers led manufacturers to begin a decisive shift from coal to oil as the predominant source of energy. A combination of low world oil prices, and the development of supertankers which greatly reduced transport costs, gave oil a decisive advantage over coal. The failure of the government's plans for coal demonstrated the limits of its control over the economy when economic circumstances dictated a course different from official wishes; and by 1970, 90% of Japan's energy requirements were met by oil (see Nakamura 1995).

The growth of the steel industry was driven, as has been mentioned, by demand for its products from other sectors of the economy, especially shipbuilding, domestic appliance manufacturers, and the infant automobile industry. In 1935, Japan had been the world's third largest shipbuilder, constructing 11% of the world's total, and most of the country's shipyards had survived the war intact. The Korean War stimulated shipbuilding, and from 1951 the government-controlled Japan Development Bank injected serious investment capital into the industry. Growing world demand for oil produced an insatiable demand for ever larger tankers, a trend accelerated by the closing of the Suez Canal in 1956, which increased the cost of shipping oil from the Middle East to Europe and North America. The Japanese shipbuilding industry bought in new technology, particularly the welding of prefabricated sections to build the new supertankers which increasingly dominated the carrying of oil. Japan swiftly became the world's largest builder of these supertankers, and by 1960 it was the world's largest shipbuilding nation, producing around 20% of world tonnage and providing an enormous market for the growing steel industry.

The growth of consumer goods manufacturing was also breathtaking. Between 1953 and 1956, the production of washing machines increased eightfold. In 1956, a mere 1% of Japanese households had televisions; by 1960 one half had sets. This explosion in the production of television sets and other electrical consumer goods was achieved by the emerging giants of Japanese electronics, such as Matsushita and Hitachi, which provided the Japanese consumer with the "three electric treasures": a washing machine, a refrigerator, and a television set. The great expansion in the Japanese automobile industry came after 1960, but, following the boost from the Korean War,

Nissan, Toyota, Isuzu and Hino not only produced cars and commercial vehicles under license from western companies, but also had begun to produce trucks and cars designed and engineered in Japan.

The economic growth of Japan between 1950 and 1972 was hardly unique, although its speed was. All industrial economies experienced substantial growth in these two decades, in part at least because of the international economic environment which existed in the quarter century after the end of World War II. The Bretton Woods structure, put in place in 1944, provided long-term monetary and currency stability, while GATT erected an unprecedentedly open trading system from which Japan benefited enormously.

The importation of technology was the factor which, above all others, contributed to postwar Japan's high-speed growth. Despite prewar industrialization, Japan was a relative "latecomer" in industrial development, and industrial growth was heavily dependent on the import of knowledge and plant from abroad. Without the liberal international economy, such large-scale purchases of the latest technology would have been far more difficult. The process began in the early 1950s and soon technology imports became a torrent. Under strict government supervision, technology worth US$281 million was imported between 1956 and 1960. Most of this money came from the large banks, which were able to recycle Japan's historically high rate of personal savings into investment capital, and "overloans" (see below) guaranteed ultimately by the government meant that unusually large amounts of money could be loaned. Also, after the Occupation ended in 1952 the Japanese government relaxed SCAP's Anti-Monopoly Law, and companies which had been members of the former *zaibatsu* began to coalesce to form *keiretsu*, conglomerates with major banks or industrial companies at their centers, which provided ample, often cheap, investment capital to other members of the group (see Tsuru).

Postwar Japan had one other unusual international advantage. The power and influence of the United States gave Japan access to raw materials and to world markets while Japan could shelter under its military umbrella (see Chapter 15). If Japan had spent 6–7% of its GNP on the military, it would have reduced overall economic growth in the 1960s and early 1970s by up to 2% a year (see Patrick and Rosovsky).

The role of government and the bureaucracy is the most controversial issue in discussions of the economic achievements of postwar Japan (see also Chapter 3). Some commentators argue that the bureaucracy had a central role in Japan's economic growth (see, for example, Johnson); others hold that the official role was either negligible or actually damaging (see, for example, Patrick and Rosovsky). The middle road is that MoF, the Ministry of International Trade and Industry (MITI), and other government bodies had differing levels of influence in different spheres at different times (see Okimoto). Japan had a historical tradition of official involvement in economic affairs which certainly continued after the Pacific War. The control and "guidance" came principally from MITI, and both the methods of influence and the institutions involved had their origins before World War II, although MITI had started life as the Ministry of Commerce and Industry in 1925, and was only named MITI in May 1949. The main instruments which the government and bureaucracy manipulated after 1950 were derived from economic institutions and controls which emerged during World War II, and had then been refined during the Occupation years. Bureaucrats exerted pressure either formally, through direct measures such as tax concessions and the allocation of scarce foreign exchange, and less formally, using personal contacts and providing "administrative guidance" (*gyosei shido*) which companies were expected to follow.

In the 1950s the bureaucrats had a powerful hand. They allocated Japan's scarce foreign currency, and therefore decided which industries and companies could import raw materials and foreign technology. During the Korean War boom officialdom acquired another tool of influence. The growth of those Japanese industries responding to US demand was threatened by a lack of investment capital. MITI persuaded MoF to introduce "overloans" which allowed the commercial banks to lend to selected firms far more money than their firms' worth or immediate prospects would justify. Since the government-controlled central banking institutions had to guarantee the risk inherent in overloans, the government and bureaucracy were thus given another lever with which to direct industry.

The less formal method of administrative guidance used the great prestige and status of the bureaucracy in Japan, and the threat of using the government's levers of influence over the economy, to guide the investment and other decisions of private companies. In the 1950s, the government and economic bureaucracy had greater opportunities to influence the development and structure of the economy than in subsequent decades. However, the bureaucracy did not always get their own way. For example, in 1956–57 Toyota challenged the Isuzu company's monopoly of diesel-powered trucks. MITI was anxious to limit competition in this sphere and "advised" Toyota not to continue. However, MITI was effectively ignored by Toyota, its chairman declaring that it had no right to interfere in Toyota's business strategy. MITI could also fail to appreciate the importance of new technology. A prime example of this occurred in 1953–54, when MITI was uncooperative with regard to Sony's plan to buy in transistor technology from the United States. On the whole, however, there were few such clashes, since there was seldom great conflict between official "advice" and the plans of private businesses.

During the 1950s, Japan experienced the so-called "management science boom". Japanese managers adopted and built upon the theoretical work of American management experts, most notably W. Edwards Deming. A particular concern was quality control, which Japanese management elevated almost to religious levels. Quality became an obsession in Japanese manufacturing industry during the 1950s and early 1960s, because the country's straitened international economic position demanded that there be no waste of valuable, imported machinery or raw materials. In 1957 the concept of "total quality control", supported by quality control circles (QCCs), was introduced into Japan (see Fruin). In addition, Japanese manufacturers innovated in logistics and production. In 1956, Toyota and its satellite companies introduced the "just in time" (*kamban*) system, requiring subcontractors to deliver components exactly when needed on the production line. This greatly reduced stocks of spares held in factories, cut the costs of warehousing, and minimized the amount of money tied up in components. Total quality control, QCCs and the just-in-time system became distinctive features of Japanese manufacturing tactics, and would be widely adopted elsewhere, including the United States, where the Japanese had first discovered these techniques (see McMillan).

Growth was rapid in the 1950s, and by 1955 Japan's industrial output exceeded the maximum reached before 1945. However, until 1967 there was a periodic brake on continuous high levels of growth. Growth sucked in substantial imports of raw materials and capital goods, causing a serious balance of payments deficit and producing a "balance of payments ceiling" on growth. The government responded with high interest rates and tight credit controls, thereby reducing domestic demand and rapidly slowing the rate of growth. These monetary restraints could be slackened once the balance of payments approached equilibrium, as the domestic economy would expand, and growth would speed up. The balance of payments ceiling arrested growth in 1954, 1957, 1961 and 1963, but after 1967 Japan did not have a balance of payments deficit and the ceiling no longer operated. Growth seemed to have no restraints (see Kosai).

The Golden Age of High-speed Growth, 1960–72

When Ikeda Hayato became Prime Minister, in July 1960, Japanese politics and society were dangerously polarized. In order to lower the political temperature, Ikeda, a former MoF official, looked for a policy which would command the support of most of the population, and hit upon the "Income-Doubling Plan", intended to double Japan's national wealth in 10 years. The announcement of the Plan has since been taken as the symbolic beginning of Japan's period of unprecedentedly swift and sustained economic growth, although, since GNP tripled between 1959 and 1971, the Plan underestimated the potential for growth. Between 1960 and 1969, GNP grew, on average, 12.1% a year and, although growth eased off between 1970 and 1973, it was still the world's highest, averaging 7.5% a year. Growth in some key industries was even more impressive. From 1961 to 1971, the manufacturing sector as a whole grew at 14% a year, while machinery manufacturing expanded by 19.6% a year. Between 1960 and 1973, labor productivity increased, on average, by 10.7% a year, but grew at a staggering 13.4% a year between 1965 and 1970, reflecting high levels of investment in new technology, as well as the generally high and improving quality of the workforce and effective management.

Industries which had taken off in the 1950s continued to grow in this golden age. Steel consolidated its growth and, using the very latest technology, the number of man-hours required to produce a ton of steel was reduced from 25 to nine between 1964 and 1973. Meanwhile, the numbers and the sizes of ships being produced in Japan carried on rising, and by 1975 the country's shipyards were building slightly more than half of the world's new ships. Industries which had only begun to develop in the 1950s soared in the 1960s. The most notable was the automobile industry. As early as September 1945, SCAP had allowed Japanese companies to manufacture trucks, but the growth of the industry, especially the manufacture of cars, was slow because of low incomes in the 1950s, the poor condition of roads, and the lack of service stations, repair facilities, and other infrastructure. In 1956, the industry produced only 100,000 trucks and cars, but the rapid rise in incomes, and therefore in demand for cars, led to rapid growth. By 1963 the industry was producing 1 million trucks and cars; by the end of the period of high-speed growth production was nearing 4 million. By 1967, Japan was the world's second largest automobile manufacturer, but this went relatively unnoticed in the outside world because only 16% of Japanese-produced cars were exported.

High-speed growth was fueled by very high levels of investment which bought and used the most advanced technology from overseas. Between 1960 and 1965, Japan paid US$684 million for new technology, and this expanded to US$1.536 billion between 1966 and 1970. In the final years of high-speed growth, US$3.205 billion was spent on importing 10,789 items of foreign technology. Between 1964 and 1971, Japan paid more for foreign know-how than any other major industrial country (see Peck and Tamura), and new technology was responsible for 45% of Japan's growth rate in the decade between 1960 and 1973 (see Ito). Heavy investment, largely by the private sector, equipped Japanese industry with the best technology that could be bought, maintaining the best of Japan's industries at the cutting edge of technology. Imported electrically powered machinery was used in developing industries, such as automobiles, and specific technologies trickled down into other, smaller-scale industries. Further, Japan's advances in research and development were especially important to the overall growth of the economy because, unlike in the United States or Britain, they were almost entirely unrelated to military research, and hence were directly useful to the civilian industrial economy.

This influx of new technology from the West had the enormous advantages of being used by an increasingly high-quality workforce, and being directed by more effective and sophisticated management methods. Education was the key to the quality of the workforce. In 1950 the bulk of the population were educated only to basic levels, and even in 1955 only 38% of those entering the workforce had been educated beyond junior high school. By 1965, however, around 58% had a high school education, while by 1975 the proportion was 91%, the world's highest (see Chapter 13). This new industrial labor force came largely from the countryside. It was the changing structure of the workforce in the decade after the Occupation which transformed Japan from a rural and agricultural society into one which was overwhelmingly urban, and employed in manufacturing and service industries. Even in the mid-1950s, more than half of the working population had been employed in agriculture. As the industrial, and essentially urban, economy grew, so the rural population was sucked into the cities at astonishing speed. In 1950, 50% of young people leaving junior and senior high schools went into agriculture; by 1960 this had dropped to 10%, and by 1965 it was only 5% (see Chapter 5).

Industrial relations following 1960 were much more peaceful than in the decade before. Enterprise unions became increasingly common and union leadership more moderate, while managements became more conciliatory. In the larger companies, management methods which had first appeared 50 years before were implemented and consolidated. Much freer from owners and shareholders than their western European and North American counterparts, Japanese managers placed less emphasis on maximizing profit, and usually demonstrated greater sensitivity to their employees' interests when framing and implementing their corporate policies. A strong strain of paternalism led the large companies to provide a whole range of welfare benefits for both employees and their immediate families, intensifying company solidarity and loyalty. In the large companies, the "three sacred treasures" – "lifetime employment", wages and salaries based largely on seniority and length of service, and enterprise labor unions – became almost universal.

By 1967 Japan was rapidly developing its own research and development capabilities. The rate of growth in expenditure on research accelerated rapidly in the second half of the 1960s, as large companies, especially electrical manufacturers, became technologically self-sufficient. Closing the technology gap had a negative side, however. It was the most important single reason for slowing down Japan's rate of economic growth in the early 1970s, even before the first "oil shock". The easy growth of the 1950s and 1960s had depended on importing and using western technology to increase production. Japan had benefited from being a relative latecomer, able to buy in and equip its factories with the latest technology from the West. In future, Japanese industry had to generate technological advance through its own research and development, and the cost of this effort moderated the rate of industrial growth.

The Slowed Economy, 1973–80

On October 16, 1973, Japan's economy was traumatized when, during the latter stages of a war between Israel and its Arab neighbors, the Organization of Petroleum Exporting Countries (OPEC) announced a two-thirds increase in the price of crude oil, from US$3 to US$5 a barrel. The Arab oil-exporting countries then imposed a complete embargo on the supply of oil to the United States, followed by progressive embargoes on other industrialized countries, including Japan. Within a month shipments of oil from the Middle East to Japan were three quarters of normal quantities, and supplies returned to normal only in March 1974 (see Lincoln). Japan was hit by a dose of "stagflation", in which the economy contracted by 1.4% in 1974 while inflation reached 31.6%. The economy quickly recovered from this malaise but in the longer term Japan's economic growth slowed appreciably. The immediate culprit for this downturn was the oil shock but the economy had, in fact, been arrested by deeper factors.

The Japanese government had been responsible for the inflation. On August 15, 1971 the strains of the Vietnam War had forced the Nixon administration to suspend the convertibility of the US dollar into gold, thereby ending the long period of international currency stability which Japan had enjoyed since 1949 and which had contributed greatly to its economic performance. By 1971 the yen was seriously undervalued, to the great advantage of Japanese exports which were relatively cheap in world markets. In December 1971, the Japanese agreed to a new exchange rate of ¥308 to the US dollar, but the yen continued to gain value. Government and big business became alarmed that the rising yen would cut exports back and cause a recession. The government began to pump money into the domestic economy, beginning an inflationary spiral which was made worse by the oil shock in November 1973 (see Uchino).

In addition to government mistakes in monetary and fiscal policy, some of the basic factors which had contributed to very rapid rates of growth in manufacturing industry had run their course by the early 1970s. The technology gap between Japan and the West had all but closed, and in many industries leading-edge technology had to be developed by Japanese companies themselves. This was expensive and led to falling rates of return on investment. Also, a greater proportion of the workforce was spending longer in full-time education, improving it qualitatively, but the rate of improvement did not match rates of the 1950s and 1960s. Improvements in labor productivity slowed, thus discouraging high rates of investment. Most of the advantages of scale which Japanese industry had been building towards in the 1950s and 1960s had been achieved by the early 1970s. Finally, the wider Japanese public was no longer willing to accept unbridled industrial growth, regardless of environmental damage and neglect of social amenities. Industry would henceforth have to meet at least some of the social costs of industrialization, and this would further arrest the rate of growth.

The crisis of 1973–74 marked a clear transition point in the performance and structure of the Japanese economy. Most obviously, the economy slowed after 1973, although it would have happened, albeit more gradually, without the oil crisis. Hesitantly at first, the growth rate crept up to 4.8% by 1976 and reached 5.3% in 1979, before the second oil shock of that year knocked it back. Only in 1984 did the economy return to an annual growth rate above 5%, at 5.1% (see Ito). As the economy recovered the structure of industry changed.

Responding to changed international circumstances, manufacturing industry quickly shifted its emphasis from the materials industries, such as metals and chemicals, to assembly and knowledge-intensive industries. Between 1975 and 1982 machining and assembly industries grew by 113.7%, while basic materials industries grew by a mere 14%. Materials industries were energy-intensive, and the rocketing of oil prices had a severe impact upon their international price competitiveness. Further, Japan was losing its competitive advantage in production chemicals, steel and other materials, along with shipbuilding, to newly industrializing countries such as South Korea and Brazil, where wage levels were far lower than in Japan. In 1973 seven of the 10 largest companies in Japan were involved in metals and heavy industry; by 1987 only one of the top 10 was (see Fruin).

With government encouragement, research and development came to be focused on numerically controlled machine tools and industrial robots, which were of immense advantage to assembly industries, and on knowledge-intensive industries, particularly information technology, biotechnology and sophisticated semiconductors (see Uekusa). As a result, even after the first oil shock giant strides were made by two industries, automobiles and information technology.

The automobile industry was the prime specimen of the assembly sector taking full advantage of the changed circumstances after 1973. The massive increase in oil prices handed a substantial competitive advantage to the Japanese automobile industry, since before 1973 it had been developing small, fuel-efficient cars as a counter to rising public hostility to pollution. In the early 1970s the Japanese government had imposed the world's strictest exhaust emission control laws, so that when North American and western European governments made their own emission regulations stricter, Japanese manufacturers were in an excellent position to meet the demand. By 1976, Japan was exporting 2.5 million vehicles a year, and in 1979 the country's automobile production overtook that of the United States for the first time. By 1980, automobiles made up 22.7% of the value of Japan's exports (see Mutoh).

The other new industries which burgeoned after 1973 were in the key areas of information and communications technology, and microprocessors. Five months after the first microprocessor was produced, in 1971, NEC produced Japan's version (see Fransman 1995). After 1973, Japan's nascent computer industry expanded rapidly. In 1972 expenditure in Japan on research and development in information technology was US$168 million, rising to US$300 million the following year, and by 1980 Japan was spending US$725 million on developing computer-related technology (see Fransman 1990). Japan rapidly became the world leader in robotic machine tools, and this gave an enormous boost to productivity in a select range of industries.

Another consequence of 1973 was a changing balance between domestic and foreign markets. Although the economy continued to be largely devoted to satisfying domestic demand, more foreign exchange had to be generated to buy expensive raw materials. Japan generated large balance of payments surpluses, as foreign demand for Japanese-made cars and consumer electrical goods exploded, and exports became more important in sustaining growth in the Japanese economy (see Chapter 9). From 1980, the surplus on the balance of payments grew very rapidly, and the contribution of exports to economic growth soared (see Lincoln). The impact of Japan's burgeoning trade surpluses was as much political as economic, as it produced dangerous friction with North America and western Europe, severely complicating Japanese foreign policy in the 1980s and 1990s.

The Recovered Economy, 1979–85

In 1979 there was another oil shock, after the oil producers doubled their prices. This gave the Japanese economy a sharp jolt, although it was nowhere near as severe as the first shock. In any case, by 1979 measures to cut oil imports had transformed Japan into the world's most efficient energy-using nation. In the 1960s, imports of energy had accounted for around 3% of GNP, but by 1984 this had been cut to 1.6%. The second oil shock did cut back the rate of growth of the economy, from 5.3% in 1979, to 3.7% in 1980, but, as we have seen, it had struggled back above 5% by 1984 (Lincoln, 1988).

Continued growth enabled Japan to overtake the United States in per-capita income in 1987, but it also led to increasingly serious friction with the outside world. After 1980 the fruits of economic growth were increasingly exported abroad as direct investment and to purchase foreign assets. Between April 1986 and March 1991, total Japanese overseas investment amounted to US$227.2 billion. Japanese-owned manufacturing bases were created in North America, western Europe and other parts of Asia. Purchases included foreign real estate, companies, shares and government bonds, notably those of the US government, which found its chronic budget deficits being partly financed with Japanese money. Japan had become the world's largest creditor nation and a true economic superpower, but there were dark clouds on the horizon. Relations with the United States deteriorated sharply during the 1980s, largely because Americans came to believe that Japan was becoming richer at their expense. Further, the rise of the newly industrialized nations in East Asia, especially South Korea, threatened competition for key Japanese industries.

Exports were to be an important, although declining, factor in Japan's recovery from the second oil shock. In 1980, exports were responsible for four fifths of Japan's economic growth, and even in 1985 they contributed 20% to growth. It was not simply the dimensions of export growth which caused trouble, it was also the products which were now central to Japan's export trade: automobiles, of which exports increased by 50% in 1979–80, and advanced electronic components competed directly with key US industries. The United States was postwar Japan's most important trading partner, while Japan was second only to Canada in US foreign trade. Since 1959 there had been trade disputes between Japan and the United States, usually over specific products, and these would usually be solved by negotiated quotas or fixed prices for Japanese exports of these products (see Tsuru). After 1980, however, trade disputes assumed much greater importance and were not really amenable to previous solutions. The immediate problem arose when influential US commentators claimed that the virtual bankruptcy of the Chrysler Corporation, an icon of US industrial prowess, was due to massive Japanese automobile imports. The automobile issue was rapidly solved by the Japanese accepting voluntary restraints on car exports to the United States, but this did not attack the heart of the problem. In many parts of the world, but particularly in Southeast Asia, US companies found it difficult to compete with the Japanese in the key semiconductor industry, in which Japan was producing 40% of world supply by 1986 (see Fransman 1995). Japan's burgeoning exports were partly the result of the serious undervaluation of the yen, which made Japanese exports relatively cheap abroad, combined with an overvalued US dollar, caused by the Reagan administration's fiscal policies, which meant that relatively cheap Japanese goods were in great demand in the United States. The Japanese cooperated in reducing the value of the dollar and increasing the value of the yen (*endaka*), but Japan's trade surpluses were only slowly arrested, and US politicians and observers increasingly claimed that Japan's economic structures meant that US businesses were not operating on a level playing field in Japan. In March 1985, the US Senate named Japan as "an unfair trading partner", and the United States increasingly pressed for fundamental structural change in the Japanese economy, through the "Structural Impediments Initiative". US fire was directed on Japanese retailing structures, which were held to obstruct the marketing of US goods in Japan; on *keiretsu* interrelationships, which were accused of providing Japanese exporters with artificially cheap credit; and even on Japanese working habits, which it was claimed denied the mass of Japanese the leisure to consume US goods. However, although criticism and threats against Japan continued in the United States, the dispute never fully caught fire. From 1993 onwards, the Clinton administration proved not to be dedicated to "Japan-bashing", while the new President's evident interest in the Pacific dimension of US interests and policies helped cool passions. Nevertheless, Japanese policy-makers remained very sensitive to foreign criticism, and the government was persuaded to begin deregulation of Japan's financial services, allowing foreign institutions to compete with Japanese banks, insurance companies and securities houses.

The "Bubble Economy", 1985–90

The Japanese economy seemed to be enjoying a second golden age in the second half of the 1980s, and seemed to be immune from the financial and general economic problems which affected much of the rest of the industrialized world. Between 1985 and 1990, GNP grew between 2.6% and 6.2% each year. When the world's stock markets were hit by a serious crisis, in October 1987, Japan's exchanges seemed to shrug off the difficulties far more quickly and completely than the other major centers. The domestic economy was awash with money, which initially financed high levels of company investment, sustaining economic growth. However, industry was soon satiated with funds, virtually unable to invest more with any serious prospect of realistic return. Interest rates were unprecedentedly low, and the mountain of cash was channeled by banks into financing the purchase of land and shares. Land prices rocketed, while between October 1987 and December 1989 share prices increased 120%. This created a "bubble economy", in which companies, corporations and individuals used the vastly inflated values of their holdings of land and shares to borrow to buy more land and shares. Banks' investment in real estate companies helped to bring about a doubling in land prices between 1986 and 1989. Japan experienced an unprecedented boom in consumer spending, much of it on credit. Domestic capital investment, heavy consumer spending, and the bubble economy sustained high levels of economic growth, conveniently sucking imports into Japan and temporarily relieving some of the international pressure to restructure international trading practices and domestic economic structures.

This could hardly last. In December 1989, the Bank of Japan, under a new President, increased interest rates. Closer analysis soon revealed that the land value of Tokyo was not really greater than that of Canada, and that stocks and shares were not truly worth nearly 400% of their value in 1985. The Tokyo Stock Exchange lost 48% of its share values in the first half of 1990, while real estate values plummeted. The bubble had burst, and banks and other financial institutions were left with a mountain of debt secured on real estate and shares by now worth a fraction of their former value. It may never be known precisely what were the true dimensions of these bad debts, euphemistically called "non-performing loans", which will never be repaid, although in January 1998 MoF estimated that the total was ¥76.71 trillion (around US$361 billion), equivalent to twice the GDP of Turkey, but this may well be an understatement of the problem. Japan's economy plunged into an unprecedented postwar recession.

The Depressed Economy, 1990 Onwards

The recession that started in 1990 was almost wholly self-inflicted. Many ordinary Japanese had been severely affected by the decline in the value of property or shares, while a large proportion of the population had financed high consumer spending with credit. The credit revolution in Japan, associated with large-scale borrowing for buying real estate and large consumer goods, and with the spread of credit cards, led to consumer debt in Japan growing from ¥9 trillion in 1979 to ¥67 trillion in 1991, giving Japanese a higher per-capita consumer debt than Americans (see Wood).

The bursting bubble led to a fierce reduction in consumer spending and an increase in savings. Since nearly 90% of Japan's manufacturing and consumer economy was concerned with meeting domestic demand, the result was a sharp recession. By 1992, growth was down to 1.3%, while in 1993 GNP actually contracted, although only by 0.2%. This downturn caused the rate of wage increases to decline from 8.3% in 1990 to 3.2% in 1992, and widely publicized fears that lifetime employment, which affected nearly 40% of the workforce, was unlikely to survive further discouraged consumer spending. Fears were heightened, and spending hit, as unemployment rose to levels unprecedented since the 1950s, although at little more than 3% of the workforce the jobless rate was much lower than in Europe or North America. One measure of the problem was that the automobile industry, which employed 10% of the industrial workforce and accounted for 13% of industrial production, experienced a 16% fall in demand between the summer of 1991 and August 1992. The fall in consumer spending was made worse by the imposition of a consumption tax, initially at 3%, but which reached 5% by 1997. The wholly unexpected Hanshin earthquake in January 1995, which damaged Kobe and neighboring cities in Japan's greatest postwar natural disaster, heaped further psychological uncertainty on Japanese consumers, who responded with another wave of saving to prepare for the unknown future, in turn reducing their spending still further. The economy was in the grip of a vicious downward spiral.

During the summer of 1993, the Liberal Democratic Party (LDP), which had monopolized government for 38 years, finally lost office, and Japan entered a period of political uncertainty which impeded strong government action to counteract the recession (see Chapter 2). The economy showed some feeble signs of recovery in 1996, but it was not sustained, and between April and September 1997 the economy actually contracted by 1.4% compared to the previous year. Large companies were able to ride out the full effects of recession by taking advantage of the "dual economy" and squeezing their subcontractors, if necessary forcing them into bankruptcy, although the human cost of this approach was reflected in the 16% increase in the number of businessmen committing suicide in 1997.

Next, Japan was hit by the backwash to the economic crisis elsewhere in East Asia. On November 17, 1997, Japan's 10th largest bank, the Hokkaido Takushoku Bank, closed, submerged beneath its debt; seven days later Yamaichi Securities collapsed. The government acted to prevent a general financial collapse and pumped money into the financial institutions. These institutions had brought their problems on themselves with deeply flawed policies during the 1980s, but the general regional uncertainties focused attention on their plight. Ominously, as mentioned above, US pressure had already prised promises that Japan's financial services would be deregulated and opened to full foreign competition, hardly an appetizing prospect for institutions which had performed so miserably (see Chapter 8). Generally, US patience with Japan showed signs of cracking, and the Clinton administration demanded that Japan stimulate its domestic economy to provide markets not only for US goods but also for imports from the wounded "Tiger" economies of Southeast Asia.

Conclusion

In recent years, some western commentators have reacted to relatively uncritical praise of postwar Japan's economic record by predicting a catastrophic collapse of the Japanese economy. The record since 1990 might seem to support their gloom, but caution would be advisable. Despite prolonged recession, at the beginning of 1998 Japan remained the world's second largest economy, with (statistically at least) among the world's richest people. Nippon Telegraph and Telephone (NTT) was probably the most valuable company in the world; Toyota was a giant in world automobiles; four out of the 10 largest companies in computer manufacture were Japanese; and in semiconductors the figure was six out of 10. Japan still has some of the world's most advanced industries, as demonstrated by the 400,000 industrial robots which represent 20% of the Japanese workforce and a substantial proportion of the world's total of 680,000 robots. The Japanese economy is certainly in serious difficulties, but it would be premature to write its obituary.

Further Reading

Allinson, Gary, *Japan's Postwar History*, London: UCL Press, and Ithaca, NY: Cornell University Press, 1997

The most recent, and best, general history of postwar Japan, covering political, economic and social themes.

Denison, E. F., and W. K. Chung, "Economic Growth and Its Sources", in Patrick and Rosovsky, cited below

A key analysis of the era of high-speed growth.

Francks, Penelope, *Japanese Economic Development: Theory and Practice*, London and New York: Routledge, 1992

An informative general account of modern Japan's economic development.

Fransman, Martin, *The Market and Beyond: Cooperation and Competition in the Creation of Advanced Computing and Electronics Technology in the Japanese System*, Cambridge and New York: Cambridge University Press, 1990; and *Japan's Computer and Communications Industry: The Evolution of Industrial Giants and Global Competitiveness*, Oxford and New York: Oxford University Press, 1995

These two works by Fransman provide comprehensive and comprehensible accounts of key industries in contemporary Japan.

Fruin, W. Mark, *The Japanese Enterprise System: Competitive Strategies and Cooperative Structures*, Oxford: Clarendon Press, and New York: Oxford University Press, 1992

This is an outstanding, although sometimes rather testing, account of the emergence of Japanese business. The appendices are an important indicator of the changing structure of the economy.

Hein, Laura E., "Growth Versus Success: Japan's Economic Policy in Historical Perspective", in Andrew Gordon, editor, *Postwar Japan as History*, Berkeley, Los Angeles: University of California Press, 1993

This excellent account of the postwar economy focuses upon the social impact of economic growth.

Horsley, William, and Roger Buckley, *Nippon, New Superpower: Japan since 1945*, London: BBC Books, 1990

This is a pioneering history of postwar Japan, written to accompany a BBC television series.

Ito Takatoshi, *The Japanese Economy*, Cambridge, MA: MIT Press, 1992

This is an excellent thematic account of the contemporary Japanese economy.

Johnson, Chalmers, *MITI and the Japanese Miracle: The Growth of Industrial Policy 1925–1975*, Stanford, CA: Stanford University Press, 1982; Rutland, VT, and Tokyo: Charles E. Tuttle, 1986

This is an influential work which arguably overstates the role of MITI in the "miracle".

Komiya Ryutaro, Okuno Masahiro and Suzumura Kotaro, editors, *Industrial Policy of Japan*, San Diego, CA, and Tokyo: Academic Press, 1988

An excellent source of case studies of the relations between individual industries and the bureaucracy.

Kosai Yutaka, "The Postwar Japanese Economy, 1945–1975", in Peter Duus, editor, *The Cambridge History of Japan*, Volume 6, *The Twentieth Century*, Cambridge and New York: Cambridge University Press, 1988

This is an excellent account of the period of economic recovery and high-speed growth.

Lincoln, Edward J., *Japan: Facing Economic Maturity*, Washington, DC: The Brookings Institution, 1988

A well-informed and rigorous analysis of Japan's economy in the 1980s.

McMillan, Charles J., *The Japanese Industrial System*, second edition, Berlin and New York: Walter de Gruyter, 1996

A sound account.

Macpherson, W. J., *The Economic Development of Japan, c. 1868–1941*, Basingstoke: Macmillan Education, 1987, and Cambridge and New York: Cambridge University Press, 1995

A clear short account of Japan's development up to Pearl Harbor.

Minami Ryoshin, *The Economic Development of Japan: A Quantitative Study*, London: Macmillan, and New York: St Martin's Press, 1986

This is an excellent account of economic change in modern Japan but it can be highly technical; it includes econometric formulae.

Mutoh Hiromichi, "The Automotive Industry", in Komiya, Okuno and Suzumura, cited above

A convenient account of a key industry.

Nakamura Takafusa, *Economic Growth in Prewar Japan*, New Haven, CT: Yale University Press, 1983

This is the basic account of Japan's economy before World War II.

Nakamura Takafusa, *The Postwar Japanese Economy: Its Development and Structure, 1937–1994*, second edition, Tokyo: University of Tokyo Press, 1995

An updated version of Nakamura's clear history of Japan's high-speed growth.

Okimoto, Daniel I., *Between MITI and the Market: Japanese Industrial Policy for High Technology*, Stanford, CA: Stanford University Press, 1989

A careful work which steers a middle course on the debate about the role of the bureaucracy in economic development.

Patrick, Hugh, and Henry Rosovsky, editors, *Asia's New Giant: How the Japanese Economy Works*, Washington, DC: The Brookings Institution, 1976

A massive collaborative work which generally takes the "no miracle" line.

Peck, Merton, and Tamura Shuji, "Technology", in Patrick and Rosovsky, cited above

An important account of the key role of technology in the era of high-speed growth.

Smith, Dennis B., *Japan since 1945: The Rise of an Economic Superpower*, London: Macmillan, and New York: St Martin's Press, 1995

This book attempts a general account of the political, economic and social development of postwar Japan.

Suzuki Yoshitaka, *Japanese Management Structures, 1920–80*, London: Macmillan, and New York: St Martin's Press, 1991

An excellent, though frequently technical, account of the emergence of Japanese management methods.

Tsuru Shigeto, *Japan's Capitalism: Creative Defeat and Beyond*, Cambridge and New York: Cambridge University Press, 1993

This is an outstanding account of the postwar economy by one of Japan's most distinguished economists.

Uchino Tatsuro, *Japan's Postwar Economy: An Insider's View of its History and its Future*, Tokyo and New York: Kodansha International, 1983

An interesting first-hand history.

Uekusa Masu, "The Oil Crisis and After", in Komiya, Okuno and Suzumura, cited above

Uekusa provides a good brief account of the economy after the oil shock.

Wood, Christopher, *The Bubble Economy: The Japanese Economic Collapse*, London: Sidgwick and Jackson, 1992; Tokyo: Charles E. Tuttle, 1993

A knowledgeable account of an almost unaccountable phase in Japan's economic development.

Yamamura Kozo and Yasuba Yasukichi, editors, *The Political Economy of Japan*, Volume 1, *The Domestic Transformation*, Stanford, CA: Stanford University Press, 1987

An excellent collection of essays on the Japanese economy in the 1970s and 1980s.

Dr Dennis B. Smith is a Senior Lecturer in History at the University of Ulster, Coleraine, Northern Ireland. His publications on modern Japan include *Japan since 1945: The Rise of an Economic Superpower* (London: Macmillan, and New York: St Martin's Press, 1995).

Chapter Two

The Political System: Stability and Change

J. A. A. Stockwin

Japan has sometimes been described as a "late-developing country" (for example, by Ronald Dore), but its experience of modern politics and government is relatively lengthy. The first written constitution on western lines was introduced in 1889 and, although it could not be described as democratic, it was continually in force until it was replaced in 1947 by a democratic constitution influenced by the Occupation authorities. The 1947 Constitution remains in force and, although in certain respects it is controversial, it has never been amended.

In other words, the running of Japan reflects a depth of political experience which is comparable to that of other major states. The nature of that experience diverges in certain important respects from that of other states, but in seeking to analyze Japanese politics it is as well to remember that Japanese political practitioners are able to call on traditions of notable pedigree and sophistication. This is so even if such traditions may differ, in certain respects, from those found widely in Europe or North America.

Change and continuity have interacted in a complex fashion in the political history of Japan. The Meiji Restoration of 1868 ushered in a period of revolutionary change, but certain features of the pre-Meiji mentality persisted and were even used to good effect by leaders in the Meiji period (1868–1912) in pursuit of their goals. Similarly, when the Americans and their allies occupied Japan shortly after August 15, 1945, and embarked upon radical reform of the political system, this did not mean that all the elements that had existed up to 1945 were eliminated. Therefore, to attain a balanced understanding of Japanese political institutions, processes and behavior since 1945, it is important to bear in mind that explanations may require a knowledge of developments which have occurred over a quite lengthy historical period.

The Occupation period (1945–52) was by any reckoning a turning point in Japanese political development, even though opinions differ about how fundamental were the changes wrought. Although it is helpful to regard the Occupation as a kind of experiment in reshaping another state's institutions and practices so as to create democracy, it could not escape the influence of international politics, since the Cold War began while it was still going on. Thus, whereas between 1945 and around 1948 the Occupation authorities were putting in place radical democratizing reforms, their priorities then changed, in the direction of coopting Japan as an ally in the struggle against "international Communism". This entailed soft-pedaling some of the reforms already initiated, giving support to relatively conservative politicians and promoting the recovery of the economy by favoring the employers rather than the unions.

In April 1952, under the terms of the San Francisco Peace Treaty, Japan regained independence and became free to run its own affairs, although linkages with the United States in particular remained close. The Occupation had affected many political institutions, often profoundly and in complex ways, but it is possible to generalize about the political system that had emerged, under two broad headings.

First, citizens were much freer than before to participate in politics, and this freedom was constitutionally guaranteed. Women had the vote for the first time; most restrictions on activities of political parties and interest groups (including labor unions) were lifted; and the apparatus of the wartime police state was abolished. The cult of the *Tenno* (Emperor) was done away with, and with it many restrictions on political expression. Attempts were made – not with complete success – to decentralize administration of some government functions, including the police and the network of state schools.

Secondly, the Occupation was concerned to finish with what had been identified as the structural irresponsibility characterizing Japanese government up to and during the war. In the old system it had been often difficult to establish where power really lay. The *Tenno* was in theory all powerful but in practice was expected to delegate nearly all his theoretical power. Political parties had occupied a central position for a few years in the 1920s, but their success had been short-lived, and they had quickly been replaced in the 1930s by the dominance of elements in the armed forces. Even so, many groups were in contention, and it was difficult to discover clear lines of responsibility. The Occupation authorities were determined that politics and government in the "new Japan" should be established on a firmer basis, with popular sovereignty as the guiding principle, and the lines of responsibility firmly stated in the Constitution.

Somewhat ironically, given that most of the occupiers were American, what emerged from these intentions was a set of relations between legislature and executive reminiscent of the British "Westminster model", rather than the American system. Thus, the electorate chooses members of Parliament in regular general elections, based on universal suffrage; groups of parliamentarians – in practice, political parties – contest elections under common labels; and the group, or coalition of groups, commanding a majority in the lower house of Parliament forms a government, which means that it produces a cabinet under a prime minister. So long as it is able to maintain its majority, the government can continue to govern, which means both administering the various ministries and agencies constituting the government bureaucracy, or civil service, and also dominating the legislative process. In other words, there is nothing to stop a government able to maintain its majority from monopolizing the legislative program and controlling the executive branch. The contrast with the American principle of separation of powers could hardly be greater.

There is no doubt that Japanese politics would have turned out very differently had something approaching the US model been followed. The rationale for preferring the Westminster model was, in part, that the bulk of liberal reformers in the interwar period had sought to enhance the roles of Parliament, the Cabinet and political parties, and reduce those of the armed forces, the elites advising the *Tenno*, and various other constitutional and extra-constitutional bodies disputing the power of Parliament.

Fundamentally, therefore, Japanese politics since the Occupation has been conducted within a parliamentary system, reminiscent of the political arrangements to be found in Britain and elsewhere. It soon became evident, however, that not all the norms and practices of politics in Britain were being followed in Japan. Whether this was a matter of cultural difference (see, for instance, Curtis), or a product of rational adaptation to current institutional rules (see, for instance, Kohno), remains controversial even today. What then was different? First of all, the party system developed some distinctive, though not exactly unique, characteristics. Following its foundation from a merger of competing conservative parties in 1955, the Liberal Democratic Party (LDP) went on to win a majority of seats in every lower house general election until 1993 and in every upper house general election until 1989. (For a comparative view, see Pempel.) Single-party dominance in turn interacted with a powerful government bureaucracy to produce, in the early postwar decades, a remarkable concentration of governmental power. This was further reinforced by the development of a network of linkages among the government bureaucracy, the dominant party, and conglomerate interests. Thus, during the period of rapid economic growth which lasted from the 1950s until the first oil crisis of 1973–74, the single-minded pursuit of economic development was undertaken by a powerful cross-sectoral elite, having the capacity broadly to accomplish its goals of economic growth. (For the "government initiative" view, see Johnson; for more skeptical approaches, see Patrick and Rosovsky, and Okimoto.)

On the other hand, it is important not to ignore another side of the picture, which is that the exercise of power appears to have been considerably more dispersed than one might imagine from the previous paragraph. The image of top-down decision-making by a tightly structured elite, in a structure characterized by clear lines of responsibility, turns out to have been at variance with political reality. Not only has the position of Prime Minister been comparatively weak, but even Cabinets appear not to have been capable of making firm decisions on key issues in such a way as the Westminster model might lead one to expect. Further, on closer inspection the closely coordinated governmental structure is seen to be divided among fiercely competing bureaucracies, out to increase their respective shares of jurisdiction and "turf". Coordination, indeed, proved a significant problem for the system to cope with, and getting things done had to be facilitated through informal networks arriving at deals involving substantial kickbacks: hence the emergence of a system needing to be lubricated by large-scale corruption or "money politics".

The Constitutional Basis of Politics and Government

The basic document underpinning the structures of politics and government is the Constitution of Japan (*Nihonkoku Kempo*), promulgated on November 3, 1946, and brought into effect from May 3, 1947. Although formally represented as an amendment of the 1889 Constitution, it is in fact a completely different document, based on the principle of popular sovereignty, replacing that of sovereignty of the *Tenno*. The subsequent history of the Constitution is full of ironies. The LDP, which has formed most of the governments since the war, has been officially committed to revising the Constitution it is pledged to uphold, while the Socialists (more recently known in English as the Social Democrats – see Appendix 4), who formed the main opposition for much of the same period, were strongly critical of US policies towards Japan, but have regarded the American-influenced Constitution as the best guarantee of peace and democracy. So far as the LDP is concerned, a formal commitment to revision has never been translated into successful action, and although at certain periods (the late 1950s, the middle 1980s) serious attempts have been made to initiate a process of revision, at other periods the party has lived with the Constitution quite comfortably, even discovering advantages from it in the exercise of power. What Watanabe Osamu has called "revision though interpretation" has in any case proved possible in certain respects.

The Constitution begins with a preamble extolling the virtues of peace and democracy. The first Chapter concerns the *Tenno*, who is defined as "the symbol of the State and of the unity of the people, deriving his position from the will of the people with whom resides sovereign power". He is specifically denied "powers related to government" and all the acts he is expected to perform have to be done "with the advice and approval of the Cabinet". Thus were removed the range of imperial prerogatives which had been a central feature of prewar government.

Chapter 2 consists of a single article (Article 9), proclaims the renunciation of war, and has been the most controversial part of the Constitution. Despite the Article, Japan has developed substantial military capacity, in the form of the Self-Defense Forces (*Jieitai* or SDF). However, Article 9 has had the important effect of greatly inhibiting, although it has not absolutely prevented, the re-emergence of military influence over government.

Chapter 3 contains an extensive array of human rights clauses (Articles 10 to 40). This is in contrast to the relatively brief list of heavily qualified clauses in the 1889 Constitution, which referred to the rights and duties of "subjects", whereas the present Constitution prefers "citizens". There is some qualification in the new Constitution as well, by reference to rights obtaining so long as they do not clash with "the public welfare" (Article 12). It is also true that in practice the notably conservative judiciary was slow to endorse protection of citizens' rights, and that the tradition of a non-litigious society was extensively used to discourage too vigorous a pursuit of human rights cases. Nevertheless, with the passage of time popular consciousness of rights has made some progress, so that Chapter 3 may be regarded as providing some degree of check on potential and actual abuse of power by authorities at various levels.

Chapter 4, which extends from Article 41 to Article 64, concerns Parliament. (This is often still called the "Diet" in English, an anachronism, since it reflects the influence of German political traditions on the 1889 Constitution.) Parliament is defined as "the highest organ of state power, and . . . the sole law-making organ of the State" (Article 41). It consists of two houses, the House of Representatives (*Shugiin*) and the House of Councillors (*Sangiin*). Both are elected – though by different systems of election, as prescribed in the electoral law – but the former is given substantially greater powers than the latter. The House of Representatives has a maximum term of four years, but may in certain circumstances be dissolved early, whereas the House of Councillors has a six-year term, with half the members being elected every three years. Ordinary bills passed by the House of Representatives but blocked in the House of Councillors become law "when passed a second time by the House of Representatives by a majority of two thirds or more of the members present" (Article 59); or a joint committee of both houses may be called, on the initiative of the lower house, in an attempt to break the deadlock. The blocking power of the upper house is, however, confined to ordinary bills, since in the case of the budget and treaties, its ability to block them is limited to 30 days. To ensure that Parliament cannot simply be bypassed, Chapter 3 contains certain stipulations regarding convoking of parliamentary sessions.

Chapter 5 concerns the Cabinet, and contrasts with the Constitution of 1889, which nowhere referred to a "Cabinet", only to individual "ministers of state" who were to "give their advice to the *Tenno* and be responsible for it". The 1947 Constitution states categorically that the "Cabinet, in the exercise of executive power, shall be collectively responsible to Parliament" (Article 66). As a safeguard against the kind of military domination which infected the polity in the prewar period, all members of the Cabinet have to be civilians. A standard list of Cabinet functions is laid down in Chapter 5, and the Cabinet is defined as consisting of the Prime Minister, as its head, and other Ministers of State. The Prime Minister appoints the Ministers, of whom a majority must be members of Parliament. (In practice, nearly all have been.) Under a clause which has the effect of removing any possibility of political interference by the *Tenno*, it is Parliament that elects the Prime Minister from among its number. If the two houses disagree, the lower house prevails.

Chapter 6, on the judiciary, was designed to ensure an independent system of courts, and provides in particular that "no extraordinary tribunal shall be established, nor shall any organ or agency of the Executive be given final judicial power" (Article 76). In what was for Japan an important innovation, the Supreme Court was made "the court of last resort with power to determine the constitutionality of any law, order, regulation or official act" (Article 81). Thus the principle of judicial review, derived from the United States, was written into the new Constitution.

In the remainder of the Constitution, two points stand out. One is that under Chapter 8, "Local Self-Government", the assemblies and chief executive officers of local public entities are elected, not appointed, this being an attempt to reverse the centralizing thrust of local administrative reform from the late 19th century. The other is that in Chapter 9 the Constitution was made extremely difficult to revise. An amendment requires "a concurring vote of two thirds or more of all the members of each House" after which it should be submitted to a referendum of the people, where a simple majority is required (Article 96). The strength of this clause is an important – though by no means the only – reason why by 1998 the Constitution remained entirely intact.

Despite criticisms from some elements – largely on the right wing of politics – that the 1947 Constitution was imposed by the Occupation authorities, it seems clear that it has shaped the political system in fundamental ways. To a considerable extent this shaping was in the directions intended. Thus in a tangible sense the sovereignty of the people replaced the sovereignty of the *Tenno*; clearer lines of responsibility were established than those under the 1889 Constitution; rights and freedoms were enhanced; and the scope for political participation was expanded. Nevertheless, the shape of politics came also to be influenced by other factors, extraneous or tangential to the Constitution, and in analyzing these we have to take into account continuities with past patterns of action, and also persistent cultural norms and values.

The Structure and Process of Politics since the 1950s

As we shall see, important changes to the way that politics worked began to emerge from the early 1990s, but the stability of structure and process was remarkable between 1955, when the LDP was formed, and 1993, when it fell from power for the first time in four decades – although by 1997 it had clawed back much of its old position of dominance. The essential stability of the system did not imply lack of evolution, but it is possible to describe it by reference to a number of fundamental factors which remained paramount throughout the period.

The Japanese system emphatically had little in common with the American system based on the principle of separation of powers, and was far closer to the British system. In the British model, however, general elections could reasonably be expected to produce changes of party in power. A peculiarity, though not a uniqueness (see Pempel), of the Japanese system under the 1947 Constitution has been that, except for brief periods, there has been a single dominant party, the LDP, which had a far more successful electoral record than any other party. Although the LDP was short of a majority between 1993 and 1997, and short of an upper house majority between 1989 and the time of writing (early 1998), it has been the largest party in Parliament, by comfortable margins, ever since it was formed.

It is widely understood – though there are some who wish to modify the assertion – that government officials exert greater influence over government policy-making, and not only over implementation, than is normal in parliamentary democracies. This has been attributed to several factors, most notably a tradition of elitist bureaucratic rule by a highly trained core of officials; the fact that the occupiers in the late 1940s needed the government officials to administer their reforms for them, and therefore could not afford to engage in much restructuring of government ministries or purging of officials; and, finally, the factionalism and venality of party politicians, who seemed more concerned with jockeying for power than with devising and promoting workable programs of policy. An important mechanism for the propagation of bureaucratic influence was the tendency from the 1950s for government officials to "colonize" the LDP by retiring early and standing for Parliament in considerable numbers under the LDP banner. At times a quarter of LDP parliamentarians and one half of Cabinet ministers were former government officials, as well as several Prime Ministers. It would, however, be wrong to conclude from this that the bureaucracy monolithically controlled the LDP: interministerial rivalries were rife, so that cross-pressures abounded.

Every year some hundreds of senior government officials on retirement have taken positions on the boards of management of companies in the private sector, as well as running government instrumentalities of various kinds. This practice, known as *amakudari* (literally "descending from Heaven"), has been lightly policed, and arguably extends bureaucratic influence to many areas of economic activity. Another device is that of establishing government advisory councils (*shingikai*) to investigate specific policy issues. By inviting selected policy specialists and interested parties onto the *shingikai*, ministerial influence could be extended through the making of policy.

The lengthy period of LDP dominance of Parliament not unexpectedly led to the formation of coteries of parliamentarians affiliated with the ruling party who were possessed of impressive expertise and networks of influential contacts in certain policy areas. From about the 1970s these came to be known as "tribes" (*zoku*), and exercised much influence over policy in their chosen areas. It is somewhat misleading, however, to regard the emergence of *zoku* as evidence of the LDP getting its own back on the bureaucracy, since for the most part these policy "tribes" were intimately linked with networks including powerful officials as well as representatives of interest groups.

In examining the structure of power, is it important not to ignore huge concentrations of economic power in the private sector, in particular as represented by *keiretsu* (conglomerate) firms, institutions such as the Agricultural Cooperative Association (*Nokyo*), groups representing small and medium-sized businesses, professional associations, and so on. The picture is extremely complex, but these groups have all had their part to play in putting pressure on government, and ultimately in decision-making.

Even though "Westminster model" arrangements, coupled with long-term domination by a single party and a strong, self-confident, government bureaucracy, suggest the likelihood of firm, bold decision-making from the center, certain factors conspire to create relatively weak central leadership. Indeed, those searching for the location of power and responsibility for its exercise are apt to become frustrated. As Karel van Wolferen has put it: "in Japan, the buck simply keeps on circulating". The prime ministerial position, for instance, is arguably less effective in the exercise of decision-making than in comparable systems such as the United Kingdom. Prime Ministers change more than twice as frequently than in Britain (where there have been 11 since 1945, compared with 24 in Japan over the same period), and although two, Sato Eisaku (1964–72) and Nakasone Yasuhiro (1982–87), have enjoyed exceptionally long tenures, a two-year term has been common. Ministers also typically stay in post for rather short periods and Cabinet reshuffles are frequent. Cabinet meetings are often less important in making political decisions than the meetings of Permanent Vice-Ministers (*jimu jikan*) which precede them. Decision-making through the painful process of forging a consensus among a large number of interests is general.

Nor has the LDP ever been a centralized or monolithic party. The phenomenon of *habatsu* (usually translated "faction", but different in some respects from factions in most European parties), is of great importance. *Habatsu* have been semi-permanent organizations within the party – almost parties within a party (though not presenting candidates in elections under *habatsu* labels) – and have been crucial in fund-raising and in the distribution of party and cabinet office. The existence of *habatsu* in part explains the rapid turnover of Cabinet positions, with each *habatsu* seeking to satisfy its members by bargaining with the Prime Minister for its share of posts. Contests among them have also determined results in elections for the party presidency, which in turn, so long as the LDP has a majority, decides the identity of the Prime Minister. Relations among *habatsu* have at times been tempestuous, as for instance in 1980, when a Prime Minister (Ohira Masayoshi) was forced to resign and dissolve the House of Representatives as a result of factional struggles, and again in 1992–93, when an acrimonious split in the party's largest faction led to large-scale defections from the LDP, the loss of its majority in the consequent general elections, and its replacement in office by a multi-party coalition government.

Between 1947 and 1994 a system of election operated for the House of Representatives, whereby each elector exercised a single non-transferable vote, but constituencies elected several members of Parliament (typically three, four or five). There was also a kind of "negative gerrymander", which had resulted from a failure to redraw electoral boundaries in such a way as to reflect the massive movements of population from rural to urban areas in the postwar period of high economic growth. The value of a single vote in the remoter countryside thus became several times the value of a vote in parts of Tokyo, Osaka or other conurbations. The combined effect of large numbers of rural seats and multi-member constituencies, in which several candidates from the same party (the LDP) were struggling against each other as well as against candidates from other parties, led to a situation in which the principal instrument of electioneering at local level was a candidate's own personal support organization (*koenkai*). This imparted to LDP politics a character of the parish pump: Members of Parliament were primarily concerned with constituency business, and in particular with the task of attracting such benefits as business investment, infrastructure spending or construction projects to their constituencies. Matters of state tended to take second place.

For various interacting reasons, the power of money within the political world became conspicuous, and in particular from the late 1980s corruption or "money politics" came to be exposed by the media in numerous scandals. The term "structural corruption" was widely employed, in several senses, but especially in the sense that a parliamentarian simply could not afford to perform all the tasks required to be re-elected without resort to "corrupt" means. The law on political corruption was a "basket law" (*zaru ho*) with so many holes in it that "money politics" continued almost unchecked.

Japan has never enjoyed an effective two-party system, and certainly not a two-party system with alternation of parties in office. The nearest approach to the former was after the 1958 general election, when the LDP won 287 seats and the Socialists (the JSP) 166 seats, accounting for all but 14 seats in the lower house, but this was often called a "one and a half party system". Opposition parties subsequently fragmented and proliferated, so that five had substantial representation at the 1976 election, and they behaved more or less as a permanent opposition, with little hope of office. Nevertheless, they exercised considerable blocking power. They could block constitutional revision, which required a two-thirds majority in both houses separately, and also exercise influence over some other kinds of legislation. For this reason, from the 1970s elements in the LDP sought to do deals with certain parties of opposition, and in 1993 most of the non-Communist opposition suddenly found itself occupying government office, though this was to be the kiss of death for the JSP. Even so, up until the 1990s it was the destiny of the opposition parties to prevent wholesale dismantling of the immediate postwar political settlement, with its conspicuous elements of democracy and pacifism.

Change and Development in an Apparently Stable System

Although the 10 items discussed above might create the impression of an unchanging set of structures and practices, in fact by the 1990s much had changed.

The argument about whether it is the bureaucrats or the politicians who exert most influence over policy-making in part boils down to a question of change over time. A plausible case can be made out that up to the ending of the postwar period of rapid economic growth – roughly, in the mid-1970s – bureaucratic power was impressive and confidently wielded. It seems significant that Chalmers Johnson's influential book *MITI and the Japanese Miracle* only takes the story up to 1975. Since the middle of the 1970s, however, a more complicated pattern has emerged, in which, broadly speaking, more or less tightly articulated coalitions of interests, often including politicians, bureaucrats and representatives of relevant interest groups, are in competition for scarce resources with other coalitions of interests, similarly constituted. This being the case, the argument about whether it is the politicians or the bureaucrats, or indeed, as is sometimes argued, the corporate sector, that really control the destinies of Japan is probably the wrong question to ask.

Since the 1950s the wealth of the corporate sector has increased many times over, so that the extent to which government can control this sector has correspondingly declined. Japan-based multinationals have become a feature of the international economy, and are obviously much more independent of government than they were after the war. Although there is no great evidence of conflict between them, this does mean that government leverage over the corporate sector is much reduced.

The wide ideological gap between government and opposition that existed in the 1950s and 1960s gradually eroded until, with the ending of the Cold War at the close of the 1980s, the ideological map of Japanese party politics had fundamentally changed. Indeed, these changes were an important causative factor explaining the political developments of the 1990s. However, the breakdown of long-standing ideological cleavages between parties has given way to new sorts of cleavage, such as the argument between large and small government, which has substantially replaced the older left/right divide over peace and war issues.

In the 1950s and 1960s the range of influential interest groups was relatively narrow, and was disproportionately concentrated in the corporate and agricultural sectors. By the 1980s and 1990s it was much broader, reflecting the far greater sophistication of the economy. In some ways, this made it more difficult to impose economic discipline, since for political reasons so many interest groups now needed to be satisfied. This in turn was an important factor explaining the financial crises of the late 1990s.

Voting behavior up to the 1980s was relatively stable and predictable, although there was a gradual slippage of support for the LDP from the mid-1950s to the mid-1970s. From the 1980s, however, it became much more unpredictable. A salient feature of the 1990s was the decline in turn-out rates at elections, marking advancing political apathy and lack of interest in a political process which more and more electors found difficult to understand.

Corruption scandals have been a constant feature of Japanese politics throughout the postwar period, but from the "bubble economy" period of the late 1980s they apparently increased in frequency and scope, while public concern with them became much more marked. This in turn led to the reformist attempts of governments in the 1990s to clean up the political scene, although towards the end of the decade it seemed evident that the structural problems leading to money politics had not been fully addressed.

Over the years between the 1950s and the 1980s the LDP had gradually come to develop a rigid system of promotion by seniority, based on the number of times a parliamentarian had been elected. In the late 1980s it was calculated that nearly all LDP parliamentarians were appointed to a Cabinet position in their fifth or sixth term in Parliament, but not before (see Stockwin et al. 1988). The system paralleled, to a remarkable degree, the seniority promotion system which existed in government ministries. Resentment at the slow rate of promotion of high-flying and ambitious younger politicians was an important element explaining defections from the LDP in 1993 and the formation of new types of government. The seniority system also meant that, although former bureaucrats still stood for election on behalf of the LDP in considerable numbers, they did so at an earlier stage of their bureaucratic careers, and arguably developed more of an identity as politicians than their predecessors had been likely to do.

It should be evident from the above that by the beginning of the 1990s there had been sufficient slow evolution of the system inaugurated with the formation of the LDP in 1955 that a major reform of that system was required. Attempts to carry out such a reform have indeed been made during the 1990s, but it has been a turbulent passage, and at the time of writing the extent of its success remains unclear.

Politics in the 1990s: Attempts to Reform the System

Opinions differ on what particular combination of factors, and with what weightings, led to the unprecedented political turbulence of the 1990s. The succession of corruption scandals, and mounting public concern about them, are one explanation; the "bubble economy" of the late 1980s, and its collapse in 1990–91, another. Some writers find the collapse of the Soviet Union, the ending of the Cold War, and the emergence of the United States as the only surviving comprehensive superpower more than coincidental with political change in Japan. The Persian Gulf crisis and the war of 1990–91, which led to international criticism of Japan for contributing only money, not troops, threw unexpected new ingredients into the Japanese political soup.

All these things no doubt had some degree of impact on what was subsequently to happen, but to discover the main cause it is necessary to concentrate upon the evolving internal dynamics of the LDP itself. The story goes well back into the 1980s, when, for reasons too complicated to go into here, the *habatsu* led by the controversial former Prime Minister Tanaka Kakuei, and subsequently by Takeshita Noboru, was much the dominant grouping in the party, playing a controlling and "king-making" role. This *habatsu* was the principal backer of the nationalistic and, to an extent, reformist Nakasone Yasuhiro, who enjoyed an exceptionally lengthy term as Prime Minister, from 1982 to 1987. Takeshita himself, a seasoned politician skilled in the murkier arts of factional politics, succeeded Nakasone, but because of unpopular policy initiatives and the impact of a major shares-for-influence affair, known as the Recruit scandal, his popularity had sunk so low in the early months of 1989 that he was forced to resign. In elections for half the seats in the House of Councillors, held a few weeks later, the JSP, led since 1986 by Doi Takako, the first woman ever to lead a Japanese political party, actually won more seats than the LDP, which it forced into a minority position for the first time in the history of that party. Considerable numbers of women were elected under progressive labels, and, to some, it appeared to be a new dawn.

These successes were a flash in the pan, however, and although the JSP also did well in general elections for the House of Representatives early in 1990, the LDP under new leadership also bounced back, and for the first two or three years of the new decade appeared to have more or less re-established the *status quo ante*. The Gulf War, as mentioned above, was an embarrassing experience, but after complex political negotiations a bill passed through Parliament in 1992, enabling contingents from the SDF to participate in at least the non-military aspects of UN peacekeeping missions (see Chapter 19).

In negotiating this bill the LDP, lacking a majority in the upper house, had to bring on side some of the smaller parties in that house, in order to forestall blocking action against the bill. Ultimately, these moves succeeded, and also helped establish linkages that were to be important in forming the coalition government that followed the fall of the LDP in 1993. The LDP, however, continued to be rocked by scandals of various kinds. In the summer of 1992 Kanemaru Shin, now the central figure of the Takeshita *habatsu* and known as Japan's political "Godfather", was forced out of politics for involvement in massive corruption. This fatefully led the Takeshita *habatsu* to split. An ambitious reformist politician called Ozawa Ichiro led a large splinter group from the former *habatsu*, and this group defected from the LDP in June 1993 to form a new party. Another small group also defected at the same time. The government of Miyazawa Ki'ichi lost a motion of no confidence and was forced to resign. In the lower house general election held in July the LDP failed to retain its majority and was unable to persuade any other party to enter a coalition with it.

On August 9, 1993, the first non-LDP government for nearly 38 years was formed, composed of seven parties and one upper house grouping. The Prime Minister was Hosokawa Morihiro, who had been governor of Kumamoto Prefecture and had an aristocratic pedigree. He had formed his own party, the Japan New Party (*Nihon Shinto*), composed largely of young political outsiders, the previous year. It did well in the elections, as did Ozawa's grouping, the New Reform Party (*Shinseito*), whose members dominated the Hosokawa Cabinet. The largest party in the coalition, however, was the JSP, which suddenly found itself in government after 45 years in opposition. Its members were demoralized after losing many seats in the July election, and found participation in the Hosokawa government an uncomfortable experience.

The government, strongly influenced by Ozawa's ideas, quickly put together a program of reforms, consisting principally of four elements: reforming the electoral system, bureaucratic deregulation, administrative decentralization and economic reforms (particularly in relation to the taxation system). Its most significant achievement was radical reform of the electoral system for the lower house, although not all the details were worked out until after another administration had come into office in 1994. Under the new system, 300 members are elected by "first past the post" in single-member constituencies (the system used in Britain, the United States and Canada), while the remaining 200 members are elected in 11 regional constituencies according to a list system of proportional representation. Each elector has two votes, one for the local single-member constituency and one for the regional constituency in which it is located. Unusually, candidates can stand for election simultaneously in both constituencies and, through a rather complex formula, a good but not winning performance in the former can favorably affect a candidate's chances in the latter. Although a key purpose of the reform was to reduce the "pork barrel" character of elections and increase the programmatic content in electoral campaigning, the first election to be held under the new system (in October 1996) showed surprisingly little change in the character of electioneering, or in the pattern of results, from those under the old system (see Stockwin 1998). In conjunction with the reform of the electoral system, the law relating to corrupt electioneering practices was substantially tightened, although loopholes still remained.

By the early months of 1994 the Hosokawa government was showing serious signs of strain, consisting as it did of diverse parties, ideological positions and interests. Hosokawa unexpectedly resigned in April and attempts to form a successor government of substantially the same composition, under Hata Tsutomu of the New Reform Party, failed when the Socialists, having been snubbed by Ozawa, pulled out. The Hata government thus took office in a minority position and survived a mere nine weeks. At the end of June, to general amazement, an alternative coalition government was launched, consisting of the LDP, the JSP and a minor party. The new Prime Minister, the 70-year-old Murayama Tomi'ichi, was the Chairman of the JSP, elected to that position a mere nine months earlier. The logic of this unlikely coalition between former political enemies was to be found both in the convergence of policy positions between left and right that had taken place over several decades, and also in the fact that both represented sets of vested interests whose concerns substantially overlapped. The Murayama government, though it had a number of important achievements and lasted much longer than most observers initially expected, ironically laid the basis for the subsequent electoral collapse of the Socialists as a political force. Shortly after becoming Prime Minister Murayama perforce abandoned, and persuaded his party reluctantly to abandon, certain time-honored policy principles, notably opposition to the SDF and to the Japan/US Mutual Security Treaty. Rather than convincing the electorate that the Socialists were intent on modernizing themselves, this threw doubt on their credibility, and also contributed to a train of events which was to lead their party to split in 1996.

Meanwhile, nearly all the other parties that had supported the Hosokawa government amalgamated in December 1994 into a single party, which called itself in English the "New Frontier Party", but in Japanese *Shinshinto* ("New Progress Party"). This party initially managed some electoral successes, and for a while seemed set fair to become the alternative government party that Ozawa had set his heart on. Nevertheless, personality clashes, differences of ideology and organizational incompatibilities soon created internal strains. Ozawa was in many ways its guiding figure, but his abrasive personality and his tendency to take radical initiatives with little consultation, ultimately helped make the party unviable.

During 1995 politics was put on hold for a while because of the Hanshin earthquake that damaged Kobe, Osaka and other cities in January, and the poison gas attacks carried out in the Tokyo subway by the Aum Supreme Truth sect (*Omu Shinrikyo*) in March. These events exposed the inadequate character of government preparations and performance in disaster relief, and added to the voices calling for wholesale review and reform of government administration.

Murayama stepped down in January 1996 and was succeeded by the LDP's President, Hashimoto Ryutaro. The LDP, though lacking a parliamentary majority, had thus first returned to government, and then recovered the prime ministerial office for one of its own. In the general election in October 1996 the LDP won about 30 more seats, but was still short of a majority. The newly formed Democratic Party (*Minshuto*), did respectably well, while the New Frontier Party fell short of expectations and the Socialists, about half of whose parliamentarians had joined the Democratic Party, were nearly wiped out. In September 1997 the LDP, profiting from defections from other parties, regained the absolute majority in the lower house that it had lost rather more than four years earlier. Then in December of the same year the New Frontier Party finally disbanded and split into no less than six successor parties, some of which were seeking to form larger parties by congenial amalgamation. The big question for the future was whether the LDP, which also suffers from serious personality and ideological divisions, would once again split, permitting a radical re-formation of the party map, or whether an amorphous large party which fails to command much enthusiasm in the electorate would be confronted by a ragbag of small parties unable to mount effective opposition.

During the period since the formation of the first Hashimoto government at the beginning of 1996, various schemes of deregulation and financial reform have been announced, and given urgency by the financial crisis which had been simmering for some years but which broke out in dramatic fashion throughout East Asia in the latter half of 1997. As 1998 began, the political games which politicians had been engaged in for several years, in face of increasing electoral apathy, seemed suddenly as if they might entail serious destabilizing consequences.

Further Reading

Abe Hitoshi, Shindo Muneyuki and Kawato Sadafumi, *The Government and Politics of Japan*, Tokyo: University of Tokyo Press, 1994

An English translation of a standard Japanese textbook on the politics of Japan, consisting of sections on the national government, local government and intergovernmental relations, government in action, political participation and political culture.

Allinson, Gary D., and Yasunori Sone, editors, *Political Dynamics in Contemporary Japan*, Ithaca, NY: Cornell University Press, 1993

A multi-authored textbook including a number of case studies. Sections concern structural features of the polity, negotiating financial reform, negotiating the role of labour, and sites and modes of negotiation, together with a concluding chapter.

Calder, Kent E., *Crisis and Compensation: Public Policy and Political Stability in Japan, 1949–1986*, Princeton, NJ: Princeton University Press, 1988

The basic argument of this book is that much Japanese public policy can be understood in terms of a "crisis-compensation cycle". There are case studies on policy towards agriculture, the regions, small business, welfare, land use and defense.

Campbell, John C., *How Policies Change: The Japanese Government and the Aging Society*, Princeton, NJ: Princeton University Press, 1992

The author analyzes the immensely complex evolution of welfare policies towards the elderly in Japan, in comparison with those of advanced western nations.

Curtis, Gerald, *The Japanese Way of Politics*, New York: Columbia University Press, 1988

The book concentrates on evolution and change in Japanese party politics, in specific disagreement with those who focus on political stability under LDP dominance, and bureaucratic power.

Dore, Ronald, *British Factory – Japanese Factory: The Origins of National Diversity in Industrial Relations*, London: Allen and Unwin, and Berkeley, Los Angeles: University of California Press, 1973; reissued with a new Afterword, 1990

A renowned study of the differences, in management, labour relations and general ethos, between Japanese and British industrial factories in the early 1970s.

Johnson, Chalmers, *MITI and the Japanese Miracle: The Growth of Industrial Policy, 1925–1975*, Stanford, CA: Stanford University Press, 1982; Rutland, VT, and Tokyo: Charles E. Tuttle, 1986

Arguably the most influential and challenging book ever written in the field, this is a highly controversial challenge to analyses of Japanese economic policy-making based on the tenets of classical economic theory.

Kohno, Masaru, *Japan's Postwar Party Politics*, Princeton, NJ: Princeton University Press, 1997

This book examines in close detail a number of key episodes in the evolution of Japanese party politics since the war. It is a stimulating study of how politicians actually behave when confronted with crisis and change.

McCormack, Gavan, *The Emptiness of Japanese Affluence*, Armonk, NY: M.E. Sharpe, 1996

A radically critical account of policy-making and its consequences in Japan. It is particularly interesting on the role of the construction industry in the structure of Japanese politics.

Okimoto, Daniel I., *Between MITI and the Market: Japanese Industrial Policy for High Technology*, Stanford, CA: Stanford University Press, 1989

The author examines Japanese industrial policy since the war, especially that promoted by MITI, from a position of some skepticism about the extent of its responsibility for rapid industrial growth.

Ozawa Ichiro, *Blueprint for a New Japan: The Rethinking of a Nation*, Tokyo, New York and London: Kodansha International, 1994

This book, by the most famous – if controversial – politician working for reform of the political system in the 1990s, sets out his proposals for a freer and more market-oriented society.

Patrick, Hugh and Henry Rosovsky, editors, *Asia's New Giant: How the Japanese Economy Works*, Washington, DC: The Brookings Institution, 1976

This multi-authored work is a comprehensive analysis of the Japanese political economy in the 1970s, underpinned by a broad belief in the efficacy of the free market, even in Japanese politicoeconomic conditions.

Pempel, T. J., editor, *Uncommon Democracies: The One-Party Dominant Regimes*, Ithaca, NY: Cornell University Press, 1990

The book is a multi-authored comparison of four political systems where regimes of single-party dominance have continued at various periods for several decades. Apart from Japan, the studies concern Italy, Sweden and Israel.

Stockwin, J. A. A., Alan Rix, Aurelia George, Daiichi Ito and Martin Collick, editors, *Dynamic and Immobilist Politics in Japan*, London: Macmillan, and Honolulu: University of Hawaii Press, 1988

The underlying argument of this book is that, while in certain areas Japanese policy making exhibits considerable dynamism, elsewhere it is startlingly immobilist.

Stockwin, J. A. A., *Governing Japan: Divided Politics in a Major Economy*, Oxford: Blackwell, 1998 (forthcoming)

This book is a fundamental updating and rewrite of the author's textbook *Japan: Divided Politics in a Growth Economy*, London: Weidenfeld and Nicolson, 1975, 1982. The chronological analysis of events is greatly expanded, and the politics of the 1990s is covered in detail.

Van Wolferen, Karel, *The Enigma of Japanese Power: People and Politics in a Stateless Nation*, London: Macmillan, 1988, and New York: Knopf, 1989

This highly controversial book had a big impact when it was published. The author maintains that Japan has no real center of accountability, that the political system is like a "headless chicken" or "truncated pyramid". It has been criticized for the excessive closure of the analysis, but contains some interesting insights.

Watanabe, Osamu, *Nihonkoku Kempo "Kaisei" Shi* [A History of 'Revising' the Japanese Constitution], Tokyo: Nihon Hyoronsha, 1987

This book presents a fascinating analysis of Japan's postwar constitutional history, and in particular of unsuccessful attempts to revise the text of the Constitution but more effective "revision by reinterpretation".

J. A. A. Stockwin is Nissan Professor of Modern Japanese Studies at the University of Oxford.

Political Economy

Chapter Three

The Economic Role of the Government

Jennifer Amyx

There has long been a debate about the role that the government has played in Japan, relative to the private sector, in bringing about the economic outcomes of the postwar period: has it been a leader, a follower, or a partner of the private sector? There is little disagreement that Japan's era of high economic growth (1955–73) derived largely from increasing inputs, but there is much controversy over the contribution of the government in mobilizing and allocating the resources comprising these inputs. Japan's prolonged economic downturn in the 1990s has heightened this debate over the government's contributions, both positive and negative, to economic outcomes.

Interpretations of the government's role range from the "developmental state" model (see Johnson) – based on a theory of growth induced from above by a strong and prescient bureaucratic state – to those interpretations which portray government intervention, in the form of industry-specific policies, as of minor consequence in achieving rapid economic growth, or even suggest that growth occurred in spite of the government's meddling. Some proponents of this latter view emphasize the leading role which private sector dynamism played in bringing about favorable economic outcomes in the era of high growth. All sides seem to agree, however, on the government's role in crafting sound macroeconomic policies during the 1950s and 1960s, and its difficulty in getting such policies right in the 1990s. Yet even here the relative weighting of government discretion and exogenous economic shifts remains a point of contention.

I shall argue here that there is an element of truth to each of the interpretations mentioned above. Confusion among different models stems largely from failure to make important temporal and sectoral distinctions. There is no single model of the Japanese government's role in the economy: examination of the government's evolving rationale for, and means of, intervention across the postwar period suggests that multiple models exist. Thus, the role of the government at the apex of the high growth era differed significantly from the role it played in the succeeding decades of slowed growth. Further, the government's role in the economy has varied, within single periods, across industries. Its interventionist role in a industry such as shipbuilding contrasts sharply with its more "hands off" approach to industries such as consumer electronics. A variation in interpretations may also exist because some scholars view the government as an exogenous variable, and relationships within the Japanese political economy as hierarchical and antagonistic; while others see it as an intervening variable in the causal forces leading to economic outcomes, and therefore as a coordinator of interests, working with the private sector through networks extending to all levels of society.

The debate over the economic role of the government initially gained prominence because developing countries sought to emulate Japan, and even those countries that were already developed hoped to learn Japan's techniques for "miraculous" growth. Given the extreme variation in Japanese economic outcomes in recent decades, however, the debate has come to take on additional meaning. Today many look to Japan to discern impediments to sustained growth, seeking to learn instead from the nation's mistakes.

The Historical Context

Up to 1945 there was considerable government intervention in all areas of society, including the economy. In wartime Japan, many of the bureaucratic institutions that would be charged with responsibility for industrial policy in the immediate postwar period were already busy with national planning, to meet strategic objectives and to finance industries critical to the war effort. These agencies thus emerged at the end of the war with a reservoir of experience in the mobilization of resources and personnel towards the achievement of national objectives.

The legal authority and the personnel of the Ministry of Finance (MoF) and the Ministry of International Trade and Industry (MITI), the critical institutions that defined sets of constraints and possibilities for economic actors in the postwar period, were left relatively unscathed from the war. Their societal position was also effectively elevated: they were empowered in relation to private sector economic actors because the institutions and political forces with which they had had to contend in wartime were effectively removed from the scene, through the purges of large numbers of private sector officials and politicians under the Occupation, and the dissolution of the industrial conglomerates, the *zaibatsu* (see Chapter 10). The absence of significant countervailing forces placed MoF and MITI in unique positions to shape both business interests and the economic landscape in succeeding decades.

The situation after the surrender also facilitated the development of the government's many network relationships with the business and political worlds, which, arguably, facilitated the coordination and implementation of industrial policy for many years, but which have been cited in more recent years as impeding economic growth. With the bureaucracy remaining the least purged of all institutions, it had a monopoly on expertise in the early postwar years. The business world was eager to obtain some of this expertise and therefore provided post-retirement (*amakudari*) posts for ministry officials. Although this expertise lost its value once private sector institutions generated leaders and expertise from within, the *amakudari* connection continued to be useful in facilitating the flow of information between government and business, offering the business world a lobbying window to government and offering the government a means of participating in the management of business, and ensuring compliance to directives, at least to some extent.

However, such informal relational networks as *amakudari* have blurred the line dividing the public sector from the private sector. In doing so, they make it especially difficult to discern where the economic role of the government ends and the economic role of the private business sector begins. While the lack of clear demarcation between the public and private sectors did not seem detrimental to growth in the first half of the postwar period, "market failures" in these types of social exchanges have been on the rise in the 1980s and 1990s. Evidence for this has been provided by a growing number of scandals involving bureaucratic officials and private sector corporations.

Japan's historical experiences also set the stage for the prominence of the financial services industry as a tool of postwar industrial policy. The nation's "latecomer" status led to the government's postwar focus on heavy industries; the capital-intensive nature of these industries, in turn, led the government to prioritize banking within the financial system. With memories of the 1927 banking crisis still vivid in institutional memory (see Chapter 8), MoF perceived its role, as ensurer of financial stability and allocator of scarce resources, to be essential in the realization of national economic goals. Regulations marking the financial landscape, until recent years, were thus direct reflections of historical experiences and enabled the government to use financial measures as a prominent tool of industrial policy.

Industrial Policy

The term "industrial policy" refers to government intervention in the market to promote selected industries. MITI has been the subject of most studies of Japanese industrial policy, because it serves as the locus of industrial policy coordination and the main actor overseeing the government's role in the economy. While MITI designates priority industries, however, it is not the lone actor: the close cooperation of other ministries is also required for the actual implementation of many policy measures. Given the financial nature of many industrial policies, cooperation between MITI and MoF has been particularly important. While discussions of industrial policy tend to focus on the role played by bureaucratic agencies such as MITI and MoF, it is important to note that a great deal of government intervention has existed in the more domestic, non-traded industries as well. This intervention, however, has been orchestrated more by elected politicians in Parliament than by the bureaucracy, because of the electoral interests often linked to such industries (for example, see Chapter 5 on agricultural interests).

Rationale

Critics of industrial policy argue that political intervention disturbs market forces, harming competition. These individuals see vigorous market competition as part of the recipe for success. In reality, however, every country has an industrial policy to some degree. In the United States, for example, the defense industry has long received preferential government treatment because it has been viewed as vital to national security. In both the United States and Japan, industrial policy has been prompted by market failures, negative externalities, anti-monopoly concerns, public procurement needs, or the desire to promote regional dispersal of jobs and industrial locations. It has also been used by both nations as a means of adjusting to shifting international comparative advantage, or to altered national priorities (see Okimoto).

However, the motivations for Japanese government intervention in the economy have been far more numerous than these. While the perception that industrial policy has been the rule rather than the exception is an exaggeration, it has nonetheless played an important role. Government intervention has come not only as a reaction to market failures, but also in anticipation of market failures. It has also been executed to ease the fall of certain declining industries through encouraging the reduction of plant capacity. For growing industries, industrial policy has been seen as a means to create a comparative advantage. Fears of excessive competition have also motivated government intervention. Through strict regulation of market entry and exit, the government has ensured a degree of industrial orderliness.

Industrial policy has also been used as a means to distribute the costs, risks, and benefits of rapid economic growth and development. Government intervention on behalf of small and medium-sized enterprises, for example, has assisted them in coping with fluctuations in business cycles and exchange rates. Further, industrial policy has been seen as a means of attaining economic security, through its focus on securing raw materials and overseas markets.

In the legal realm, the government has exempted corporations in certain industries from anti-monopoly regulations. This, in turn, has given rise to industry cartels. Relaxed environmental and safety laws for designated industries have been another means of lowering industry costs and granting competitive advantage in the world markets. Weak control over the export of dual-purpose technology has been an additional legal component of industrial policy.

The trade component of industrial policy measures has included export tax credits, import tariffs and quotas on foreign goods, non-tariff barriers, exchange rate manipulation, tolerance for dual pricing and dumping, and information gathering by the government on foreign markets. Research and development have also been areas of government intervention on behalf of the economy. Here, industrial policy has taken the form of government contracts for research and development, the establishment of national laboratories, subsidies for national research consortiums, and patent policies for intellectual property and technology diffusion. Human resource development is a further aspect of industrial policy, and often takes the form of graduate research fellowships or government-funded training at national or overseas laboratories.

The Progression of Industrial Policy

The main goal of industrial policy in the immediate postwar period was catching up with the West. To carry this out, the government removed bottlenecks in the economy by supporting the reconstruction of critical industries such as electric power, shipping, steel and coal (see Nakamura). Government policies which prohibited the development of attractive alternative financial instruments also stimulated savings in this period. At a time when the demand for investment funds exceeded supply, the government intervened as well to allocate scarce capital to preferred industries and thereby facilitate investments viewed as falling within national interests. During the early part of this high growth era, the government both promoted infant industries with potential for rapid growth, and sheltered less efficient industries with a high capacity to absorb labor (see Raphael and Rohlen). The net outcome was significant aggregate growth. Industrial policy helped reduce risks and uncertainties, without totally removing the competitive element of the market.

The Law for the Rationalization of Industries, promoted by MITI and enacted in 1952, is a prime example of early legislation on industrial policy. It targeted iron, steel, steel rolling, oil refining, metals, chemical fertilizers, soda, and dyes; provided for subsidies for the upgrading of technology, and for loans of government-owned machinery and equipment; and shortened depreciation periods for certain types of equipment and facilities in these industries (see Nakamura). The Law was typical of many other general financial measures of government intervention, in that it also included favorable tax treatment and subsidies for designated industries. The Law also involved the injection of public funds into the building of infrastructure, such as improved road and port facilities. The government's role here was not simply to promote growth, but rather to promote equitably distributed growth. By building up appropriate industrial infrastructure in more rural areas, the government was able to attract industries to these regions, and thereby lessen the demographic shifts that might have otherwise taken place.

In the 1950s, several public finance corporations were established by the government to issue loans to designated industries. The Export-Import Bank, for example, provided long-term credit for exports, and the Japan Development Bank (JDB) provided funds for capital investment. An examination of the composition of the JDB's lending from 1951 to 1955 reveals that four industries received the largest proportion of loans, while all other industries combined accounted for less than 25% (see Nakamura). Banks such as the JDB and Export-Import Bank, in turn, were funded by the savings of individual Japanese through the postal savings accounts, which were funneled through the Fiscal Investment and Loan Program (FILP) in MoF's Finance Bureau.

The creation of these government financial institutions not only permitted the government to direct public money in the direction of targeted industries, it also facilitated the diversion of private sector financial institutional funds to these industries. This came about through indicative lending, wherein loans issued by government financial institutions to targeted industries indicated to the private sector which industries would be favored by the government in succeeding years. Such government signaling indicated that private sector investments into such industries were implicitly government-guaranteed and thereby virtually risk-free.

In the mid-1950s, indicative lending was a critical component of industrial policy. Since then, however, indicative lending has been of declining importance. The proportion of funds available to industry through the FILP has declined steadily, and leading corporations in targeted industries today are capable of meeting their financing needs through retained earnings, bank loans (both domestic and overseas), and the issuance of corporate stock and bonds (see Okimoto).

From 1955 to 1970 the Japanese economy had grown at an average rate of 9.7% a year. By the time that the high growth era ended, the economic role of the government no longer resembled the developmental state model, since, although industrial policy was still being used, it had only a marginal impact at best and a negative impact at worst (see Raphael and Rohlen). The economy then entered an era of what could be called "medium" growth, running at an annual average rate of 4.24% from 1971 to 1990, before falling to an average of just 1.42% from 1991 to 1995. Against this background industrial policy increasingly dealt with the unintended consequences of growth, such as industrial pollution, rather than with attaining growth itself. In this era, industrial policy also reflected a shift to targeting technologies rather than industries (see Vestal).

In the 1980s and 1990s, the development of human resources, and the facilitation of linkages between research institutions and industry, have gained in prominence, at the same time that many other measures have been phased out. In the latter half of the 1990s, the government has been especially keen to develop the kinds of links among government, industry and research institutions seen in "Silicon Valley" and elsewhere in the United States. Initial steps towards developing these kinds of cooperative arrangements have been made in Japan, at the government-owned Tsukuba University and other sites.

Just as the salience of particular industrial policy measures has altered over time, so have the industries targeted changed, as Japan has progressed through different stages of development. The textile industry, for example, was the first postwar target of government industrial policy, and the choice of this industry reflected Japan's learning from the experiences of other nations which had engaged in this industry with success at a similar point in their economic development. When newly industrialized countries, such as Taiwan or India, began to enter the textile industry as competitors, the Japanese government shifted its targeted export industries to steel and machines, such as cars, motorcycles, and trucks (see Reischauer).

Administrative Guidance

Although industrial policy may involve explicit legislation, clearly defining targeted industries or specific policy measures, the bulk of industrial policy in Japan has been carried out through administrative guidance (*gyosei shido*), a practice which involves government direction or inducement of certain private sector behavior to promote economic growth and development. The practice is made possible by the general policy-making pattern in Japan, wherein a ministry with authority over specific policy areas proposes legislation to establish a large policy framework. Once this legislation is passed, the ministry fills it with bureaucratic details through ministerial ordinances (*shorei*), official orders, and "notifications" (*tsutatsu*) which largely circumvent elected politicians and formal parliamentary proceedings. This process gives the bureaucratic agencies a significant amount of latitude in ensuring private sector compliance and permits flexible enforcement, a concept in striking contrast to the stringent application of a more limited industrial policy in the United States and some other western countries.

Administrative guidance in the postwar period encouraged targeted industries to compete at home but simultaneously lent them significant support in their competition abroad. Thus, the government created a distinctive dynamic of domestic competition and aggressive export promotion in these industries. Some, such as sewing machines, automobiles, and ships, were assisted virtually every step of the way. However, not all targeted industries were necessarily successful, and not all successful industries were necessarily targeted ones (see Okimoto).

Variation in Outcomes

Successful cases of Japanese industrial policy include shipbuilding, steel, electrical power, telecommunications, and semiconductors. These industries tended to be capital-intensive, involving heavy sunk investments, mass manufacturing, and standard technology. Further, they tended to be strategic, "upstream" industries, exerting significant effects on other parts of the economy. Industrial policy success in these industries may have benefited as well from Japan's latecomer status, and its opportunity to observe the successes and failures of other nations.

Among cases of mixed success and failure can be counted the computer, pharmaceuticals, and aircraft industries. Their functional requisites vary from those found in the industries that were successfully targeted. Computers, for example, require state-of-the-art technology, while pharmaceuticals require intensive research. For various reasons, the Japanese government has had less success fulfilling these needs. Manufacturing, which the government has proved adept at facilitating, plays a less critical role in the aircraft industry than it does in some of the more successful industries.

Those industries where significant government intervention did not result in enhanced competitiveness or efficiency include petrochemicals, software, agriculture, construction and distribution. Government targeting of the petrochemicals industry can be seen as a major miscalculation in retrospect. The industry's reliance on oil, combined with Japan's lack of domestic oil supplies, and the "oil shocks" which hit in the early and late 1970s, doomed this industry to failure. Unsuccessful policies in the software industry largely stemmed from the rigid regulation of the nation's financial services, which inhibited venture capital formation. As seen elsewhere, successful software industries are inextricably tied to start-up firms requiring such capital. Policy failures were also evident in efforts at industrial consolidation in the 1960s.

Agriculture, construction, and distribution are examples of industries where notorious diversions from public funds were not simply misjudgments, but had the acknowledged purpose of coddling industry, not for the enhancement of aggregate economic growth but for the equitable distribution of growth and the provision of employment, particularly to rural regions. These industries have survived on government subsidies and public procurement, as well as protection from international markets. The ministries in charge of these industries, such as the Ministry of Agriculture or the Ministry of Construction, tend to be deeply penetrated by political actors and interest groups. Unsurprisingly, the principle of equitable distribution of growth or employment tends to go hand in hand with electoral interests, and the cultivation of constituency support for elected politicians. The coexistence of these policies propping up inefficient industries with the nation's aggregate growth, at least through the 1980s, suggests to some that growth has been realized despite the state rather than because of it.

There have also been many untargeted industries which nonetheless became world competitors. Thus, we find that industrial policy has most definitely not been a prerequisite to success in the Japanese economy. Consumer electronics, precision equipment, and office automation have performed extraordinarily well on their own. The characteristics of these industries include market-driven demand, volume production, and price-sensitive, incremental innovation. In consumer electronics and automobiles, which emerged in the 1960s, state intervention may have actually hindered exports.

Impediments to Effective Industrial Policy in the 1970s, 1980s and 1990s

The rise in the value of the yen beginning in the 1980s, the maturing of the Japanese economy, the over-regulated domestic market, and the lure of the global marketplace have driven much of the production operations of large Japanese corporations overseas in recent years. For example, in 1995 Japanese companies manufactured more overseas than they exported from Japan, for the first time (see Hirsch and Henry). Many of these companies served as the pillars of national growth during much of the postwar period and were targets of industrial policy. Although they remain the best in the world at what they do, the Japanese economy does not benefit as it once did from their prosperity. Because much of their income today is attained overseas, it no longer flows into government coffers or contributes to domestic employment.

These are important developments, as government intervention in the economy in the earlier era of high economic growth was based on the assumption that the interests of the private sector and those of the government were congruent. Rising points of divergence between corporate and government actors have served as major impediments to the government's playing as effective a role in the economy as it did in earlier periods. Extraordinarily high levels of government debt, both in aggregate numbers and as proportions of GDP, as well as the 40-year record level of unemployment reached in the 1990s, are in striking contrast to Japan in the mid-1960s when the nation was at the apex of its growth, and government and private business interests were much more aligned.

Many of the industrial policy measures involving trade have also come under fire from Japan's trading partners, particularly the United States, since the 1980s, when Japan's trade deficit soared. For example, the talks between the United States and Japan under the Structural Impediments Initiative focused on non-tariff barriers, such as collusive bidding practices in the Japanese construction industry. Many of the trade measures of industrial policy have also been increasingly inhibited by Japan's participation in international institutions such as GATT or the OECD, where such practices are not permitted.

Inter-bureaucratic competition and conflict, divergence in goals and interests between MITI and companies, and inter-company competition have also made many of the industrial policy measures related to research and development relatively ineffective since the latter half of the 1970s (see Callon). In addition, political instability has affected the nature and effectiveness of government intervention in the economy. While the stability of the bureaucracy and its prominent role in industrial policy-making might suggest that instability in Parliament should matter little, history has shown that the Japanese policy-making dynamic often shifts in periods of crisis to one where politicians come to play a more conspicuous role (see Calder). The number of decisions of a magnitude requiring involvement by politicians has increased in the 1990s, just as stability within the political arena has decreased.

Political stability was indeed an important factor in allowing the government to implement long-term planning and make industrial policies more effective in an earlier period. The Liberal Democratic Party (LDP), which controlled the government from 1955 to 1993 (and again since 1996), made high growth the central component of its party platform, and provided a context within which policies encompassing long time horizons could be taken as reliable signals to markets of future trends (see Raphael and Rohlen). Although Japan's economic difficulties in the 1990s preceded the interruption of LDP rule, political instability since has not boded well for the economy. While the costs of proceeding with the status quo in the 1990s exceed those of change, a shifting government role in the context of a stagnant economy nonetheless requires decisions to be made about who will win and lose, decisions with electoral repercussions. The "positive sum" interaction among bureaucrats, politicians, and the private sector which was prevalent during the high growth era is no longer possible in the era of slowed or zero growth.

Finally, to the extent that industrial policy worked, it would not have worked so well if the government had not simultaneously gotten macroeconomic policy right and maintained financial stability. Of course, during the period of fixed exchange rates and closed markets, success in macroeconomic policy was a less formidable feat than in the decades since. In addition to industrial policy, many more general policies also helped the nation move along the road to economic growth. By 1951, for example, a tax system was in place that rewarded plant and equipment investment across industries with reduced taxes, thereby favoring the accumulation of capital (see Nakamura). Tax incentives for household savings were also in place by this time, subsidizing banks and industry throughout the postwar period.

Administrative Reform

Many of the industrial policies discussed above have provided further support for the claim (mentioned above) that there has been a blurring of lines between the government and other, private economic actors. A recognition of the potential dangers arising from this lack of clear demarcation between the public and private sectors is reflected in the prominent position that proposals for administrative reform (*gyosei kaikaku*) have occupied on Japan's political agenda in the 1980s and 1990s. The administrative reform movement ushered in by Prime Minister Nakasone Yasuhiro's government in 1982 represented an overt attempt by the private sector to limit the government's interference in their activities (see Allinson 1997). The movement then had a resurgence in 1993, when Hosokawa Morihiro formed the first non-LDP government since 1955, promising to "break the collusive relationships between politicians, bureaucrats and business".

Indeed, in the wake of the bursting of the speculative asset "bubble" in 1990–91, the negative effects upon the economy of such intimate relations among politicians, bureaucrats, and business were self-evident, as the nation entered its longest period of economic stagnation since 1945. The concept of administrative reform – the improvement of the organization, structure, and functioning of the executive branch of government – initially was revived by the Harbinger Party (*Shinto Sakigake*) in 1993, under the pretext of bringing about more democratic accountability in government. However, the LDP (from which the Harbinger Party broke away, only to enter a coalition government with it in 1994) soon shifted the emphasis to cost-cutting and reorganization to reflect new administrative needs (see Nakano in *Asian Survey*).

Deregulation

Deregulation measures were advanced significantly under Nakasone's government (1982–87), which used deregulation as a means to counter a "hawkish" foreign policy image by emphasizing domestic politics (see Otake). The pattern of government retreat from some policy areas in the 1980s and 1990s was assisted by the government's growing difficulty in regulating these areas. An increasing conflict of interests among the support groups comprising the LDP constituency put the government in the line of fire between interests, and therefore made it desirable for the government to reduce its regulatory visibility, to avoid being seen as favoring one supporting interest over another (see McKean).

One area affected in the 1980s was that of land use, when Prime Minister Nakasone used deregulation, in combination with fiscal policies, to stimulate investment in the construction industry. A number of problems eventually arose out of this deregulation effort, however, when land prices skyrocketed. Deregulation was thus effectively reversed in the late 1980s.

While the deregulation championed in 1980s implied a reduction in the government's role in the economy, it was often really re-regulation, or a different type of government role (see Vogel). It is important to note in this regard that in some cases, a heightened government role is actually necessary in one area to counter deregulation in another. A case in point is the deregulation of financial services. Giving private sector financial institutions greater freedom to act simultaneously heightened the incentives for imprudent lending decisions; thus, while regulations were relaxed, heightened monitoring by the government was a desirable and necessary accompaniment to deregulation, but was largely absent.

The exclusive nature of the relational networks linking government with industry in Japan also serves in the 1990s as an impediment to deregulation and the liberalization of markets. Outsiders, both domestic and foreign, find it difficult to penetrate this closed system, even after formal regulations are dismantled. While such network links arguably reduced the costs of coordination and information, and lent power to regulative efforts in an earlier period, instances of "market failure" in such relationships are more prominent today, suggesting that these networks needed to be altered or loosened for industries to adapt to new international exigencies.

Privatization, Public Corporations, and Government Ownership

Industrial policy in its most extreme form involves direct government ownership of the means of production. One area where one might expect to see such direct commercial participation by the government is in energy markets. However, although Japan's resource constraints are second to none and energy markets are an area in which government ownership remains the norm in many areas of the world, the Japanese government's active role in this sector is not as an owner but as a banker (see Chapter 4).

Japan is notable, however, for its relatively large number of public corporations (*kodan*), and other government-related "special legal entities" (*tokushu hojin*). These entities (listed in Table 3.1) were each established by the government through specific legislation, and each comes under the jurisdiction of one or more of the government's ministries or agencies. In total, they numbered 92 as of July 1, 1995. The Japanese government – which, confusingly, refers to all of them as "public corporations" in its English-language publications – defines a *tokushu hojin* as a body which:

> . . . is established primarily when particular activities are better managed in the form of a profit-making enterprise, when efficiency in performance is more likely to be achieved than under direct operation by the national government agencies, or when more flexibility in financial or personnel management is required than is normally possible under the laws and regulations pertaining [to] government agencies (quoted from Government of Japan).

Included in the list of 92 special legal entities are a number of companies which have recently been partially privatized, but are still at least 50% government-owned: Nippon Telegraph and Telephone (NTT), the eight companies formed out of Japan National Railways (JNR), and Japan Tobacco.

NTT used to have a monopoly on domestic telecommunications (while another government-owned company, KDD, was responsible for international services). In 1985, NTT was partially privatized and the government began to implement the gradual liberalization of telecommunications. The company remained 65.6% government-owned 10 years later (see Nakano in *Journal of Japanese Studies*), but a reduction in this proportion is expected to accompany the splitting of the company into three, planned for 1999.

Until 1987, JNR was also wholly publicly owned, but differed from most other state-owned rail networks in having to face competition from private railways within and between the major cities. Although its assets were liquidated, following its division into six Japan Railway (JR) companies responsible for regional passenger services, a freight company, and a separate corporation to settle JNR's debts, the transition to private ownership has not been smooth. The payment of the huge amounts of debt left behind continues to be an issue contested by the JR companies and the government. Much of the problem with privatization lies in the integration of the low-premium Railway Workers' Pension Plan into the higher-premium welfare pension (*kosei nenkin*) plan for corporate workers.

In 1994, Japan Tobacco followed NTT and the former JNR into partial privatization. Again, more than half of its shares remain in the hands of MoF, although further public offers are planned.

The total number of special legal entities increased in the high growth era and has remained much the same since. However, some corporations have been wholly privatized, notably Japan Air Lines (JAL), which was sold in 1987. In addition, as of July 1994 seven other corporations were designated as "privatized" (*minkan hojinka sareta*) but retained their status as *tokushu hojin* (and are therefore listed in Table 3.1). They are now more akin to private corporations, with less government regulatory involvement or financial support than upon their establishment.

The privatization or dismantling of some of the many public corporations and other *tokushu hojin* has became an issue for serious debate in the 1990s. Many, such as the JDB, are viewed as having served their purpose; many are also losing money. The FILP, which still funnels postal savings to these public corporations, no longer generates enough returns from these loans to keep pace with what it owes to postal savings depositors or pensioners, and the privatization of some of these entities would help pay off the debt owed to the FILP. In 1997, government advisory councils (*shingikai*) submitted proposals to privatize state insurance companies and the Post Office savings system. As of 1998, however, the government had not yet acted upon these proposals. Privatization of postal services, postal savings, and postal insurance would involve a spectacular political battle if it was ever seriously pushed forward, because many powerful LDP politicians find continued government ownership to their electoral benefit. Among the 24,000 post offices across the nation, 75% are "special" post offices, where the postmaster is appointed by those in the area with considerable financial resources. These appointed postmasters comprise critical components of many politicians' rural political machines. In addition, a weak stock market might not be able to absorb the issuance of new shares, thus making this a difficult task.

Conclusion

Clearly, there are difficulties in arriving at a single model of the Japanese political economy and in attributing economic outcomes to a monocausal explanation of the government's role in the economy. Particular measures, industries targeted, and degrees of government intervention have varied, depending on the stage of development reached at particular points in time. Discussions of Japan's rapid economic growth and subsequent slowdown must also take into account factors other than the government, such as Japan's latecomer status, the international system, the global economic climate, market forces, culture, the organizational strength and business structures of Japanese firms, labor/management relations, and the nation's high savings rates.

A proper understanding of the economic role of the Japanese government, however, must also be situated in a cross-national comparative context. Some aspects of the Japanese government's role in the economy are commonly found elsewhere, while other aspects appear distinct. Those countries sharing the largest number of characteristics similar to Japan's appear to be nations in East Asia, such as South Korea, and particular nations in western Europe, such as France. A strong role for the central bureaucracy in the economy has been found not only in Japan, but also in France, the United Kingdom, and many countries in Asia. Extensive financial regulation, and the use of financial regulation as a tool of industrial policy, are also common to France and Asian nations such as South Korea. The blurring of boundaries between the public and private sectors is common as well throughout Asia, and in France and Italy.

Indeed, the presence of industrial policy or even some of the primary characteristics of its supporting government institutions are not what make the economic role of the Japanese government distinctive. Distinctions are found in some of the particulars surrounding implementation, however, and in the blend of all of the various characteristics (see Okimoto). French governments have tended to choose national champions with guaranteed national and government markets, for example, while Japanese governments promoted competition among a restricted number of companies. This aspect of controlled competition in Japan may be one reason why Japan attained higher levels of economic performance during its high growth era than France did. Japanese industrial policy has been distinctive in that it has involved mechanisms for cooperation between government and industry, while at the same time promoting competition within industries. Thus, mechanisms for competition have been integrated into government policies, reflecting a commitment by the political elite to "market-conforming" methods of intervention in the economy.

The Japanese government faces difficulty in the 1990s in disentangling itself from industrial policy, now that many of the initial rationales outlined above are no longer present. The need for the temporary allocation of resources in the 1950s led to government practices which then became ingrained in the political culture, making it difficult to halt them once the external environment had changed and the efficiency of such policies had declined. The extent to which the government's historical role in the economy has become institutionalized and remains in place today is perhaps another distinctive characteristic, viewed from a cross-national perspective.

In conclusion, the causal link between Japan's industrial policy and economic growth is difficult to prove unequivocally: there has clearly been an important economic role played by the government, but scholars have yet to quantify this contribution. The qualitative evidence from industrial policy in Japan provides us with a mixed picture and suggests that Japan's period of rapid economic growth was the product of a complex set of factors. By helping targeted industries, industrial policy can simultaneously end up hurting others. In most cases of industrial policy in Japan, the government influenced the industrial structure and the pattern of economic growth, but did not control it.

Two analogies may prove illustrative here. First, we might liken the Japanese state to a referee in an athletic contest, being particularly careful not to confuse it with the winning team or the owners (see Okimoto, for example). We might also, or instead, liken the government's role to that of the conductor of a symphony orchestra, carefully distinguishing its role from that of the composer: the government has brokered, but not determined, economic outcomes (see McKean, for example).

Japan is experiencing in the 1990s its most prolonged period of economic stagnation since World War II, following the bursting of the speculative asset bubble in 1990–91. This suggests that a different economic role is needed for the government than that being fulfilled at present. Ironically, as Japanese domestic forces push for further deregulation, privatization and a shifting role for government, foreign governments call upon Japan to intervene more, in order to stimulate the domestic economy in light of the financial crisis in East Asia. It seems that a consensus on the proper economic role of the government has yet to be found.

Further Reading

Allinson, Gary, *Japan's Postwar History*, London: UCL Press, and Ithaca, NY: Cornell University Press, 1997

An integrated analysis of Japan's social, economic, and political changes since 1945. Chapter 3, "Growth, 1955–1974", outlines the preconditions for growth, processes of growth, and actual economic performance of the period 1955–74, and contains a number of fresh insights.

Allinson, Gary D., and Yasunori Sone, editors, *Political Dynamics in Contemporary Japan*, Ithaca, NY: Cornell University Press, 1993

Allinson and Sone bring together six Japanese and six US scholars to examine Japanese domestic political issues which have become controversial since the 1970s.

Calder, Kent, *Strategic Capitalism: Private Business and Public Purpose in Japanese Industrial Finance*, Princeton, NJ: Princeton University Press, 1993

An analysis of Japanese industrial finance which concludes that the dynamism of the private sector was integral to the achievement of industrial transformation.

Callon, Scott, *Divided Sun: MITI and the Breakdown of Japanese High-Tech Industrial Policy, 1975–1993*, Stanford, CA: Stanford University Press, 1995

An analysis of the relationship between MITI and Japanese high-technology firms which finds widespread breakdown in MITI's industrial policy for the sector.

Government of Japan, *Organization of the Government of Japan*, Government of Japan, 1996

An overview of the organization and function of the executive branch of the Japanese government, public corporations and other government-related bodies, and advisory councils.

Hirsch, Michael, and E. Keith Henry, "The Unraveling of Japan Inc.: Multinationals as Agents of Change", in *Foreign Affairs*, March/April 1997

Hirsch and Henry argue for a disjuncture in interests in the 1990s between the Japanese government and Japanese multinational corporations.

Ikuta Tadahide, *Kanryo: Japan's Hidden Government*, Tokyo: NHK Publishing, 1995

Insights into the network of relationships extending from the central government bureaucracy.

Johnson, Chalmers, *MITI and the Japanese Miracle: The Growth of Industrial Policy 1925–1975*, Stanford, CA: Stanford University Press, 1982; Rutland, VT, and Tokyo: Charles E. Tuttle, 1986

A groundbreaking work on Japan's economic bureaucracy which develops the concept of the "developmental state".

Kaplan, Eugene, *Japan: The Government-Business Relationship*, Washington, DC: US Bureau of International Commerce, 1972

A classic articulation of the concept of "Japan, Incorporated".

McKean, Margaret, "State Strength and Public Interest", in Allinson and Sone, cited above

This chapter argues that the mobilization of new domestic interests and the coordinating role of the state often lead to policies serving Japanese public interests.

Nakamura Takafusa, *Lectures on Modern Japanese Economic History, 1926–1994*, Tokyo: LTCB International Library Foundation, 1994

This book traces the evolution of the Japanese economy through the Showa period (1926–89) and into the 1990s, arguing that the economic institutions developed in the immediate prewar and wartime periods laid the groundwork for the economic success which followed.

Nakano Koichi, "Becoming a 'Policy' Ministry: The Organization and *Amakudari* of the Ministry of Posts and Telecommunications", in *Journal of Japanese Studies*, Volume 24, number 1, Winter 1998

Nakano examines the system and role of *amakudari* in this important ministry.

Nakano Koichi, "The Politics of Administrative Reform in Japan, 1993–8: Toward a More Accountable Government?", in *Asian Survey*, March 1998

Nakano here stresses the role played by the Harbinger Party in providing the initial definition of administrative reform as more democratic accountability in government, and placing it on the political agenda, following the interruption of LDP one-party rule in 1993.

Okimoto, Daniel I., *Between MITI and the Market: Japanese Industrial Policy for High Technology*, Stanford, CA: Stanford University Press, 1989

A very accessible work which casts MITI within the broader context of the system in which it operates, highlighting its skillful management of political and business relationships, and offering a typology of political exchanges in the Japanese political economy, but at the same time underscoring the costs of industrial policy in Japan.

Otake Hideo, "The Rise and Retreat of a Neoliberal Reform: Controversy Over Land Use Policy", in Allinson and Sone, cited above

Otake argues that neoliberal reforms surprisingly contributed to the escalation of land prices in Japan, thus producing a less than desired outcome.

Raphael, James, and Thomas Rohlen, "How Many Models of Japanese Growth do we Want or Need?", in Henry Rowen, editor, *Behind East Asian Growth: The Political and Social Foundations of Prosperity*, London and New York: Routledge, 1998

The authors consider some of the issues encountered when attempting to extrapolate a model for economic growth based on Japan's experiences.

Reed, Steven, *Making Common Sense of Japan*, Pittsburgh, PA: University of Pittsburgh Press, 1993

A section of this book argues that administrative guidance is effective in Japan, even though the bureaucracy is weak, because the bureaucracy plays the role of mediator in "prisoner's dilemma" situations, such as excessive competition, where a mediator is needed in order to achieve the optimal outcome.

Reischauer, Edwin O., *The Japanese Today*, Cambridge, MA: Belknap Press, 1988

An overview of various areas of Japanese society, including the state's role in the economy and Japan's political institutions.

Samuels, Richard J., *The Business of the Japanese State: Energy Markets in Comparative and Historical Perspective*, Ithaca, NY: Cornell University Press, 1987

Samuels examines the industrial policy debate and Japanese state intervention in markets in the context of energy policies, arguing that a "reciprocal consent" dynamic best represents the relationship between government and business.

Vestal, James, *Planning for Change: Industrial Policy and Japanese Economic Development, 1945–1990*, Oxford: Clarendon Press, and New York: Oxford University Press, 1993

A major contribution to a better understanding of the character and evolution of industrial policy. Vestal notes the coupling of pro-growth measures, aimed at stimulating selected industries, with anti-growth measures, aimed at blocking competition, rationalization and growth in other designated industries.

Vogel, Steven K., *Freer Markets, More Rules: Regulatory Reform in Advanced Industrial Countries*, Ithaca, NY: Cornell University Press, 1996

Vogel contributes to an understanding of the dynamics behind regulatory policy changes in Japan, the United States, and the United Kingdom, by identifying the forces for change, actual changes made, and the politics driving regulatory reform.

Woodall, Brian, *Japan Under Construction: Corruption, Politics, and Public Works*, Berkeley, Los Angeles: University of California Press, 1996

An examination of the political dynamics underlying Japan's construction industry, which pays particular attention to the *dango* system of collective bidding in the public works market in Japan.

Zysman, John, *Governments, Markets, and Growth: Financial Systems and the Politics of Industrial Change*, Ithaca, NY: Cornell University Press, 1983

An analysis of the role that financial systems play in the conduct of industrial policy in Japan, France, Germany, Britain, and the United States.

Jennifer Amyx is a doctoral candidate in the Department of Political Science at Stanford University.

Table 3.1 Special Legal Entities (*Tokushu Hojin*), as of July 1, 1995

MITI
Japan National Oil Corporation,[1] Metal Mining Agency of Japan,[2] Coal Mine Damage Corporation,[2] Japan Small Business Corporation,[2] Electric Power Development Company, Limited,[3] High Pressure Gas Maintenance Association,[4] Japan Electric Meter Standardization Center,[4] Japan Bicycle Racing Association, Japan External Trade Organization (JETRO), Institute of Developing Economies, Japan Motorcycle Racing Association, New Energy and Industrial Technology Development Organization

MITI and MoF
Tokyo Small Business Investment Promotion Company, Limited,[4] Nagoya Small Business Investment Promotion Company, Limited,[4] Osaka Small Business Investment Promotion Company, Limited,[4] Small Business Credit Insurance Corporation,[5] Japan Finance Corporation for Small Business[5]

MITI and ministries/agencies other than MoF
Japan Regional Development Corporation,[1] Water Resources Development Corporation,[1] Japan International Cooperation Agency (JICA),[2] Japan National Railways Settlement Corporation,[2] Pollution-Related Health Damage Compensation and Prevention Association

MoF
Japan Tobacco (JT), Incorporated,[3] People's Finance Corporation,[5] Housing Loan Corporation,[5] Japan Development Bank (JDB),[6] Export-Import Bank of Japan,[6] Central Bank for Commercial and Industrial Cooperatives[6]

MoF and other ministries or agencies
Central Cooperative Bank for Agriculture and Forestry,[4] Agriculture, Forestry and Fisheries Finance Corporation,[5] Environmental Sanitation Business Corporation,[5] Japan Finance Corporation for Municipal Enterprises,[5] Hokkaido-Tohoku Development Corporation,[5] Okinawa Development Finance Corporation,[5] Fund for the Promotion and Development of the Amami Islands

Ministry of Transport (MoT)
Japan Railway Construction Corporation,[1] New Tokyo International Airport Authority,[1] Japan Shipbuilding Industry Foundation,[1] Kansai International Airport Company, Limited,[3] Hokkaido Railway Company,[3] East Japan Railway Company,[3] Central Japan Railway Company,[3] West Japan Railway Company,[3] Shikoku Railway Company,[3] Kyushu Railway Company,[3] Japan Freight Railway Company,[3] Railway Development Fund, Maritime Credit Corporation, Japan National Tourist Organization

Ministry of Construction
Japan Highway Public Corporation,[1] Metropolitan Expressway Public Corporation,[1] Hanshin Expressway Public Corporation,[1] Japan Workers' Housing Association

Ministries of Transport and Construction
Honshu-Shikoku Bridge Authority,[1] Housing and Urban Development Corporation,[1] Teito Rapid Transit Authority[7]

Ministry of Posts and Telecommunications (MPT)
Postal Life Insurance Welfare Corporation,[2] Kokusai Denshin Denwa Corporation (KDD), Limited,[3,8] Nippon Telegraph and Telephone Company,[3] Nippon Hoso Kyokai (NHK)[9]

Ministry of Agriculture, Forestry and Fisheries (MAFF)
Japan Agricultural Land Development Agency,[1] Forest Development Corporation,[1] Livestock Industry Promotion Corporation,[2] Japan Raw Silk and Sugar Price Stabilization Agency,[2] Japan Racing Association, National Association of Racing, Mutual Aid Association for Personnel of Agriculture, Forestry and Fishery Organizations

Table 3.1 Special Legal Entities (*Tokushu Hojin*), as of July 1, 1995 (Continued)

MAFF and Prime Minister's Office
Northern Territories Issue Association

Ministry of Health and Welfare
Social Welfare and Medical Services Corporation,[2] Pension Welfare Service Public Corporation,[2] Social Insurance Medical Fee Payment Fund, Social Development Research Institute, Association for the Welfare of the Mentally and Physically Handicapped, Farmer Pension Fund

Ministry of Labor
Employment Protection Corporation,[2] Smaller Enterprise Retirement Allowance Mutual Aid Corporation,[2] Labor Welfare Corporation,[2] Japan Institute of Labor, Mutual Aid Association for Construction, Sake Brewing and Forestry Retirement Allowances

Ministry of Education
Japan Scholarship Foundation, Mutual Aid Association of Private School Personnel, National Education Center, Japan Arts Council, Japan Society for the Promotion of Science, Japan Private School Promotion Foundation, National Stadium and School Health Center of Japan

Ministry of Education and MPT
University of the Air Foundation

Ministry of Foreign Affairs
Japan Foundation

Ministry of Home Affairs
Japan Firefighting Standards Association,[4] Mutual Aid Fund for Official Casualties and Retirement of Volunteer Firemen

Economic Planning Agency
Overseas Economic Cooperation Fund (OECF), Japan Consumer Information Center

Science and Technology Agency
Power Reactor and Nuclear Fuel Development Corporation,[2] Research Development Corporation of Japan,[2] Japan Atomic Research Institute, Japan Information Center of Science and Technology, Institute of Physical and Chemical Research

Science and Technology Agency, MoT, and MPT
National Space Development Agency of Japan[2]

Environment Agency
Japan Environment Corporation[2]

1 A "public corporation" (*kodan*)
2 A "business corporation" (*jigyodan*)
3 A "special corporation" (*tokushu gaisha*)
4 A "privatized special legal entity" (*minkan hojin sareta tokushu hojin*)
5 A "finance corporation" (*koko*)
6 A "special bank" (*tokushu ginko*)
7 The only remaining example of the government's "management foundations" (*eidan*), created before and during World War II; in charge of part of the Tokyo subway system
8 Literally, "International Telegraph and Telephone"
9 Literally, "Japan Broadcasting Association"; responsible for public service radio and television

Sources: Adapted from Government of Japan, *Organization of the Government of Japan*, 1996, pp.100–3; and Gyosei Kanricho [Management and Coordination Agency], *Gyosei Kiko-zu* [Map of the Administrative Structure], Tokyo: Sorifu [Prime Minister's Office], 1995, p. 248

Chapter Four

Sources and Uses of Energy

Richard J. Samuels

The Japanese have had to devise ways of coping with their relative lack of domestic sources of energy ever since industrialization began. The problem has intensified in the postwar period, as rapid economic development has brought exponential increases in demand for energy, just as the long-established hydroelectric and coalmining industries have come to play ever smaller roles in the economy. Hydroelectric power was already fully installed by the 1950s, before the economic "miracle" of the 1960s, and today accounts for less than 4% of the country's primary energy supply. Domestic supplies of coal, which were already inadequate by the 1960s, covered barely 8.5% of consumption. Other, newer sources of energy have been tapped, but only to a strictly limited extent. Thus, Japanese oilfields do not produce enough petroleum to meet even 1% of total demand, and such recently developed alternatives as geothermal power or synthetic fuels cannot yet provide more than 2% of energy consumption.

Accordingly, Japan's dependence upon imported supplies, which account for around four fifths of its consumption of primary energy, is significantly greater than that of any other industrialized country. In particular, it imports more petroleum than any nation except the United States. This dependence on foreign sources of energy is compounded, in the eyes of Japanese policy-makers, by their longstanding perception that there is great uncertainty in the global resources markets. Thus, the economy itself has become a major consideration in the making of policy on national security, within which energy stands alongside international trade as a leading concern. Indeed, national security has come to be defined more broadly than in some other industrialized countries, since Japanese policymakers perceive a pressing need to guarantee access both to advanced technology and to raw materials, in order to secure what they refer to as "comprehensive security". This requires continuing creativity in acquiring and maintaining stable primary supplies. In addition, policy-makers have ensured that Japanese industries and individuals alike pay a premium for their consumption of energy, which encourages them to be as efficient as possible in their uses of these relatively scarce resources. As a result, the economy is extremely energy-efficient, producing 16% of global wealth by consuming just 6% of global energy.

Another notable feature of Japan's energy use is that, to a greater extent than in any other country, the consumption of energy has been dominated by large industrial consumers rather than by households, which account for a remarkably small proportion of total national consumption. As recently as 1973, a quarter of Japanese energy use was for electricity generation, nearly half was used in the manufacturing sector, and transportation accounted for little more than 10%. By the 1990s, however, the share for manufacturing was still below 50%, while the share for transportation had increased to 20%.

Patterns of Energy Use

The ways in which energy is used in Japan, and the sources from which it is derived, have changed dramatically over the decades since World War II, but especially following the first global oil crisis that began in October 1973, in connection with the Third Arab/Israeli War. This is widely referred to in Japan as the first oil "shock" (*shokku*), reflecting the impact on the economy of two key events in that month (see Sampson). First, the 13 member countries of the Organization of Petroleum Exporting Countries (OPEC) initiated a series of sharp rises in the price of crude oil. Second, the 10 member countries of the Organization of Arab Petroleum Exporting Countries (OAPEC) – seven of which were then also members of OPEC – began making cuts in their production of oil for supply to selected countries, including Japan.

Before the crisis, the use of energy had increased at the dizzying rate of nearly 25% a year. Consumption grew from 41 million tonnes of oil equivalent in 1950 to 340 million tonnes in 1973. Soaring consumption was reflected in the changing market shares of different fuels. In the 1950s, Japan tried to achieve at least a degree of self-sufficiency by relying on domestic coal and hydroelectricity, but the availability of hydroelectricity was limited by nature, and the usefulness of domestic coal was limited by the high cost of extracting it. Japan had had a significant coal industry long before World War II, but by the 1950s relatively cheap imported crude oil threatened to overwhelm domestic coal. Responding to political pressure from the coal industry and from the coalminers' unions, politicians and the Ministry of International Trade and Industry (MITI) attempted to restrict petroleum imports. However, the industries using oil were reluctant to take part in a program that would raise fuel prices, and MITI proved unable, even with the substantial influence that it wielded at that time, to impose its preference for coal. Thus, between 1950 and 1973 oil increased its share of total primary energy consumption from less than 5% to 75%. By the time that the OPEC and OAPEC decisions began to take effect, around 90% of Japan's oil came from sources in the Middle East, notably including supplies through the Arabian Oil Company, which had been created in 1957 under an agreement with Saudi Arabia and Kuwait (see Sampson). The sheer volume of energy consumption, and the crucial role of oil supplies in Japan's postwar growth, are strikingly indicated by the fact that Japan's consumption of oil in an average week in 1973 was around the same as its consumption for the whole of 1941.

Meanwhile, the share of coal had dropped from nearly three quarters of the total to less than one quarter between 1950 and 1973 (see Table 4.1). Once the political obstacles had been overcome, typically by fiscal measures to provide generous compensation of displaced workers and firms, Japan's postwar industrialization and rapid economic growth proceeded with easy access to relatively cheap energy, particularly imported oil.

The proportions of total energy use supplied by different fuels changed soon after the oil shock of 1973–74. Yet the change came about less as a result of the oil shock itself than in response to successful efforts – motivated by the longstanding concern for "comprehensive security" in general – to diversify energy supplies and reduce dependence on oil. A leading role was played in these efforts by the Agency for Natural Resources and Energy. This agency had been established under MITI in 1973, before oil prices were increased, but it soon became the main vehicle for the government's effort to organize the national response to the oil shock, especially after the OECD had rejected Japan's proposal for a multilateral body that could share the dwindling oil supplies among that organization's member countries (see Sampson). Through the new agency, MITI collaborated with the leading energy interest to develop nuclear power, import liquefied natural gas (LNG), and promote the use of liquefied petroleum gas (LPG). Nuclear power was developed in part to provide a degree of independence from foreign energy resources, while the development of LNG and LPG was motivated by a desire to reduce the pollution caused by the burning of oil for power generation. Nuclear power and LNG together accounted for only 7% of energy consumption in 1973, but by 1995 they supplied more than 20% of total demand, split almost evenly between them. The increasing use of nuclear power and gas came about largely at the expense of oil, which accounted for 75% of the market in 1973 but for only 56% in 1995. Over the same period, the market share supplied by coal fell slightly, to 17%, and that supplied by hydroelectricity declined to 4% of total energy demand.

Given the changes that were occurring even before the first oil shock, Japan was in a position to respond with considerable success to the crises of the 1970s. The short recession and high inflation of 1974 led to some significant political uncertainty, and the second oil shock, beginning after the Iranian revolution in 1979, saw a tripling of the price of crude oil. Yet the rate of economic growth rapidly returned to the highest level in the industrialized world, at least until the economic "bubble" burst in the 1990s (see Chapter 1). Although economic output more than doubled between 1973 and 1996, the "energy intensity" of the economy, as measured by representing energy as a percentage of GDP, declined considerably. Japan's energy intensity is now the fifth lowest among OECD members (following Austria, Denmark, Italy, and Switzerland). A range of measures was taken between 1973 and 1996, by MITI and the major corporations, to achieve greater efficiency and security. Japan moved much of its energy-intensive industrial capacity offshore, stepped up the development of electronics production and other industries with relatively low energy requirements, and diversified its sources of primary energy. In particular, Japan reduced its dependence upon petroleum, from more than three quarters of its primary energy supply in 1973 to just 54% in 1996. The Middle Eastern sources that had supplied 90% of Japan's petroleum up to 1973 were supplying around 74% by 1996, and more oil was being bought from the United States, Mexico, Nigeria and other alternative suppliers. Between 1973 and 1996, the proportion of primary energy supplies accounted for by the consumption of coal actually increased slightly, while the use of LNG and nuclear power both increased more than 10-fold. Private firms and public agencies together maintain the world's largest stockpile of petroleum (at 142 days), and, despite slack oil prices in recent years, Japan has maintained a vigorous program of research and development. Conservation has succeeded better in Japan than in any other industrial democracy: in 1989 Japan needed only 63% as much energy to produce the same GNP as in 1973.

Energy-producing Industries

Japan's oil, coal, gas, and electric power industries have virtually always been privately owned, in striking contrast with European countries, where state-owned energy monopolies were routinely established as "national champions". Government-owned entities, such as the Electric Power Development Company or the Japan National Oil Corporation, are each limited by law to assisting or subsidizing the private energy firms. Each energy-producing industry has distinctive structural features.

Oil

The oil industry in Japan has never been fully integrated from "upstream" (extraction of crude oil) to "downstream" (refining and distribution of oil and oil products). As a result of prewar agreements and conflicts among refiners, suppliers, and foreign interests, it has been vertically truncated instead. Few Japanese refiners process their own crude oil, and, while their number increased in the 1990s, there are few Japanese-controlled oilfields, at home or abroad. Refiners downstream, organized as the Petroleum Association of Japan, have economic and political interests which differ from those of the oil-producing firms upstream, organized as the Petroleum Development Association. These two distinct industrial groups often have failed to coordinate their political support for oil policy.

Even though considerable efforts have been made to consolidate the refining industry, with 29 separate firms it remains the most fragmented heavy industry in the Japanese economy. The industry is divided into firms that are affiliated with foreign capital and those that are wholly domestic. Only two Japanese refiners are wholly foreign-owned, and firms with at least minority foreign equity account for half of Japan's refining capacity. The struggle to domesticate this industry has been a central theme of Japanese energy policy in the 20th century, and has divided the industry politically. Consequently, the sale by Caltex of its shares in Nippon Petroleum Refining to Nippon Oil, Japan's largest refiner/distributor, for more than US$2 billion in 1996 represented a watershed in the history of the industry.

In the interests of improving efficiency, MITI has encouraged some deregulation in the petroleum industry (see Choy, February 1997). For most of Japan's industrial history, government policy required that refining be done "at the doorstep of the consumer", a euphemism for a strategy of capturing maximum value added and security simultaneously. In 1986, MITI, under pressure from foreign oil-producers which wanted to integrate downstream, agreed to change this policy and to license the import of refined petroleum products. It granted licenses, however, only to firms that already possessed capability for refining and storage. Effectively, this meant that licenses were granted only to members of the existing oil oligopoly. In 1996, MITI relaxed that requirement and stipulated instead that firms must have the capability of quality control testing and at least 70 days' worth of storage. This change in the regulations opened the Japanese market a little further, intensifying competition among refining firms and cutting deeply into their earnings. Once again, they felt significant pressure to restructure, but they have continued to ignore it.

Coal

The coal industry has also undergone major structural adjustment. For years its division into large and small firms frustrated the implementation of a coordinated energy policy. The large mines were controlled by the *zaibatsu*, the conglomerates that dominated the Japanese economy until their dissolution under the Occupation. The most prominent of the *zaibatsu* with coalmining interests were Mitsui and Mitsubishi, which used the profits from mining to fund the development of their industrial empires.

From 1948 to 1950, the coalmines that had belonged to the former *zaibatsu* were managed directly by MITI (see Johnson), but after the reconstitution of Mitsui, Mitsubishi and others in the 1950s, in the form of looser conglomerates known as *keiretsu* (see Chapter 10), the mines were returned to their control. By then, however, these highly diversified groups had begun to look for ways to depart from coalmining, since they had competing and more attractive interests in oil and nuclear power. For many years their wishes were frustrated by the owners of smaller mines, who often succeeded in nurturing local political support to block bureaucratic efforts to consolidate the industry, and, later, in securing the creation of subsidy programs for coal as a declining industry. Their actions retarded the "energy revolution" sought by Japan's policymakers, and kept mines open longer than could have been justified on economic grounds alone. The policy of supporting the domestic coal industry also led to the imposition of tariffs on imports of oil and restrictions on its use. While coal consumption increased by more than half, more than one third of the increase was captured by imported coal, which was often cheaper and/or of higher quality. Additionally, the use of oil grew 15-fold during the decade, demonstrating that restrictions on its use were largely honored in the breach. Public funds that might otherwise have been designated for Japan's high-growth industries were diverted politically to prop up coal and other "sunset" industries instead. By 1959, the new era of cheap oil was finally recognized, and efforts to prop up the domestic coal industry began to fade. The domestic output of coal fell from about 60 million tonnes in 1960 to 9 million tonnes in 1988. Enormous subsidies were paid out, but the government has been phasing them out in recent years. Finally, in 1996–97 the electric power companies, which had been persuaded by MITI to buy at least 5% of their coal supplies from the domestic industry, refused to go on paying significantly more for domestic coal than for foreign coal; while the decision by Mitsui Coal and Mining to close its Mi'ike coalmine, Japan's oldest and largest, in March 1997 left Japan with just two mines still producing coal (see Dawkins).

Electricity

Japan's 10 electric power companies are among the largest utilities in the world. They are privately owned but have long been closely supervised and regulated by MITI and its predecessors. Indeed, they owe their existence in their present form to the Electric Power Control Law of 1938, implemented in 1941, which secured the merger of 103 companies into (at that time) nine large companies, combining generation and distribution functions (see Johnson). In particular, the structures for the rates that they charge to their customers, their decisions about the siting of power stations and other facilities, and their choices among different fuels, have long been subject to MITI oversight and intervention. Their preferences for coal in the 1950s, oil in the 1960s and nuclear power in the 1970s were secured through careful negotiations with suppliers and with MITI officials, and have been very influential in determining the balance of energy supply and demand in Japan (see Table 4.2).

Like the natural gas utilities, the electric utilities enjoy regional monopolies and are more uniform in size than the coal or oil companies are. In the past the only competition among these utilities involved efforts to attract industry to one region over another. Prices were static and, when compared to the costs of electric power in other countries, were relatively high. They remained high even when the cost of fuel imports rapidly declined because of the strengthening of the yen (see Lesbirel). However, this all changed in the 1990s. A growing awareness that Japan was paying comparatively large sums for electric power helped generate a coalition in support of regulatory reform. Consumers, joined by independent suppliers and industrial firms, gained MITI's support for regulatory change, and in 1995 the Japanese Parliament passed the Electric Utility Industry Law, designed to promote competition in the electric power market. The first legislation aimed at restructuring the energy industry in decades, the Law promoted competition by loosening licensing restrictions for suppliers, and by increasing price competition. Deregulation in the electric power industry had long been opposed by those who claimed that energy was too important strategically to be left to the vagaries of market competition, but those who pointed to new energy sources in Central Asia and elsewhere prevailed. The government became convinced that another oil embargo, like those experienced in the 1970s, was only a remote possibility, and that Japan had more to gain by promoting the development of cheaper, more stable, and more diverse sources of energy (see Evans and Samuels 1998). Industrial users responded to the expensive monopolies by securing permission from MITI to expand capacity for auto-generation. In addition, an entirely new industry was spawned as independent power producers began to establish market share in the 1990s. These firms provide power directly to customers and sell their excess supply to the power utilities.

Energy Policy

As indicated above, Japanese energy policy has reflected not only the changing international environment, as mediated through the central concern for "comprehensive security", but also the often incompatible interests of diverse constituents. Policy-makers have had to tread a fine line between pursuing national security and promoting market efficiency. Like any other energy-consuming nation, Japan has had to acquire reliable supplies of energy at reasonable cost. Japanese energy policy has been more focused and sustained than energy policy in other industrial nations in the postwar period, yet energy security has remained an elusive goal throughout.

The authoritative forum for the coordination of energy policy is MITI's Agency for Natural Resources and Energy (discussed above), which was established in response to the prevailing view that the pursuit of "comprehensive security" would require considerable coordination, both within the government and among the powerful but diverse industrial consumers of energy. As the ministry responsible for the healthy development of industry, MITI serves both as trade gatekeeper and economic coordinator. MITI has access to detailed information about domestic market activities through its interactions with the Federation of Economic Organizations (*Keidanren*), the leading business interest group in Japan, and with the various industry associations. Through its formal advisory councils (*shingikai*), notably the Comprehensive Energy Policy Advisory Council, which deliberate on matters of production, imports, prices, and market structures, MITI is also able to supervise negotiations among stakeholders in Japan's complex and trade-dependent energy economy. Less formally, MITI enjoys powers of administrative guidance (*gyosei shido*), which it uses to secure cooperation with regulated parties, often accepting Keidanren compromises, in order to find ways of minimizing the dislocations of "excessive competition" through coordinated action. Negotiations among different segments of Japanese capital are thus institutionalized in the process of making energy policy. For example, steel firms could be enticed to accept more domestic coal than they would prefer, and electric utilities could be induced to pay more for nuclear fuel, in part because MITI is able to authorize the bargaining necessary for effective public policy.

The case of Lion's Oil is instructive when considering MITI's ability to influence the energy industry through administrative guidance. When Lion's Oil, a gasoline retailer attempted to import gasoline purchased on the open market in the mid-1980s, it was doing something that by then was legal, but that still presented a potential threat to the Japanese refiners' control of the market, and MITI attempted to prevent the transaction. The bank which Lion's Oil had sought support from withdrew its financing, and the owner of the company was forced to capitulate and sell the gasoline, relabeled as naphtha, a petrochemical feed stock, to Nippon Oil. MITI was able to have its way in the end, but having its way required lengthy and strenuous efforts involving several branches of government, banks, and petroleum refiners in a move against one small firm (see Sato, and Upham).

Aside from negotiations with private firms, bargains must be struck within the government as well. MITI must compete with the Science and Technology Agency for control of the agenda in nuclear power, and with the Economic Planning Agency to determine forecasts of supply of, and demand for, energy. As in other areas of policy-making, MITI must also contend with the Ministry of Finance (MoF) for budget allocations.

All policy bargains are facilitated and enforced by politicians. From 1955 to 1993, when Japan was, as T. J. Pempel has put it, a "one party dominant" democracy (see Pempel), the ruling Liberal Democratic Party (LDP) was required to adjust and coordinate disputes among its constituents, the most powerful of which were industrial and financial interests. The LDP worked diligently to preserve stable solutions to the wrenching effects of energy crises and industrial restructuring. LDP politicians played an important and very active role in the negotiations among energy producers and consumers, in which they were more likely to become advocates than jurists. For example, virtually every Prime Minister in the postwar period had to impose costs on coal-consuming heavy industry, in order to placate the smaller mineowners and the coalminers, who mobilized legions of local politicians. After the LDP lost power in July 1993, and even after it returned to office, initially as the largest element in a coalition government from July 1994, policy bargaining within the conservative political elite became far more complicated. Now several conservative parties, both inside and outside the governing coalition, as well as the Social Democrats who were part of that coalition, also had to be satisfied. It has become quite difficult for politicians to reconcile fundamental differences about the desirability and pace of deregulation in the various energy industries.

Keidanren is the primary forum for coordination within the private sector for the reconciliation of energy policy preferences, and for the adjustment of demands among vendors, suppliers, producers, and consumers. It was Keidanren that helped the LDP government of the day to develop its original "Comprehensive Energy Program," which blocked many attempts to nationalize the energy industries in the 1960s. Keidanren also coordinated allocative decisions about energy supply and production throughout the postwar period, and in the 1990s it has been prominent among those calling for deregulation and liberalization.

Energy Alternatives

The development of policy to achieve "comprehensive security" in energy supplies has generally been focused on two main issues. Policy-makers have been concerned to maximize the diversity of sources of energy, both by geographic origin (supplier) and by type of source (supply), in the light of their changing perceptions of global instability and Japanese vulnerability. However, they have also been concerned to promote research and development of alternative sources of energy, such as solar, geothermal or biomass.

Diversifying Suppliers and Supplies

Policies towards suppliers have included the acquisition of foreign energy reserves, as well as accommodation with foreign oil companies and oil-producing nations, through long-term supply contracts and through "direct deal" purchases from state-owned oil firms, bypassing the major oil companies. Japanese governments have also sought, with some success, to reduce dependence on foreign oil supplies, especially after the first oil crisis. Oil imports accounted for 80% of Japanese energy consumption in 1973, but declined to only 56% in 1995.

"Energy diplomacy" has also played an important part in Japanese strategies for energy diversification. Thus, for example, in 1973 the Japanese government responded to the OPEC price rises and the OAPEC embargo by moving immediately and vigorously to disengage from Israel. Japan joined the growing list of nations recognizing the Palestine Liberation Organization, Japanese firms began observing the Arab "blacklist" of firms doing business in Israel, and investment by Japanese interests in the Middle Eastern oil-producing states (both Arab and non-Arab) increased dramatically. This investment included the decision by the Mitsui group, with government support, to build a major petrochemical complex at Bandar Shahpur (now Bandar Khomeini) in Iran. This project proved politically and economically costly when the complex was destroyed in 1979, during the Iranian revolution.

Developing Renewable Sources of Energy

Renewable sources of energy, such as solar or geothermal power, have been actively subsidized and promoted since 1974, when MITI's Agency for Industrial Science and Technology began collaborating with the private sector on the "Sunshine Project" to promote research and development in this area. However, renewable sources still account for only marginal amounts of Japan's energy, amounting to just 1.3% in the financial year 1995–96.

After 1973, as energy prices rose, economic growth fell, and government policies encouraging conservation of energy began to have some impact, the rate of growth in the total consumption of energy declined to just over 1% a year until the mid-1980s, when it once again began to increase faster than economic growth. After the Iranian revolution and the second oil shock in 1979, Japanese policymakers decided to accelerate research and development of alternative energy. As in other industrialized countries, officials involved in making economic policy preferred to create a new public entity to research and develop synthetic fuels. Accordingly, MITI submitted proposals for legislation to create a state-owned "alternative energy public corporation" in 1979. However, the MoF objected on fiscal grounds, while Keidanren objected to this new form of state intervention in the economy. By the time that the opposition from these two powerful sources had been taken into account, Parliament was being asked to pass the legislation which created the New Energy Development Organization, since expanded and renamed the New Energy and Industrial Technology Development Organization. To placate the MoF, new taxes were placed on petroleum and electricity consumption, and earmarked to support this new institution; and, to satisfy Keidanren, its Chairman was made Chairman of the new body.

The Nuclear Alternative

Initiated as long ago as 1955, Japan's nuclear power program has been far more ambitious than any other program for alternative sources of energy. For many Japanese officials, nuclear power has two major potential benefits. At the same time that it has held the promise of ameliorating vulnerabilities to variations in energy supplies, it also promises to provide commercially viable technologies that might simultaneously enhance Japan's international competitiveness. To date, however, this promise has not been fulfilled.

The decision to accelerate the development of nuclear power was taken in the early 1970s, after two decades in which Japan had first imported foreign reactor designs, and then indigenized them. Nuclear power has since become the source of 12% of Japan's primary energy supplies (as of 1995), and of 20.5% of its electric power (as of 1996). In Japan, as, for example, in France, nuclear power has been promoted by public officials, utility executives, and vendors as the most promising avenue for achieving long-sought "energy independence", and an end to vulnerability to international political developments (see Samuels 1994).

The Japanese companies that build "indigenized" Japanese reactors belong to the same industrial and financial groups (*keiretsu*) that had dominated the coal industry just decades earlier and that have long dominated the Japanese financial markets. Mitsubishi Electric obtained its pressurized-water reactor technology under license from the US producer Westinghouse, and Toshiba obtained its boiling-water reactor technology under license from Westinghouse's rival, General Electric. Mitsubishi Electric and Toshiba have since become so proficient in their respective technologies that they have been able to become fully fledged partners with both of these US corporations. The Japanese reactor companies also collaborate with MITI to encourage the regional electric power companies to expand their nuclear capacity, and all parts of this "nuclear energy iron triangle" collaborate to dampen opposition from local and national environmental pressure groups. One part of this strategy allows the electric power companies to pay more than the world market price for nuclear fuel and to treat stockpiled fuel as a capital expense. This expense generates a risk-free return above market rates, facilitates investment in new nuclear capacity, and, since the costs are borne disproportionately by households and commercial establishments, it subsidizes industrial consumption (see Samuels 1989).

Japan's nuclear program has implications beyond energy supply that affect the country's comprehensive security. In particular, there is Japan's ambitious plutonium program which, if it succeeds, will close the nuclear fuel cycle and enable Japan to achieve full energy independence. The plan is to reprocess spent fuel from light-water reactors, in order to recover plutonium that would be recycled in mixed-oxide fuel in light-water reactors and eventually (by 2030) in commercial fast-breeder reactors. It was thought that closing the fuel cycle would contribute to energy security while reducing costs and ameliorating Japan's substantial problems with waste management. However, the program has prompted expressions of serious concern from other countries (see Skolnikoff et al.). Some are concerned that non-nuclear powers will cite Japan as a precedent and justification for pursuing similar reprocessing and breeder programs. They fear that other nations may not pay the same degree of attention to safety and proliferation issues that Japan has displayed. Second, the stockpile of plutonium that Japan's program has generated (about 10.8 tons, some in Japan and some in Europe) has led to concerns that other countries in Asia will pursue clandestine or overt nuclear weapons programs, and will justify them by the real or fabricated fear that Japan's plutonium stocks could be converted into a nuclear weapons capability directed at them. Third, some nations are concerned that such fears are justified, that Japan itself will decide to pursue a nuclear weapons program, and that it will be able to do so because of its plutonium stockpile. Fourth, some countries worry that increasing plutonium stocks and international transfers of fissile materials will increase the difficulty of managing Japan's plutonium program, and increase the risk of possible diversion by terrorists. Finally, many experts disagree with the rationales that the Japanese government has put forth in defense of its plutonium program. They are unconvinced that it is indeed the best way to improve energy security, promote economic benefits, and create economic advantages, and they note that both France and the United States have abandoned plans for developing breeder reactors.

Domestic reactions to Japan's nuclear program also influence energy options. The Japanese people have been profoundly ambivalent about the introduction of nuclear power, especially as it is on a relatively large scale. Many recognize and accept the prevailing wisdom among opinion-formers, that nuclear power has the potential to provide Japan with its elusive energy security, but as the preoccupation with energy security has faded, fears about the safety and wisdom of nuclear power slowly have tipped the scales of the public debate against nuclear power (see Choy, August 1997). In particular, nuclear accidents in the United States, in the former Soviet Union, and in Japan itself have stimulated an influential movement of opposition.

This opposition has been bolstered by a series of relatively minor incidents at nuclear power plants during the 1990s. Public confidence in nuclear power was badly shaken in December 1995, when a coolant leak at the Monju experimental fast-breeder reactor sprayed 1,500 tons of highly reactive material into an equipment room. Although no radioactivity was released, and the reactor remained under full control, plant operators were found to have tried to cover up their actions during the mishap. Officials at the government-owned Power Reactor and Nuclear Fuel Development Corporation (*Donen*) were fined by local governments and by its supervisory authority, the Science and Technology Agency. The public was outraged, and the future of Japan's nuclear power program was threatened. Donen was again roundly criticized when a fire and explosion occurred at its nuclear fuel reprocessing center at Tokaimura in March 1997. Several dozen workers were exposed to small amounts of radiation, and it was later revealed that plant managers required workers to relay a falsified chronology of events in order to cover up their mismanagement. Once again there were public calls for criminal prosecutions of Donen officials, for the closure of the plant, and for the reorganization of Donen itself. The combination of these two incidents has empowered local citizens' groups fighting the construction of nuclear power plants in their regions. In 1996, the citizens of one town voted against the construction of a nuclear power plant in their area, and other groups have followed suit. More than 90% of the public in Japan "feel uneasy" about nuclear power, according to opinion surveys, and public confidence in nuclear power has never been so low. As a result, more funds are today made available to purchase "public acceptance" through funding unrelated projects in localities than are actually spent designing and constructing reactors. These side payments notwithstanding, Japan's nuclear program has never been as threatened as it is today.

Conclusion

Japan's system for obtaining energy supplies is characterized today, as in the past, by relentless competition among bureaucratic interests and rapidly shifting market forces. In this sense, it constitutes just one of the many ways in which Japanese market players and government officials develop economic priorities and strategies, channeling competition into cooperation. However, precisely because energy is so crucial to the economy, the system's specific functions, its successes, and its occasional failures cannot be explained without recurring reference to Japan's unusually high level of energy dependency, or an appreciation of how Japanese elites define the country's "comprehensive security". In particular, the system would not work as successfully as it has if individual consumers did not tolerate relatively high domestic prices for goods and services. In the belief that value added "at the doorstep of the consumer" enhances economic security and, indeed, technological development, Japanese companies have invested in relatively expensive energy supplies, for the same reason that they invested in a wide range of other inefficient but import-substituting projects. They have proceeded, with encouragement and support from politicians and officials, in the belief that when consumers deny themselves foreign versions of products that can be made domestically, and/or effectively subsidize domestic production through higher prices, healthy Japanese industry will contribute over the longer term to creating a more secure, less dependent, and more technologically sophisticated economy, that serves the interests of producers and consumers alike.

Portions of this chapter appeared in Richard J. Samuels, "Energy in Japan", in Richard Bowring and Peter Kornicki, editors, *The Cambridge Encyclopedia of Japan* (Cambridge and New York: Cambridge University Press, 1993).

Further Reading

Choy, John, "Japan's Energy Market Getting Shocked", in *JEI Report* number 6B, Japan Economic Institute, February 14, 1997; and "Deregulation Jolting Japan's Energy Sector", in *JEI Report*, number 33A, August 29, 1997

Dawkins, William, "Light at the End of the Tunnel", in *Financial Times*, May 12, 1997

Energy in Japan, Tokyo: Institute of Energy Economics, bimonthly publication

This is the English-language version of the Institute's regular report.

Evans, Peter C., and Richard J. Samuels, "Kisei Kanwa to Enerugii Anzen Hosho wa Ryoritsu Dekiru" [Electric Power Deregulation versus Energy Security: Is There a Trade-off?], in *Ekonomisuto*, February 24, 1998

Harrison, Selig, *Japan's Nuclear Future: The Plutonium Debate and East Asian Security*, Washington, DC: Carnegie Endowment for International Peace, 1996

Harrison examines Japan's plans to expand its nuclear power industry from the perspective of concerns about nuclear proliferation.

Hein, Laura E., *Fueling Growth: The Energy Revolution and Economic Policy in Postwar Japan*, Cambridge, MA: Harvard University Press, 1990

Hein traces the early postwar history of the Japanese coalmining industry, with special reference to the role of the labor unions.

Japan Review of International Affairs, quarterly publication

The Fall 1997 edition of this journal, published in English by a think tank affiliated with the Japanese Ministry of Foreign Affairs, includes two important articles: "Japan's Energy Policy: Current Status and Issues" by Fujime Kazuya; and "The Future of Nuclear Power in Japan" by Kono Mitsuo.

Johnson, Chalmers, *MITI and the Japanese Miracle: The Growth of Industrial Policy, 1925–1975*, Stanford, CA: Stanford University Press, 1982; Rutland, VT, and Tokyo: Charles E. Tuttle, 1986

Lesbirel, S. Hayden, "Wheeling and Dealing: Reforming Electricity Markets in Japan", presentation to a "Roundtable" on The Politics of Economic Reform in Japan, Australia Japan Research Center, Australian National University, July 26, 1996

Morse, Ronald, *Turning Crisis to Advantage: The Politics of Japan's Energy Strategy*, New York: The Asia Society, 1990

This "think piece" reviews Japan's policy options at the time of the Gulf War.

Pempel, T. J., editor, *Uncommon Democracies: The One-Party Dominant Regimes*, Ithaca, NY: Cornell University Press, 1990

Samuels, Richard J., *The Business of the Japanese State: Energy Markets in Comparative and Historical Perspective*, Ithaca, NY: Cornell University Press, 1987

Samuels, Richard J., "Consuming for Production: Japanese National Security, Nuclear Fuel Procurement, and the Domestic Economy", in *International Organization*, Autumn 1989

Samuels, Richard J., *"Rich Nation, Strong Army": National Security and the Technological Transformation of Japan*, Ithaca, NY: Cornell University Press, 1994

Sampson, Anthony, *The Seven Sisters: The Great Oil Companies and the World They Made*, London: Hodder and Stoughton, and New York: Viking Press, 1975

Sato Taiji, *Ore wa Tsusansho ni Barasareta* [I Was Murdered by MITI], Tokyo: 1986

Skolnikoff, Eugene, Suzuki Tatsujiro, and Kenneth Oye, *International Responses to Japanese Plutonium Programs*, Cambridge, MA: Center for International Studies, Massachusetts Institute of Technology, August 1995

Skolnikoff, Suzuki and Oye review Japan's highly controversial plan for a plutonium-based nuclear power industry.

Suetsugu Katsuhiko, editor, *East Asian Electricity Restructuring Forum: Final Report*, Cambridge, MA: Center for Business and Government, Kennedy School of Government, Harvard University, June 1996

This report examines the prospects for international cooperation on the development of electric power in East Asia, with special reference to the changing role of electric utilities and deregulation.

Upham, Frank, "The Man Who Would Import: A Cautionary Tale About Bucking the System in Japan", in *Journal of Japanese Studies*, Volume 17, number 2, Summer 1991

Richard J. Samuels is Ford International Professor of Political Science at the Massachusetts Institute of Technology, and Director of the MIT Japan Program. The author would like to thank *Jennifer Lind* for her research assistance in the preparation of this chapter.

Table 4.1 Shares of Energy Sources in Japan's Primary Energy Market, selected years 1950–95 (%)

	Coal	*Oil*	*Gas*	*Hydroelectricity and geothermal power*	*Nuclear*
1950	74.1	4.6	0.0	21.3	0.0
1960	52.4	31.6	0.9	15.1	0.0
1970	22.5	68.7	1.1	7.2	0.4
1980	17.4	70.2	6.4	6.7	5.9
1988	18.5	56.6	9.4	5.4	10.0
1995	16.5	55.8	10.8	3.5	12.0

Source: Ministry of International Trade and Industry

Table 4.2 Shares of Energy Sources in Japan's Electric Power Market, selected years 1960–96 (%)

	Coal	*Oil*	*Gas*	*Hydroelectricity*	*Nuclear power*
1960	42.1	15.7	0.1	42.1	0.0
1970	19.9	53.6	1.5	23.7	1.4
1980	8.1	42.6	12.2	17.5	15.6
1996	9.8	23.5	25.3	20.7	20.5

Source: Ministry of International Trade and Industry

Chapter Five

Agriculture and Fisheries

Aurelia George Mulgan

Agriculture in the Economy

Japanese agriculture is small-scale, inefficient, and gradually declining. Although farm output and productivity have been raised by technological improvements, the economic and social significance of farming – as an occupation, as a way of life, as a form of land use and as an industry contributing to national income and national output – has continued to diminish. In 1994 agriculture accounted for only 1.6% of GDP, down from 7.4% in 1965. Over approximately the same period, the number of people employed in agriculture – who had accounted for around half the total workforce immediately after World War II – fell from 9.8 million, or 20% of the total workforce, in 1965, to 3.4 million, or 5%, in 1995 (according to Norinsuisansho Tokei Johobu).

A densely populated mountainous country, Japan has less than 15% of land under cultivation. Productivity in agriculture is around one third that in manufacturing and non-agricultural industries, and behind that of agriculture in many other industrialized countries. Even with considerable government assistance, very few Japanese farms are economically viable, in the sense of being able to sustain full-time farmers at a standard of living comparable with the rest of the community. The average income of those employed in agriculture in 1995 was less than 40% the average wage of those employed in manufacturing. The basic unit of agricultural production is the family farm, often with scattered plots amounting to little more than one hectare in total size. Of Japan's 2.6 million commercial farm households (households marketing agricultural products), the great majority derive most of their income from sources other than agriculture. Farming is predominantly a part-time enterprise, with members of farm households employed in a range of other occupations. Very few young people enter directly into farming careers. Most new entrants to farming are between 45 and 60 years old, having left non-farm jobs in order to take over family farms. As a result, more than half the total farm population is now aged over 65 years.

Japan's food self-sufficiency rate has declined along with farming. Japan is now the world's largest importer of food products, both processed and unprocessed. The most important source of agricultural imports for the Japanese market is the United States; other important suppliers include Australia, Canada, New Zealand and, to a lesser extent, EU and Southeast Asian countries.

The most important agricultural commodity produced in Japan is rice. The gross output of rice exceeds that of every other single crop, at 10.3 million tonnes in 1996. The rice crop occupies 42% of land dedicated to agricultural production and yields income of about ¥3.875 trillion (34% of gross output value in 1994). Land-extensive farming, such as of wheat and barley, occupies only one tenth of the area planted in rice. Rice is harvested by more than three quarters of all Japanese commercial farm households. Other major products are livestock commodities (23% of gross output value), vegetables (22%) and fruit (8%).

Fisheries

Fisheries play an important part in Japan's food industry since, in spite of the growing consumption of meat, fish remains the largest source of animal protein for the Japanese people. In recent times Japan has dominated global fisheries, taking around 15% of the world's catch throughout most of the 1970s and early 1980s. Since the mid-1980s, however, its annual catch has declined significantly, from 12.6 million tonnes in 1984 to 7.3 million tonnes in 1996. Since 1989, its catch has been surpassed by both China and Russia (ABARE p. 267; *MAFF Update* May 30, 1997; Bergin and Haward p. 1n.).

A number of reasons account for this decline: the effects of over-fishing in Japan's coastal waters; increasingly stringent international restrictions on fishing in the high seas, as well as in the exclusive economic zones (EEZs) of other countries; the impact of the rising value of the yen on the economic competitiveness of fish exports; and increasing preference among consumers for imported "gourmet" varieties. Japan remains a major world fishing power, however, as well as being by far the world's largest producer in aquaculture.

Japan's fishing industry is divided into inland water fisheries and marine fisheries. The former has a relatively insignificant share of the total, producing 94,000 tonnes, or just over 1% of total production, in 1995 (*MAFF Update* May 30, 1997). Marine fisheries are further classified into three categories based on different catching areas.

The first category comprises coastal fisheries within 10 nautical miles of the coastline, operating either from small boats of less than 10 gross tonnes or from the shore and under licenses issued by local fisheries cooperatives. In 1995, coastal fisheries accounted for 1.883 million tonnes, or 26% of total production (*MAFF Update* May 30, 1997). In 1993, coastal fisheries provided work for 275,000 people, 85% of the fisheries workforce, mainly employed in small, family-based fishing enterprises (JIIA p. 129). Often, like small-scale farming, these provide part-time work for family members whose primary employment is elsewhere.

The second category covers offshore fisheries within 10 to 200 nautical miles from the shore (thus, within Japan's EEZ), operating from boats of between 10 and 100 gross tonnes. In 1995, offshore fisheries accounted for 3.173 million tonnes, or 43% of total production (*MAFF Update* May 30, 1997).

The third category involves distant water fishing beyond Japan's EEZ, both on the high seas and in the EEZs of other countries, usually from boats of more than 100 gross tonnes. In 1995, distant water fisheries accounted for 812,000 tonnes, or 11% of total production (*MAFF Update* May 30, 1997).

Licenses for both offshore and distant fisheries are issued by government agencies, at prefectural and national level respectively, taking into account previous catches and a perceived need to maintain stability and harmony in the industry, meaning the respective shares of its various participants. The main species of fish caught in marine fisheries are mackerel (728,000 tonnes), flying squid (410,000 tonnes), sardines (339,000 tonnes), anchovies (357,000 tonnes) and tuna (281,000) tonnes (*MAFF Update* May 30, 1997). Japan also continues a small whaling industry, catching 126 whales in 1994 – a minute catch in comparison with catches of more than 3,000 in 1985 and 27,000 in 1965, but still internationally significant (OECD 1996b p. 165; ABARE p. 268).

Offshore and distant water fisheries in 1993 together employed 50,000 people, 15% of the total fisheries workforce. As in agriculture, the profile of the fishing workforce is generally ageing, and total numbers engaged in fisheries are falling along with the declining catch (down 27% between 1983 and 1993). In 1993, 40% of all fisheries workers were above the age of 60 (JIIA p. 129).

Besides fish caught in open waters, aquaculture is also an important part of fishing production. Marine aquaculture yielded 1.272 million tonnes in 1995, 17% of total production, while inland aquaculture produced 73,000 tonnes (*MAFF Update* May 30, 1997). The main aquaculture products are edible seaweeds, oysters, pearl oysters and scallops, and certain varieties of fish, mainly yellowtail and sea bream.

Japan's trade in fish products shows a heavy preponderance of imports over exports. Declining Japanese fish production, combined with the stronger yen and growing demand for gourmet fish products, has led to a steady increase in the amounts of fish products imported during the 1980s and 1990s. In 1994, fish imports totalled 3.296 million tonnes, the equivalent of 40% of total domestic production, while exports were only 296,000 tonnes, or 4% of domestic production) (OECD 1996b p. 161). The main imported species were shrimps and prawns, tuna, salmon, mackerel, and cod.

The Policy Framework

By a range of economic measures, Japan's agricultural producers are the second most protected in the OECD, after Switzerland's (OECD 1996a p. 187). As a result, consumers are forced to pay food prices that are, on average, twice world market prices (OECD 1996a p. 194). The degree of protection enjoyed by Japanese farmers reflects the scale of government intervention in the agricultural economy. A host of laws, policies, financial measures and institutions have been specifically designed to provide support and protection for agriculture by aiding farm production, regulating agricultural marketing and commodity distribution, and promoting the farmers' welfare.

This extensive and complex framework of agricultural assistance is justified as maintaining self-sufficiency in food, raising agricultural productivity, promoting the living standards of farm households, fostering the vitality of rural communities and providing stable prices for consumers. Many of these goals are expounded in the Agricultural Basic Law of June 1961, otherwise known as the "Charter of Japanese Agriculture", which affirms the contribution of farmers to the nation.

One of the major targets of government intervention is producer prices. Most agricultural commodities are subject to support and stabilization schemes, linked to control of imports. The most comprehensive and rigid program of price support and distribution control has applied to rice under the Food Control Law. Farmers have been permitted to sell their rice only through government-controlled distribution channels, known as the Food Control system, in which government-sponsored agricultural cooperatives (*nokyo*) have been dominant as rice collectors and distributors. The Food Agency, a branch of the Ministry of Agriculture, Forestry and Fisheries (MAFF; *Norinsuisansho*), has purchased most of this rice from the farmers at a price decided by the Minister. The government's buying price, or "producer rice price", has been the most important item on the annual agricultural policy agenda.

The Food Control system was partially liberalized in 1969, when the government established an alternative channel, known as "voluntarily marketed rice", which permitted farmers to bypass the government and sell directly to wholesalers, although the new distribution channel remained dominated by the *nokyo*, and prices were still subject to administrative oversight. Since 1990, a small amount of voluntarily marketed rice (about 20%) has been sold at auction to rice wholesalers administered by a semi-governmental agricultural agency. The introduction of this system was designed to introduce a modicum of market competition into domestic rice sales at the wholesale level. In 1995, with the passage of a new Staple Food Law to replace the Food Control Law, farmers were additionally permitted to sell rice directly to retailers and consumers, provided that they notified the government. This enabled them completely to bypass the *nokyo* as distribution agents. Under the new Law, the Food Agency purchases only 1.5 million tonnes, or 15% of total production, for stockpiling purposes, with the aim of guarding against shortages and ensuring a stable supply of rice for consumers.

Until 1995 state control of rice marketing effectively prohibited the import of foreign rice. Since then, in response to international pressure for liberalization through GATT, Japan has agreed to limited imports of rice, equivalent to 4% of domestic consumption in 1986–88, increasing to 8% by 2000. This commitment was part of a package of agricultural policy measures to which Japan acceded under GATT's Uruguay Round Agreement on Agriculture, in December 1993. However, the use of import quotas rather than tariffs has enabled the Food Agency to control the sale of imported rice, offering it on the Japanese market at price levels comparable to those for domestic rice and using the profits generated to help support the domestic market. The combination of managed imports and residual controls on domestic distribution ensures that Japanese farmers continue to receive around seven times the world price for their rice (EAAU p. 359).

In the case of wheat and barley, the government offers producers a guaranteed price within a structure of relatively high import tariffs. Until 1995, the Food Agency controlled all imports of wheat and barley under an import quota system. Since then, licensed companies have been allowed to import wheat and barley, provided that certain tariff equivalents are paid. However, the Food Agency remains the sole purchaser within the tariff quota.

Prices of livestock products have been maintained by a semi-government agency, the Livestock Industry Promotion Corporation (LIPC), which was renamed the Agriculture and Livestock Industries Corporation (ALIC) in October 1996. Price support has been achieved partly through direct subsidies or deficiency payments for manufacturing milk and beef calves, and partly through price stabilization systems for designated dairy products, pork and beef. The Corporation controls market prices within a predetermined range through purchasing operations that limit the volume of sales on the domestic market. Imports have been regulated by Corporation trading, import quotas and tariffs, or combinations of these approaches. Import liberalization has brought some changes to this system. The Corporation's import rights have been eliminated either partially, for designated dairy products, or totally, for beef. In the latter case, import quotas were abolished in 1991 and replaced by a tariff, which is being reduced in stages, from 70% initially to 50% in 1991–94, then, through further annual reductions, to 38.5% in 2000. In 1995, however, MAFF imposed regulations allowing increases in tariffs if beef imports exceeded certain limits. The tariff system for beef has made it more difficult for ALIC to stabilize beef prices on the domestic market.

Price stabilization schemes, funded by producers and the government, also operate for a variety of other products, such as designated vegetables, raw fruit for processing, pulses, eggs, and silk. Sugar beet and cane, potatoes for processing, and some other products are supported by minimum price guarantee systems, while soybeans and rapeseed receive a direct subsidy, in the form of deficiency payments. Import liberalization for these products has occurred progressively over a period of some decades: soybeans and rapeseed in 1961, refined beet and cane sugar in 1972, fruit purée and paste, fruit pulp and canned pineapple in 1988, raw silk in 1995, and so on. Liberalization, however, has not affected domestic price support arrangements, and in some cases volume control, by state trading, has been retained.

While price supports and import restrictions have been among the main instruments of government assistance to farmers, public subsidies have also been made available for other purposes. For example, large sums are allocated in the budget for agricultural land and infrastructure development, under expenditure headings such as "agricultural production basis consolidation" or "agricultural land preservation management", as well as for general economic and social infrastructure in rural areas, under the headings "rural consolidation" and "agricultural road consolidation". These comprehensive public works programs in the countryside benefit farmers by subsidizing input costs, and rural dwellers generally by raising the standard of local amenities and infrastructure. At the same time, they generate lucrative contracts for local construction companies and provide employment opportunities for members of farm households. Expenditure on agricultural and rural public works accounted in 1996 for 56% of MAFF's general budget, more than double the proportion in 1975, which was 23%. Government-subsidized loans are also made available to farmers to assist with other capital costs, such as the erection of livestock facilities.

In addition, farmers benefit from a range of other government concessions. Agricultural income is subject to preferential taxation treatment in relation to the incomes of wage and salary earners. Land tax rates on most agricultural land are one tenth of the tax rates on residential land, and often effectively much less.

Farm land is also free of inheritance tax if it continues as agricultural land. Farmers are charged lower rates for electricity than urban consumers and they have access to supplementary old age pensions which are not available to non-farmers.

With respect to fisheries, government policy is aimed at maintaining the industry and maximizing fish supplies. The onshore sale of fish is conducted partly through wholesale markets operated by fisheries cooperatives, and partly, and increasingly, through direct purchasing agreements between large buyers, such as supermarket chains, and large suppliers, such as fisheries cooperatives or vertically integrated fishing companies. As in agriculture, the government seeks to maintain prices at a level sufficient to guarantee reasonable returns to producers, using a variety of mechanisms, including tariffs, which have averaged 4% since 1995 (OECD 1996b p. 23), as well as quotas and price stabilization schemes. The Fisheries Agency of MAFF, for instance, operates a stockpile for tuna similar to that operated for rice by the Food Agency.

In relation to offshore and distant fisheries, the government plays a crucial role in securing access to foreign fisheries and negotiating international fishing agreements. Annual meetings are also held with representatives of neighboring countries, particularly Taiwan and South Korea, to regulate the supply of imported fish to Japanese markets. In addition to price support measures, considerable government financial assistance is also furnished for the stated purpose of encouraging structural adjustment in the fishing industry, in line with international obligations, but in effect to subsidize fishing enterprises and fishing communities. For instance, funds are allocated to the upgrading of port facilities and the "invigoration" of fishing villages, the improvement of the marine environment, the reduction of vessel numbers, the modernization of plant and equipment, improvements to fish-processing plants, research and development of new fisheries and products, and the promotion of fish consumption (OECD 1996b p. 162; *MAFF Update* September 6, 1996; ABARE pp. 287–8). As with agriculture, the generous provision of public works to fishing communities not only reduces costs to producers but also provides valuable employment opportunities for members of fishing households not engaged in full-time fishing.

Representation of Agricultural and Fishery Interests

Farmers in Japan, like farmers in many other developed economies that protect agriculture, benefit from strong organizations with an established voice in government. In the Japanese case, one farmers' group has dominated the economic and political life of the countryside: the nationwide Agricultural Cooperative Association (*Nogyo Kyodo Kumiai*), better known as *Nokyo* or JA (short for "Japan Agriculture"),which was established by law in 1947. Both farmers and non-farming "associate" members may belong. In 1994, they totalled just under 9 million people (Nihon Nogyo Nenkan Kankokai pp. 369 and 613). The activities of Nokyo and its constituent agricultural cooperatives – which, as mentioned, are also called *nokyo* – are all-encompassing, extending from basic farm-related functions, such as the marketing of agricultural produce and supply of farm inputs, to every conceivable type of social, cultural, medical and consumer service, such as banking and credit; purchasing and supplying members with goods for their business or daily living; installation of joint-use facilities, such as rice mills, rice centers, country elevators, spayers, breeding facilities, pastures or milk collection facilities; insurance; facilities for medical use, transportation, processing, storage and repairs; land improvement or development; and management of agricultural lands through contract farming.

Nokyo in its totality constitutes a massive and highly complex grouping with a multitude of organizational offshoots, including women's and youth organizations. It brings together a collection of several thousand separately constituted *nokyo* that are independent in organization and internal decision-making structures, but interdependent in the flow of goods, services and finance. Nokyo thus forms a pyramid-shaped structure, with a base line made up of unit *nokyo* operating at city, town or village level. These join together to form prefectural federations which are specialized according to function, which in turn form national federations. Within this horizontal structure a vertical division separates multi-purpose *nokyo*, which conduct the full range of economic and other services for members at the local level, from special-purpose *nokyo*, which perform a more limited range of functions in relation to particular farm products or economic activities. Another special type of *nokyo*, the central union, both prefectural and national, is formally charged with representing the interests of the *nokyo* and their members to government.

Almost all Japanese farm households, no matter what they produce or their level of their engagement in agriculture, belong to Nokyo. Its presence in the Japanese countryside is ubiquitous and its functions comprehensive and diverse. Unlike most farm cooperatives elsewhere, it is not merely an economic group with self-help functions. It is a social institution, a social movement, a group that encapsulates, expresses and reinforces the social and cultural mores of the countryside; a vast bureaucracy with a multitude of officials extending the organization's reach into the remotest areas of Japan; an electoral support organization and pressure group, with policy interests that range over the entire agricultural economy; and an interlinked enterprise network (*keiretsu*) that competes with other giant financial and trading corporations on equal terms. Nokyo has long been regarded as one of the nation's most politically and economically powerful organizations, as influential in lobbying as the large federations of manufacturing and service businesses. The best-known of Nokyo's national organizations are the Central Bank for Agriculture and Forestry (*Norinchukin*), the marketing and purchasing federation *JA Zenno*, and the insurance federation *JA Zenkyoren*. Each has financial and business operations running into astronomical figures.

Nokyo's primary policy concerns relating to farmers center on matters that affect producers' incomes, such as producer prices, the degree of border protection, or budget subsidies for farm assistance programs. Nokyo submits "demand" prices for different agricultural commodities and backs these up with both public and behind-the-scenes lobbying, as well as direct negotiations with the government. The scale of organizational mobilization has traditionally been greatest in the case of the producer rice price, which has occasioned annual rituals of Nokyo-led public demonstrations and marches by farmers and *nokyo* leaders.

Nokyo also pursues its own independent organizational interests. In particular, it is concerned with the management viability of individual agricultural cooperatives and the levels of profits generated by its different businesses. Nokyo's economic activities include not only the importation and distribution of farm inputs, such as farm machinery, equipment and agro-chemicals, but also their manufacture, in areas such as stockfeed and fertilizers, mainly through subsidiary companies. Constituent *nokyo* and their subsidiary companies are also engaged in agricultural product processing, such as milk, fruit juice and livestock products of all kinds. In addition, Nokyo's financial activities extend to stock and bond purchases, and large quantities of loans to other financial institutions, not all directly related to agriculture.

Throughout their postwar history, Nokyo's businesses have been underpinned by a number of government concessions, such as relatively low corporate tax rates, and by their administratively sanctioned monopolies over rice collection and distribution and other activities, for which Nokyo is paid commissions and various other service fees and subsidies by the government. Nokyo also receives budget subsidies for carrying out agriculture-related projects, programs and functions on behalf of the government. In this way, Nokyo has been inextricably interwoven into the whole fabric of agricultural administration, fusing the roles of government agent and farmers' organization.

Nokyo has also been extensively involved in electoral activities such as vote mobilization, campaign assistance and political funding to candidates seeking political office at all levels of government – national, prefectural and municipal. Nokyo supports both its own officials and other sympathetic candidates in exchange for sponsorship of Nokyo's and the farmers' interests in politics. Its electoral prominence is due to a particular combination of factors. First, the *nokyo* incorporate the strategically important farm electorate within their membership. Secondly, the electoral power of the agricultural cooperatives has been enhanced by the over-representation of the more sparsely populated rural districts, which magnified the political significance of the farm vote until the electoral reforms of 1993–94 (see below, and Chapter 2). Thirdly, Nokyo has functioned as a "rice roots" electoral infrastructure in the countryside for Japan's dominant political party, the LDP, providing the principal organizational link between LDP politicians in Parliament and supporting votes in rural and semi-rural electorates.

Besides Nokyo, other government-sponsored farmers' organizations have also been created to perform designated functions under law, although they do not have the broad functional scope or universal membership of the *nokyo*. They are the land improvement groups (*tochi kairyo dantai*); the agricultural mutual aid associations (*nogyo kyosai kumiai*); and the agricultural committees (*nogyo i'inkai*). Each of these is specialized according to function and each is involved in petitioning the government for policy benefits in relation to their activities. Other farmers' associations without government sponsorship are the left-wing farmers' unions (*nomin kumiai*); innumerable groupings representing the interests of specialist farm producers, such as dairy farmers, tobacco growers, or tea growers; and a broad category of farmers' political leagues (*nomin seiji renmei*), which act as the spearheads of the *nokyo* and other farmers' organizations in elections and agricultural policy activities (*nosei katsudo*).

Within the fisheries industry, organized interest groups, particularly producers' cooperatives, also play a key role in representing the interests of producers, particularly in relation to the government's management of fishing stocks and fishing rights. The cooperatives are based in the main fishing ports, where they are in charge of fishing rights within their designated coastal waters. While the prefectural governments establish overall quotas for particular species, the cooperatives decide which rights are to be allocated to which individual enterprises. The fisheries cooperatives also conduct a range of economic and technical activities supporting fishing enterprises and communities, including finance, marketing and training. The individual cooperatives are federated into prefectural associations and the National Federation of Fisheries Cooperative Unions (*Zengyoren*), which, like Nokyo, acts as a powerful lobby group.

Other associations represent the interests of particular fishing sectors, for instance the various tuna industry associations, including the Federation of Japanese Tuna Fisheries Cooperative Unions (*Nikkatsuren*), the National Offshore Tuna Fisheries Association (*Kinkatsukyo*), and the Japanese Overseas Purse Seine Fishing Association (*Kaimaki*). The influential Japanese Fisheries Association represents the interests of all parts of the industry (Bergin and Haward p. 61).

Representation of agricultural and fisheries interests to government is also fostered by consultative processes centering around the activities of formally established advisory councils (*shingikai*) attached to MAFF. These bodies, appointed by the Minister, contain representatives of relevant interest groups, such as the cooperatives, as well as consumer organizations, business groups, academia, the media and so on. In agriculture, the most prestigious grouping is the Agricultural Policy Advisory Council, which issues reports outlining basic goals for agriculture and serving as policy guides for administrators and the farming industry in general. The Rice Price Advisory Council, on the other hand, has traditionally attracted considerable publicity through its annual deliberations on the producer rice price. The Central Fisheries Adjustment Advisory Council gives advice on the contentious issue of allocating national fishing licenses, and is paralleled by prefectural councils.

Ad hoc councils are also established from time to time. In 1997, for instance, the Fisheries Agency established a Council on Basic Policy Concerning Fisheries to consider policy changes required to meet Japan's obligations under the UN Convention on the Law of the Sea (*MAFF Update* October 24, 1997). Though theoretically independent, these councils generally support MAFF policies and serve to legitimize new policy directions conceived by its officials.

Administrative Institutions

The extensive control and regulation of agricultural and fisheries production in Japan depend on a number of key public and semi-public institutions, each with its own interests to safeguard and enhance. The chief instrument of state intervention is MAFF and its associated agencies, particularly the powerful Food Agency and Fisheries Agency.

MAFF's basic rights of intervention in the agricultural economy are embedded in its founding legislation, the 1949 Law Establishing the Ministry of Agriculture, Forestry and Fisheries (*Norinsuisansho Setchiho*). MAFF is charged with administering around 125 other laws as well as ministerial ordinances governing all aspects of farm production, distribution, marketing and trade. It is also responsible for drafting relevant new legislation and compiling annual agricultural budgets, as well as negotiating agricultural, forestry and fisheries policies with the ruling party or parties.

Although MAFF does not command the prestige and power of either MoF or MITI, it is still a major force within the Japanese bureaucracy. In personnel numbers it is four times the size of MITI, but smaller than MoF. A large proportion of MAFF personnel are technical and engineering officials (*gikan*) specializing in public works. MAFF consumes around 4–5% of the total General Account budget and is the fourth largest allocator of budgetary subsidies after the Ministries of Health and Welfare, Education, and Construction. Its fortunes have inevitably suffered to some degree with the economic decline of the activities that it administers, but its standing has been preserved for the most part because of its largely untrammelled regulatory powers, the relative size of its budgetary outlays, the political "weight" of agricultural interests in national politics, and the importance of farm connections to the LDP. MAFF also constitutes a powerful vested interest in its own right, with a considerable stake in the maintenance of agricultural support and protection. It has little incentive to advocate deregulation of the agricultural economy when relaxation of controls would inevitably undermine its fundamental rationale and remove much of the need for government subsidies to agriculture.

MAFF has established multiple organizational linkages to assist in the process of administering the agricultural economy. These auxiliary organs are called government-affiliated agencies (*gaikaku dantai*) and number in the hundreds. They operate under varying degrees of MAFF control, with funding derived in differing proportions from MAFF's budget and other sources. In conducting public policy functions they provide a range of private services to group members, principally other agricultural organizations, but in many cases they also form an important channel for the distribution of agricultural subsidies and, in some instances, for funding generated by state trades in farm products.

The agricultural *gaikaku dantai* provide a lucrative source of employment for officials who have retired from MAFF, while some invite farm politicians into leadership positions as a means of expanding their influence over the allocation of subsidies. These intermediary agencies, like MAFF itself, have developed a vested interest in the maintenance of government support to agriculture, both as a basis for group functioning and as a source of financial benefits. Collectively, they form a substantial organizational obstacle to the abolition of regulatory controls on the agricultural economy and the largesse flowing from high levels of government intervention.

Political Institutions

Another vital element in the system of agricultural support and protection has been the relationship between agricultural interests and party politicians, particularly politicians from the LDP, which has relied on the votes of farmers and rural dwellers to win seats crucial to the maintenance of its parliamentary majority. Lacking its own organization in the countryside, the LDP has also turned to the organized power of the *nokyo* to help ensure electoral victories for its candidates in rural and semi-rural constituencies. Protection of rural interests has, therefore, been deemed essential to the party's survival in government.

The LDP has effectively hijacked the more vote-productive components of agricultural assistance and protection, such as price support, opposition to agricultural trade liberalization and subsidies for land improvement and rural infrastructure, as deliberate mechanisms for gathering votes. The electoral system in use for elections to the House of Representatives up to 1993–94, in which multi-member constituencies required competition among individual LDP candidates, encouraged a system of patronage politics linking individual parliamentarians to localized agricultural and fisheries interests, particularly through the delivery of geographically specific public works projects. The new electoral system, used for the first time in 1996, has retained some of the features of the old, notably enforced intra-party competition.

The LDP has had a strong incentive to adopt policies that would preserve not only the loyalty but also the size of its rural support base. Income support and the other forms of agricultural regulation and assistance have encouraged large numbers of people to stay in farming, cultivating small plots of land even if only on a part-time basis, thus preserving the fundamental structure of agriculture and keeping large numbers of LDP supporters in the countryside. If defined as persons 20 years old and over residing in farm households, the agricultural electorate has remained a relatively significant component of the national voting population. Although the number of farm household voters declined from 20.1 million to 15.9 million between 1960 and 1980, and fell even further to 12.1 million in 1996, this still represented 12.4% of the national electorate in that year, in spite of the fact that the farm vote more than halved in percentage terms between 1955 and 1979 and almost halved again between 1979 and 1996. Considering the period of time covered (1950 to 1996), the decline has not been particularly dramatic. In spite of the fact that the LDP's main supporting groups have diversified over time and its support rates in urban areas have risen, farmers have remained the LDP's traditional constituency and the core of its electoral support base. Indeed, it could be argued that the defection of LDP politicians to other more urban-orientated parties in 1993–94 merely served to underscore the LDP's identity as the farmers' party (George Mulgan 1997b pp. 878–90).

Electoral malapportionment, meanwhile, guaranteed that farmers' votes always continued to be more significant than their absolute numbers. The effect of malapportionment of voters among parliamentary seats was to over-represent the least populated districts and under-represent the most densely populated constituencies. The electoral reforms of November 1994, which eliminated the most extreme cases of malapportionment, have made it more difficult, but not impossible, for politicians to cater selectively to agricultural interests (George Mulgan 1997b pp. 893–95).

In the future Japan's voting population will continue to urbanize, necessitating a greater orientation among all parties towards attracting the votes of city dwellers. Farm household population will continue to fall, and as a result the farm vote will continue to decrease. Lack of successors for family farms may produce an even more rapid reduction in farm household numbers. Nonetheless, farmers in the short to medium term will remain a valuable component of the electoral support base of the LDP, and, if the party remains in government, either ruling on its own or in coalition, it will no doubt seek to preserve this rural connection by appropriate policies.

Pressures for Change

The Japanese system of agricultural protection, publicly justified as sustaining food security and rural values, and materially cemented by the mutually supporting interests of farmers, agricultural organizations, bureaucrats and politicians, has not been without its opponents, both domestic and external. Of prime importance has been international pressure (*gaiatsu*) from foreign governments seeking access to the Japanese food market (George Mulgan 1997a pp. 174–205). Much of this pressure has come from the United States in direct bilateral negotiations, as well as from other agricultural exporting countries, such as EU members, Australia and New Zealand. Multilateral trade forums, notably GATT, have also been influential.

Conflict has usually centered on the issue of access for particular products to the Japanese market and has often lasted over several decades. For instance the disputes between Japan and the United States over citrus and beef began in the early 1960s and were not finally resolved until the late 1980s, while imports of apples were liberalized in 1971, but banned until 1994 on quarantine grounds. The rice market issue arose later, in the mid-1980s, becoming pivotal in the Uruguay Round of GATT negotiations (1986–93). It was during the mid- to late 1980s, however, that the most serious escalation of trade pressure on Japan took place. The United States reacted negatively to the exponential growth in Japan's bilateral trade surplus during that period, while the Uruguay Round, by placing agriculture at the center of negotiations on freeing world trade, also greatly increased the level of international scrutiny of Japan's general system of agricultural support and protection.

Criticism of this system has also been articulated by a number of domestic institutions and interests. Within the bureaucracy, MAFF has faced opposition from three major ministries: from MoF, eager to reduce budgetary outlays to agriculture in the face of a growing fiscal deficit; from MITI, aiming to encourage moves towards trade liberalization in the interestsof Japan's export-oriented manufacturers; and from the Ministry of Foreign Affairs, seeking to maintain good relations with the United States and to foster Japan's reputation as a good international citizen.

Business groups, too, particularly those representing Japan's major export industries, have been opposed to the high levels of agricultural support and protection, because of the burden on taxpayers in general and therefore on corporate tax rates in particular, and because they threatened Japanese access to other markets, particularly in the United States. Among these groups, the most vocal has been the influential Federation of Economic Organizations (*Keidanren*), representing the interests not only of Japanese industrial exporters but also of food processors using relatively expensive domestic inputs.

The mass media have also acted as forums and mouthpieces for the perceived interests of consumers and taxpayers. Most major newspapers have consistently supported a liberalization agenda, reminding their readers of the high cost of food in Japan compared with world prices, a comparison that steadily became more invidious as the yen appreciated against the US dollar from 1985 onwards.

These domestic interests, however, would not have been sufficient on their own to force any degree of liberalization on the powerful rural sector, or its allies in MAFF and the LDP. No political party was prepared to take up the cause of consumers against the farmers. Parliament periodically resolved unanimously to retain agricultural import barriers for the sake of maintaining food self-sufficiency. The crucial, determining factor in favor of liberalization was, therefore, international pressure, which domestic forces favoring liberalization were able to harness as a means of encouraging a process of reform (George Mulgan 1997a pp. 200–5].

Even then, the extent of reform has been limited and its timing drawn out, following a familiar pattern of small concessions, reluctantly conceded and gradually implemented. The Uruguay Round Agreement on Agriculture eliminated all quantitative barriers to agricultural imports, except for rice, yet many of the supports and controls remain in place, including production subsidies, price support and stabilization schemes, state trading and, last but not least, relatively high import tariffs.

Since the Uruguay Round was completed, most administrative prices for farm products have been frozen at 1991 levels, only showing a declining trend in 1997. Actual pricing systems have also remained substantially the same, in spite of new arrangements for imports in some cases. ALIC retains some import rights for designated dairy products and silk, while, in the case of beef, high tariffs and domestic stock purchases by ALIC continue to protect producers, while quantitative limits remain on rice imports in a state trade administered by the Food Agency, which also trades in imported wheat and barley. Although the present minimum access arrangement for rice will expire in the year 2000, Japan has not agreed to liberalize rice beyond that date. It has merely agreed to negotiate the matter in 1999–2000 under the auspices of the World Trade Organization. In the meantime, the Food Agency's continuing controls on the marketing of both domestic and imported rice enable it to frustrate the sale of foreign rice.

In addition, large quantities of budgetary subsidies continue to be poured into agriculture. As a political gesture to compensate farmers, MAFF and the LDP worked out a comprehensive package of subsidies amounting to ¥6.01 trillion in 1994, including ¥3.55 trillion for agricultural public works projects and ¥2.46 trillion for non-public works projects. The main pillars of agricultural support and protection thus remain comparatively intact.

In its fisheries policies too, Japan has faced international pressure from other governments, in this case seeking to regulate and constrain Japan's fishing activities outside its own coastal waters, rather than to open up its market to greater competition. Japan has been forced to make a series of concessions, both in the interests of other countries' coastal fishing industries – notably, its reluctant acceptance of 200-mile EEZs in the 1970s – and for the purposes of conserving stocks on the high seas, such as under the UN-sponsored moratoriums on driftnet fishing in the South Pacific (1990) and globally (1992); the multilateral agreements among North Pacific fishing nations restricting fishing for salmon (1992) and pollock (1993); and the convention signed with Australia and New Zealand for the conservation of southern bluefin tuna (1994)). Japan also came under pressure from the International Whaling Commission in the 1980s to accept the international moratorium on commercial whaling.

While conservation of international fishing stocks may be in the long-term interests of the Japanese fishing industry, the government's aim has consistently been to maximize its own harvest, with little immediate concern for conservation. In this, it may be seen as responding to domestic political pressure, from both the industry and consumers. In contrast to agriculture, no significant internal constituency supports external demands for conservation. Japan's willingness, if somewhat reluctant and belated, to accept the restrictions of international treaties and international law, even against the predominant perception of its own interests, nonetheless underlines its readiness to play the role of a responsible and law-abiding international citizen.

The Future

Fisheries policies are likely to center around maintaining supplies for consumers and managing the effects of long-term reduction in the scale of the domestic fishing industry. Agricultural policies will be faced with the so-called "crisis in agriculture" occasioned by the aging of the farming population, a lack of successors to farms, and the growing incidence of farmers abandoning arable land.

At the same time, pressures will mount on MAFF to reduce agricultural subsidies, because of budget deficits; to find a workable solution to the perennial problem of expensive small-scale farming; and to create agricultural markets that better reflect supply and demand conditions. Continuing regulation of rice production and distribution are causing a massive surplus which undermines the government's stockpiling policy and incurs large losses for the agricultural cooperatives.

For all its much vaunted economic and political clout, Nokyo is also facing various organizational difficulties. It is proving vulnerable to the decline of agriculture and to changes taking place in domestic markets. In the new era of deregulation and intensified competition, Nokyo's protected financial and economic institutions appear inflexible, anachronistic and slow to adapt to fundamental changes in their economic and policy environment.

For all these reasons, it is now widely accepted that agricultural policy has reached a crossroads. In late 1996, an advisory panel to MAFF submitted a report on its review of the Agricultural Basic Law. It was critical of the failure of the nation's agricultural policies to raise agricultural productivity sufficiently through expansion of the scale of farming operations. It called for further study of the feasibility of formulating a new Basic Law, emphasizing the need to ensure a public consensus around such issues as a stable food supply as well as the stability of farm management and the revitalization of rural areas. In April 1997, an Investigative Council on Basic Problems Concerning Food, Agriculture and Rural Areas was established as a new advisory committee to the Prime Minister. Its main task is to formulate a new Basic Law to replace the existing one. The Council will hand down its report in 1998, perhaps presaging fundamental change in the basic philosophy of agricultural support in Japan.

Further Reading

ABARE (Australian Bureau of Agricultural and Resource Economics), *Japanese Agricultural Policies: A Time of Change*, Policy Monograph No. 3, Canberra: Australian Government Publishing Service, 1988

A good general source for gaining an understanding of basic Japanese agricultural policy, but now somewhat out of date.

Bergin, Anthony, and Marcus Haward, *Japan's Tuna Fishing Industry: Setting Sun or New Dawn*, New York: Nova Science Publishers, 1996

Contains a general summary of fisheries facts and policies, as well as more specific discussion of tuna fishing.

East Asia Analytical Unit (EAAU), *A New Japan?: Change in Asia's Megamarket*, Canberra, ACT: Department of Foreign Affairs and Trade, 1997

A general overview of the Japanese economy, with a chapter on agriculture.

George Mulgan, Aurelia (1997a), "The Role of Foreign Pressure (*Gaiatsu*) in Japan's Agricultural Trade Liberalization", in *The Pacific Review*, Volume 10, number 2, 1997

Provides a considered assessment of the impact of foreign as opposed to domestic pressure on the opening of Japan's agricultural markets.

George Mulgan, Aurelia (1997b), "Electoral Determinants of Agrarian Power: Measuring Rural Decline in Japan", in *Political Studies*, Volume 45, number 5, December 1997

Examines how the Japanese farm sector illustrates the key measures of rural electoral influence in advanced industrialized societies and the principal causes of decline in the electoral basis of agrarian power.

George Mulgan, Aurelia, *Farmers in Politics: Rice-Roots in Japan*, forthcoming

Japan Agricultural Cooperatives (JA) Homepage, www.rim.or.jp/ci/ja/

Mostly in Japanese, but includes selected articles in English, giving major speeches of leaders, some up-to-date statistics on agriculture and agricultural employment, and discussion of agricultural policy issues.

Japan Institute of International Affairs (JIIA), *White Papers of Japan*, Tokyo: The Japan Institute of International Affairs, 1994–95 edition of annual publication

Useful compilation of annual abstracts of official reports and statistics of the Japanese government.

MAFF Update, www.MAFF.go.jp/, various dates

Mostly in Japanese, but includes selected articles in English; a good up-to-date source for all major statistics on agricultural production and prices, as well as explanations of major policy initiatives and discussions.

Nihon Nogyo Nenkan Kankokai, editors, *Nihon Nogyo Nenkan* [Japan Agricultural Yearbook], Tokyo, Ie no Hikari Kyokai, 1997 edition of annual publication

A comprehensive outline of all major developments in agriculture, agricultural organizations (especially Nokyo) and agricultural policy for the year, plus relevant statistics.

Norinsuisansho Tokei Johobu, *Poketto Norinsuisan Tokei* [Pocket Agriculture, Forestry and Fisheries Statistics], Tokyo: Norin Tokei Kyokai, annual publication

Useful for up-to-date agricultural production statistics.

OECD (1996a), *Agricultural Policies, Markets and Trade in OECD Countries: Monitoring and Evaluation*, Paris: OECD, 1996 edition of annual publication

Useful for comparing agricultural policies in OECD countries, although analysis tends to be superficial for any one country; contains useful statistics for rates of protection.

OECD (1996b), *Review of Fisheries*, Paris: OECD, 1996 edition of annual publication

Contains comparative data on fisheries in OECD countries and a chapter summarizing trends in Japan.

Dr Aurelia George Mulgan is a Senior Lecturer at the University of New South Wales, Australian Defence Force Academy.

Chapter Six

The Production Revolution in Manufacturing

Charles McMillan and Tom Wesson

In the ebb and flow of the industrial revolution and its aftermath, it seems perverse to speak of "Japanese" manufacturing, let alone American, British or European. In and of itself, manufacturing, like medical science or the weather, is hardly culture-bound, nation-specific, or unique. In medicine, research on the circulation of the blood, heart disease, or cancer is not greatly enhanced by comparisons between countries – or is it? Medical researchers and public policy analysts alike have become acutely aware of national differences. Japan, for example, spends vastly less money on health care than the United States, yet the differences in measures of longevity, both male and female, and in measures of child and maternal deaths at birth are stunning. On measures of basic educational learning and achievement on test scores in science and mathematics, similar and dramatic differences exist.

Clearly, institutional and historic factors are at play. Cultural differences, so allegedly important in earlier writings on the spread of multinational firms, are now recognized as less relevant. Of more relevance are organizational and managerial approaches to the production of economic output and of knowledge. In this context, the Japanese manufacturing approach has three unique features: the origins of its philosophy; its application to manufacturing and services companies alike; and its fundamentally revolutionary character. In this chapter we address these issues in a comparative context, and discuss the historical context of Japanese manufacturing and the underlying striving for ever higher benchmarks towards global leadership. We take as a framework Joan Woodward's classification of production systems, namely "craft", "mass production", and "continuous process", and consider how state-of-the-art Japanese manufacturing selectively draws on the cost and quality advantages of each type.

An Overview of the Japanese Production System

Despite the stereotypes related to Japanese processes, from low-paid sweatshops to the glistening rhythm of the automated factory manned by industrial robots, there can be little doubt about the powerful influence of modern forms of scientific management, originating in the ideas of Frederick Taylor. This production emphasis is clear in the tactical policies of Japanese factories. Products exported to the world market have earned unrivaled reputations for quality and reliability. Comparative studies show that plant and equipment in Japan are replaced twice as rapidly as in Britain, and faster than in the United States. Statistics show that stock turn, that is, inventory turnover, is substantially higher in Japan than in other OECD countries. Waste reduction policies and energy efficiency are unrivaled. Are these results simply a function of resource scarcity? Are there cultural issues at work here, especially the receptivity to new ideas? Or is the issue of production emphasis more fundamental to Japan's entire strategy on new industries and technologies?

Japan's emphasis on manufacturing and production processes must be understood as part of the country's obsession with catching up with the West. From the Meiji period (1868–1912) to the present, Japanese government and industrial leaders have recognized that, as a resource-poor and technology-dependent country, Japan had the most to gain from improving manufacturing output and productivity. Further, the costs of many resource inputs for Japanese industry have been much higher than in Europe or the United States. Until sharp increases in oil prices changed the world energy situation in 1973, North American managers had never faced up to these constraints, and even in the 1990s they understand them only imperfectly.

Japan's advantages thus include a century's experience of managing resource scarcity. Despite this, Japanese management is influenced by the dynamics of internal organization. When employees are recruited from rural villages or the best universities, they bring to the organization a desire for security and long-term careers. The "lifetime" employment system in large companies reduces managerial flexibility by making wages essentially a fixed cost, but it also serves to promote experimentation and innovation with technology (see also Chapter 13). The relative absence of a basic conflict between worker and machine reduces the social backlash of technological change or even disputes about job demarcation.

Today, Japan is at the center of the major technological changes associated with automation. Japan is revolutionizing western concepts of mass production, including the use of robotics, computers, and management software in relation to monitoring quality; the transformation of craft building to mass production; inventory management; and distribution. Japanese institutions and individuals together hold about 50% of current global patents, a key to subsequent commercialization. The Japanese dominate foreign patent listings in the United States, at 21% versus 9% for Germany, and 3% each for France and Britain. The Japanese know the challenge of becoming world-class scientific innovators, but what is less well appreciated by foreigners is how well Japan can afford to pay for a science-based economy. Japan outspends all other OECD countries in civilian research; 20,000 Japanese study in US universities; Japan rivals the United States in research employment, at about 65 for every 10,000 workers. By the end of the next decade, Japan will be a world leader in the next century's technologies: unmanned space systems, artificial intelligence, biotechnology, advanced materials, high-energy physics, media power, and many commercial applications of advanced pharmaceuticals. Leading-edge Japanese corporations approach strategic planning in a different way than North American companies, while the operating imperatives are focused on such key areas as brand names, market share, competitive cost position, and the like. Yet these issues are not central. Production remains the key focus of Japanese corporations – a reflection, perhaps, of the engineering profile of corporate demography, but also one of the new realities of technological development.

Japanese corporations spend far less time than western companies do on the "strategizing preoccupation" with such issues as market share, product positioning and product-market mix. Indeed, for most Japanese managers these are symptoms or outcomes of more fundamental variables, embedded in the production system of the organization. Yamaha's strengths in "synthetic sounds", Sony's skills in "production miniaturization", and Toyota's approach as a "systems assembler" illustrate the exceedingly sophisticated approach to gaining competitive advantage in global markets.

Japanese companies must balance a range of factors on a day-to-day basis not readily appreciated in the West, including the appreciation of the yen, domestic versus Asian sourcing, and external contracting of components. Japanese technological leaders recognize these trends, hence the need to innovate using standardized, off-the-shelf components, developing truly world-scale benchmarking in both production and marketing. The diffuse decision-making patterns of Japanese companies and the relative absence of rational norms to "control" future events in the long term make operational decisions supportive of long-term strategy. S. C. Wheelwright has emphasized this point as follows:

In Japan, the integrity of the production system and strategic purpose comes first. But Japanese manufacturers also realize that decisions at the level of operations can, if handled in a wise and consistent manner, have a useful cumulative effect at the level of strategy. Experience has taught the Japanese the value of placing even short term manufacturing decisions at the service of long term strategy – a lesson that American companies have learned only imperfectly.

The Origins of Japan's Production Strategy

The phenomenal success of Japanese manufacturing in industrial products, from motors and ships to consumer goods such as electronics of every description, is usually dated from World War II. The influence of US management theorists in industrial engineering, statistics, and quality control is often credited with shaping Japanese managerial practices. Indeed, the historic roots of Japan's postwar recovery stem from this very strength. Japan's centuries-old skills in crafts of all sorts – pottery, ceramics, papermaking, printing, carpentry, textiles, all still carefully nurtured – have contributed enormously to the supply of highly literate and dexterous workers and to an appreciation of craftsmanship and such spin-off capabilities as product design, miniaturization, and product simplicity (see Patchell). Historically, leading-edge Japanese manufacturers have emphasized engineering processes through time and motion studies, simplification of work, elimination of waste, and improvement of plant layout; and their skills and manufacturing outlook have played their part, as shown by documented studies of companies as diverse as Toyota, Kawasaki, Honda, and Hitachi. They have also benefited from having the marketing function externalized to Japan's specialist and diverse trading companies (*shosha* – see Chapter 7). Certainly, the desire to learn from western technology has placed production – especially issues of process technology – at the forefront of post-war redevelopment (see McMillan).

Foreign experts have had a lasting impact on Japanese management. Fifty years separate the two waves of foreign intellectual ideas. The first, inspired by the time and motion studies developed by Frederick Taylor and his disciples, such as Henry Gantt, came at the turn of the 19th and 20th centuries. The second, characterized by the statistical quality control theories of W. Edwards Deming and W. Juran, arrived after the Pacific War. Both waves involved the core activities of industrial engineering, or, to use its historic name, "scientific management". Indeed, scientific management became vastly more developed in Japan even than in the United States, in part because of the more central role of engineers in management strategies, and because in electronics and robotics the core topics of Taylor's early studies are seen as critical to global marketing success.

Taylor's influence in Japan came about with the translation of *Principles of Scientific Management*, first published in the United States in 1911. The Japanese version, published under a title which may be translated as *The Secret of Saving Lost Motion*, sold almost 2 million copies. As early as 1908, Iwatare Junihoko of Nippon Electric, Japan's first joint venture, studied scientific management at Western Electric in the United States, and subsequently introduced it to his company. Similarly, Yasukawa Daigoro visited Westinghouse Electric during 1913–14, and went on to introduced the bonus system at Yasukawa Electric Equipment in 1915. By 1919, the government was subsidizing the Labor Management Cooperation Society (*Kyocho Kai*), which had a section on scientific management, the Industrial Efficiency Institute (see Nakase).

In Japan, then, scientific management gained an immediate and impressive foothold, and it has influenced management thinking for generations. In Britain, by contrast, Taylorism was totally ignored until World War I, and, once it had become known, it was greeted with universal hostility from management and labor unions alike. Why was this so, and what lessons does the introduction of scientific management suggest about management in these two countries?

In contrast to the hostility to Taylorism in Britain, Japan's reception was open and widespread. General economic conditions obviously were favorable. There was a fear of serious conflict between labor and management, a perception of inefficient business and government practices relating to imperialist conflicts, the rise of the *zaibatsu* holding companies (see Chapter 10), and a felt need to reform personnel practices. It was left to Japanese managers to take the initiative. Two important books on Taylor were published, one by Hoshino Yukinori, who was in the United States when Taylor brought out his *Principles of Scientific Management* – itself published in Japan in 1912 as *Kengaku Yoroku* (A Report on Observations) – and the other by Ikeda Toshiro, with a title that may be translated as *Secrets For Eliminating Futile Work and Increasing Production*, of which more than 1 million copies were sold. The President of Mitsubishi Goshi Kaisha distributed 20,000 copies of it to his employees, and the head of Kawasaki Shipbuilding Yard gave 50,000 copies to his workers. Numerous articles were printed on various aspects of Taylorism, and books appeared on Taylor's theories of differential piece rates, and on the bonus system developed by Taylor's colleague, Henry Gantt.

This legacy of Taylorism and scientific management in Japan is significant, because the historic origins of Japan's strong orientation towards engineering production help to explain its present manufacturing strengths. Inventory management, reduction of waste, miniaturization, quality control (QC) circles, robotics, rapid cycle times and automated factories – all these elements of "lean production" have become today's agenda for industrial engineering, but their practice is more advanced in Japan than anywhere, even the United States.

As outlined in Table 6.1, traditional mass production engineering as applied to assembly line industries – automotive plants are the classic example, but there are numerous others – has favored economies of scale, relatively long production runs, and minimal product diversity, except through add-ons, special features and interchangeable components. Economies of scale result from declines in unit costs as absolute volume increases over a given time period. Long production runs decrease costs because of learning and sequential improvements with cumulative volume. Production specialization adds to productivity because of the decreased need to change equipment and to incur set-up costs as a result of product diversity. Inherent in this kind of automation is a system that manages centralized data flow, while still directing and controlling material flow and production conversion.

Consider the traditional automobile assembly line, which uses capital-intensive processes for extrusion and pressing. Increasing capital intensity may lead to higher productivity, mainly by replacing workers and reducing direct operating costs. However, novel management systems have turned this traditional approach on its ear. Instead of having large-scale, capital-intensive investment geared to relatively inflexible outputs, flexible manufacturing systems, operating with such processes as quick die change and just-in-time inventory, turn the old assumptions upside down, allowing dramatic improvements in productivity. In traditional manufacturing, production diversity forces up unit costs because of the downtime incurred in setting up equipment. Quick die change technology permits set-up time in minutes or even seconds, rather than weeks, days or hours. Quick set-up times allow more small lot production, hence greater product diversity. An indirect benefit is provided by lower in-process inventories and instantaneous lead times for suppliers. Learning effects disappear, since the system is as smart on the first unit as the thousandth: for production purposes, the learning curve is flat. Indeed, the integrated production organization, in its extreme form, is characterized by production changeover costs that approach zero. Economic order quantities change from large lots to one unit.

The cost and quality advantages of Japanese manufacturing originate in painstaking strategic management of people, materials and equipment – that is, in superior manufacturing performance. Japan's manufacturing strengths cannot be reduced to simple formulas or slogans, but one real advantage, widely understood in today's enormously competitive market, is product-specific cost. As with the philosophy of total quality management, where the aim is to eliminate defects in the early stages of design and production planning, the concept of design to cost (or target cost) is central: assess the potential, but realistic, selling price and work backwards through each stage of the value chain. The goal is to assess the appropriate level of value added, but in reverse (which in fact is a value-deduction chain); and real cost, less profit, less mark-up for each stage of the value chain. Thus, for example, the various contributions to the cost of a car may be assessed as follows:

- 50% on purchasing components;
- 20% on assembly;
- 2% on distribution logistics;
- 13% on sales and marketing; and
- 15% on dealers' costs and margins.

This approach is driven by markets and by customers, which means that managers and workers must understand external relationships and what the market will accept. This concept is price-driven costing: defining what the customer will pay, and then working backwards – to costs, product design, engineering and manufacturing – all the costs that enter the barriers to lowering prices to customers.

Design to cost illustrates a recognition that companies and factories succeed as a team or as a network of strengths, not as an isolated unit. This is the power of Japan's *keiretsu* model of industrial groups (see Chapter 10), perhaps Japan's unique contribution to organizational theory (see Ferguson).

Like total quality control, the Japanese focus on dynamic cost analysis has two features: first, an assessment of costs for the main organization and for the network of suppliers; second, a dynamic approach that illustrates the constant pressure to reduce costs, enhance productivity, and improve product quality. Nobody is immune, neither workers, nor suppliers, nor subcontractors. Consider one simple case study. When Sony produced its new camcorder, developed in great secrecy and with considerable skill – involving a reduction in the number of moving parts, enhancement of quality, and reduction in production costs – not only did Sony itself achieve all its production goals, an almost unbelievable accomplishment, but Sony's record, once achieved, was copied by eight other electronics companies within six to 12 months, each firm adding its own innovations and improvements. This type of innovation involves high risks but leads to high payoffs from which winners reap huge gains.

To take another example: with the rise in the external value of the yen in 1993, where each unit rise in the yen against the US dollar meant a loss or profit of ¥10 billion (US$100 million), every detail came under scrutiny at Toyota and other car manufacturers. More than 1,600 detailed cost savings were introduced, from reducing the size of the bracket to hold an electric wire (costing ¥20 for each car) to reducing the size of the curve in the windscreen. At Toyota and Honda, fully 80% of cost savings come from their suppliers.

The emphasis on scientific management in Japan has led to a variety of widely used production tools in industry. These hardware systems are not well-known in the West, although companies in the United States and Europe are adopting them in whole or in part. The tools are not novel in themselves, but their extensive application and their relationship to other strategic areas, such as marketing and personnel, make them very significant. In particular, progress cost curves, *kamban* inventory techniques, and robotization are major production tactics which reinforce company-level marketing strategies, and form part of the broad mosaic of Japan's societal strategies to gain technological superiority.

Experience Curves

Experience curves have a variety of other names – learning curves, progress cost curves, progress functions – and a history that can be traced back to 1938 (see Yelle), with initial reference to the observation that, as the quantity of units manufactured doubles, the number of direct labor hours it takes to produce an individual unit decreases at a uniform rate. Such phenomena are related to, but distinct from, economies of scale or technological advances; and it has been found that they tend to vary by product and industry. For example, US research has revealed that the experience effect was about twice as great in assembly (26%) as in machining (14%), because assembly work involved greater complexity and uniqueness. Studies carried out in a number of

Japanese industries point to the fact that various firms, in such industries as color televisions, motorcycles, jeweled watches, cameras, small cars, and steel, have superior records of accumulating experience and lower cost. The actual rate of decline is product-specific, from about 12% in automobiles, and 15% in color televisions, to as much as 40–50% in semi-conductors and integrated circuits. Put differently, compare automobiles and auto parts. The productivity of the Japanese automobile industry exceeds that of its rivals in France, Germany, Italy, Spain and the United Kingdom by around 30%, while in the components industry Japanese productivity is around 2.5 times greater. Toyota's target for new model development is a stunning 12 months; Detroit's average is around 28 months.

As a management tool, the experience curve is not an automatic or purely engineering phenomenon: experience effects must be planned. For one thing, there is a "learning by doing" effect, whereby tasks carried out for the first time take longer than with experience, such that unskilled workers take more time than skilled and experienced workers. Feedback and scientific work study are parts of this process. There must be a major managerial effort to reduce costs by improvements in work layout, better supervision, simplification of product design, and more rationalized purchasing, inventory control, and scheduling. There also must be a recognition that experience curve effects may not apply to all products in the same way (for instance, semiconductors versus television assembly); in all situations, for instance, when raw materials account for a significant percentage of value added; or where technology changes, as when transistors replaced vacuum tubes, or when micro-circuits replaced mechanical parts in adding machines and calculators.

The analysis of experience curves has many implications for management. In the first place, management must recognize their planning use, and this means that good engineering data are essential. As noted below, the use of experience curves in design to cost is meaningless without a solid data base. The experience curve also provides an effective means of coordinating production strategy with research and development, and with marketing. Consider the case of pricing. Cost-conscious producers may develop a strategy of using different pricing levels, depending on the degree of competition and the stage that the life cycle of the product has reached. A firm can adopt a strategy of price exploitation in order to build profit levels but not necessarily to gain market share, assuming competitive conditions. Pricing for market share, in contrast, suggests a strategy of sacrificing profits, and medium-term cash flow, to gain longer-term market share. When competition intensifies, as new entrants are attracted by price exploitation, higher cumulative production allows for a shift to production based on lowest-cost performance.

This analysis corresponds precisely with the practices of Japanese manufacturers in scores of products. However, there are other factors that need to be considered. For one thing, a high growth rate requires increases in capital investment for the newest, best-practice equipment and processes. Without them, a firm may drop off the experience curve. Another related issue is the strategic time horizon. Firms emphasizing short time horizons and quick paybacks are likely to forgo long-term market share and cumulative learning. Such cumulative learning may well mean that, barring some technological breakthrough, a firm which falls behind can never catch up.

The acceptance of the importance of experience curves is indicated in the technique of design to cost. The particular approach of design to cost is aided by another technique, namely reverse engineering. Instead of designing a product on paper with engineering specification laid out in blueprints, companies take an existing product foreign to the firm, and break it down into component parts and processes; learn the basic design but improve on it; and reassemble the product on the basis of creative improvements. Reverse engineering and design to cost mean that a particular cost becomes the production constraint, and the manufacturing process is designed to meet that target. Today, this approach is not uniquely Japanese, but it is better understood and more widely adopted in Japan's advanced manufacturing industries than it is anywhere else.

The *Kamban* Inventory System

The experience curve is a technique which relates to the design and technology of a particular product. However, the most radical innovation in manufacturing, virtually unknown in western countries until recently and even less widely practiced, is the *kamban* (or *kanban*) system, often referred to in English, and even, sometimes, in Japanese, as the "just in time" system, or "lean production". *Kamban* is an approach to inventory and parts delivery that combines just-in-time delivery with a rigid scheduling flow to reduce waste and inventory costs, from incoming parts, through work in process, to the finished product.

Bruce Henderson, formerly head of the Boston Consulting Group and a leading analyst of corporate strategy, has described this approach as follows:

> Just In Time is a colloquial expression for describing a philosophy and technique that may be the ace in the game of manufacturing productivity. In the history of the industrial revolution, it may rank with interchangeable parts, precision gauge blocks, assembly lines, time and motion studies, and powered conveyors.

The genesis of the *kamban* system owes much to the brilliant thinking of Ono Tai'ichi (often also referred to in English as Ohno Taiichi), a former Vice-President of Toyota and one of the great manufacturing geniuses of this century. A production concept which had an especially strong impact on Ono was the supermarket as it had developed in the United States. A supermarket is not only a vast storehouse of numerous products, it is a point where customers are directly confronted with the last link in the distribution chain, and therefore with any damaged goods, perishables, or products in short supply. Empty shelf space or gaps become the trigger mechanism for staff to replace products. Additionally, large supermarkets usually mean high rents, placing a premium on layout. The ideal is some optimum space which is available for large inventories (warehousing), but also adaptable for quick turnover of and easy replacement of stock.

These issues in supermarket management impressed Ono and the management group at Toyota, who began asking themselves whether a system applicable in a service institution such as a supermarket could have any relevance to the mass production style typical of the car industry. After all, the entire car industry is extremely complex. It consists of thousands of individual parts assembled in a stepwise sequential fashion. There are several different models and permutations for specific components. The models themselves change every four to five years. Further, the car industry, like many other manufacturing industries – consumer electronics, aerospace, telecommunications – is at once labor-intensive, involving skilled and semi-skilled workers, and capital-intensive, involving high investment in land, plant equipment and distribution.

Conventional production planning combines a sales forecast projected over a given time period with a production schedule for each stage of the work flow, in an ordered sequence. Holding buffer stocks (inventory) where potential bottlenecks in delivery can occur, or where small lot production of components is uneconomical, achieves managerial flexibility in coordination and information. Standard production planning is a critical aspect of organization design, where information on demand scheduling leads to "uncoupling" of production scheduling and product delivery decisions by inventory. The benefits of such an approach stem from the reduction of uncertainty caused by fluctuating demand for products. In essence, Toyota and other manufacturers have switched from a system of estimated production, based on final demand, to job order production, based on minimum inventory of finished product. The key to this approach is the availability of flexible and short set-up times for assembly line machinery.

Most precepts of organization design are consistent with this classic approach to production. The basic aim is one of managerial control, even if inefficiencies or trade-offs are not explicitly stated. Many costs, for example, are not known or even calculated. Imbalances between high-turnover stocks and low-turnover or dead stocks are often not corrected. Managers may prefer to cushion the risks of their own errors by adding to buffer stocks, usually to reduce exposure to disruption, such as workers' absenteeism, breakdowns of machines, forecasting errors, or worse, to pinpointing directly the production line fault. Additionally, increased management control and the desire for flexibility decrease the need to rely on direct labor for anything short of the physical job requirements of the worker (see Argyris).

The Toyota system – now widely emulated as the world standard by other carmakers and manufacturers – incorporates the costs and benefits of conventional product planning. The specifically Japanese environmental need to husband raw materials and to reduce factory waste led to Toyota's first principle – namely, to abandon or eliminate waste by supplying only the necessary parts, as and when they are needed. The key to this principle was to define, control, and minimize the lead time from initial entry of parts and raw materials to their actual use. As much as possible, in-process stocks would be eliminated, or reduced to the absolute minimum. Both inventory costs and labor time expended on handling materials are considered as surplus and thus waste – the key to lean production.

As a means to implement this policy, the Toyota team turned to Ono's idea, based on his understanding of supermarkets: parts would be supplied, as they are needed, just in time. Instead of projecting materials requirements from the beginning stage of production to each subsequent stage, they do the reverse. They start from the final stage of output and work backwards; instead of "push out", they use "pull out". Information flows in a reverse order, with workers in the final stage dictating the flow of parts needed from the prior stage, right through to the initial production point. In this way, each step in production is carefully synchronized, and parts flow from one step to the next just in time. The elimination of buffer stocks and the tight coupling of the line increase the inflexibility of the process. Interruptions in the flow of parts at any one stage break the continuity of the whole line. Stable flows and close integration of the whole line depend both on internal coordination of production stages, and orderly delivery of supply and parts. The design of specific jobs, and relations with suppliers, must reflect this on-line flow. In the 1990s, Toyota's production system is the benchmark model for manufacturing and service firms alike (see also Chapter 7).

Japanese automobile manufacturers operate with annual production plans, broken down into two-step monthly plans. The first step consists of production estimates for all types and models; the second step is an actual plan for production. Both are provided to subcontractors, which then determine their own requirements. It is at this stage that "production smoothing" takes place, that is, the translation of the monthly plan into a daily schedule, outlining production by shift and by model. By tying the average cycle time needed to produce one unit of any model, the overall daily production schedule, and with it, the daily parts required, can be determined. This overall scheduling is assisted by sophisticated computer programs, which have features similar to those used for planning material requirements.

On the assembly line itself there is a computer tape readout listing the requirements for each model as it moves along the line. The tape indicates the cycle time, and provides information on parts, accessories, colors, and the like. The availability of this information allows the worker to operate the *kamban* to control excess inventory.

The word *kamban* itself literally refers to a small plate or card, now plastic but originally paper, and measuring about three and a half inches by nine inches, which provides the information to workers at each production station. There are, in fact, two types of *kamban* card that each worker takes from the parts container of existing stock to the stock point of the previous stage. The production *kamban* is then left at the stock point, as a dispatch signal to replace the inventory used. The two *kamban* cards together form a real-time information system, indicating production capacity, stock usage, and manpower allocation.

In the early years at Toyota, *kamban* were used only in specific areas of the production system, and it took more than three years to introduce them throughout the company.

Today, the *kamban* system is not only applied in all Toyota's factories but has also been extended to its major suppliers. In this sense, the internal operations of a Toyota plant operate as an integrated, synchronized system, linked to an "invisible conveyor" extending to supplier lines.

To balance out production across all stages, so that no surplus stocks are built up and no production stage is short of components, Toyota and other manufacturers have succeeded in reducing the average lot size, usually by shortening set-up time from hours to minutes. A corollary, however, is that the maintenance of equipment must be very effective. Flexibility in production output is assisted in several ways. Because most workers operate several machines, reductions in job rotation and the hiring of temporary workers can increase output. Temporary workers are the first to go if demand slackens. Altering work shift times, by requiring early attendance or overtime, or by changing the cycle time, also allows flexibility in output. If necessary, regular workers, who are normally not laid off, may be shifted to other work, such as maintenance, training for set-up activities, or making parts normally purchased from suppliers.

A critical element in the operation and practice of the *kamban* system is the network of human relationships among managers, workers, and supervisors, such as Japan's software management systems of quality control, and *jishu kanri* – literally "autonomous management" – conducted through voluntary groups of workers contributing to managerial functions. Japan's system of using work groups as the basic structure of the labor force is well-suited for *kamban* scheduling. QC circles are assisted by groups of staff specialized in, for instance, engineering, accounting, or production control, who can be on call to help work groups solve problems. Perhaps the area of greatest success has been in the set-up times for particular processes, which make short lot production feasible and thereby greatly reduce the lead times for production planning. In addition, unlike in most western practice in organization design, Japanese companies use engineers directly with blue-collar workers. Indeed, they usually wear the same clothing and participate in the semi-annual bonuses. More to the point, such practices break down differences in status and other artificial barriers to open communications.

The differences in attitudes towards the relative autonomy of the worker are seen in Toyota's control of the assembly line. In the conventional assembly line, only an extreme event would merit stopping the entire line, and any unwarranted stoppage could lead to the dismissal of workers. Toyota and other Japanese companies have challenged the need for a constantly moving assembly line. At Toyota, a red cord runs the length of the assembly line and any worker can stop it. Stoppages are usually caused by equipment problems, shortages of parts, or discoveries of defects. In the center of the line in a Toyota factory is an enormous number board, something like a stadium score board, which indicates the specific spot where the line has been stopped, the area of the problem, and the time involved. The system reinforces the basic aims of eliminating waste and reducing inventory. In this approach, assembly production *per se* represents less than 10% of costs.

This explains why allowing workers to stop the line serves as a method of pinpointing the source of errors and directing resources to the exact trouble spot. Since all workers and foremen are fully aware of the impact of a stoppage, the strong possibility of detection acts as a deterrent to slipshod work or careless behavior. The origins of the idea can be traced to the textile weaving equipment at Toyota Automatic Loom Works, the company from which Toyota Motors was originally spun off: Ono Tai'ichi and several of his colleagues began their careers there. In the same way that the loom stops when wool runs out from a spindle, the line stops when a defect is discovered, or when an excess or shortage of parts occurs.

The managerial consequences of this system should not be misunderstood. Where the conventional production system, with its buffer stocks and decoupled production stages, gives a premium to managerial flexibility and control, the lean production system operates with different trade-offs. The line is virtually inflexible, with minimal buffer stocks and greater reliance on workers for inspection and quality control. Yet the high risk which at first glance appears to be a feature of the Japanese system also raises the awareness of individual workers, who come to understand the need to perform to ever higher standards.

The capacity of the *kamban* system for its internal operations is greatly fostered by its open consultation across functions, its decentralized planning among work groups, and senior management's willingness to challenge conventional assumptions of hierarchical control. In this respect, the general application of the *kamban* system is limited only by the willingness of management and workers to make it function. Japanese companies investing abroad have the advantage of designing their factories and floor layouts with the *kamban* principle in mind.

The more extensive form of *kamban*, however, involves the application of inventory control and delivery to parts suppliers. Here the principles of *kamban* are applicable to any country, although, in practice, there are import obstacles. The linkages between suppliers and parent companies go far beyond the simple relationship of buyers and sellers that remains common in North America and Europe. The parent company is not just a buyer of parts and components: it provides credit, equipment, machinery, technical knowledge, and managerial experience, often by transferring "retired" senior managers to become directors of supplier firms. More directly, parent firms also hold shares in the suppliers, even when the suppliers are also quoted separately on stock exchanges. This kind of organization explains why Toyota, for example, can contract out around 70% of its components, while General Motors makes around 67% of components internally. The impact on the cost structure of the automobile industry, and other similar industries, is revolutionary.

In North America, competition among large enterprises is generally confined to legal corporate entities. In Japan, by contrast, subcontracting relations add to the complexity of the corporate planning process, and make oligopolistic competition severe. Informal corporate planning linkages extend to major and minor suppliers, and to trading firms and banks (see also Chapter 10). Not only does this system strengthen the competitive edge of Japanese manufactures, by providing the parent firms with virtually guaranteed delivery schedules and assured product quality, it acts as a clear barrier to supplier firms, such as those from overseas, that want to break into the close family circle. It also explains the traditional reluctance of Japanese manufacturers to enter foreign markets through takeovers, rather than new start-ups. The reason is simple: their family group structure would be difficult to replicate. The forward planning systems of the manufacturers require close cooperation with the major and minor suppliers – Toyota alone, for example, has more than 200 – and this planning relationship is nurtured by close personal ties, consultative committees, and exchange of information. Technology, production techniques, and improvements in productivity are the main areas of cooperative effort, especially since the car industry is facing so many challenges. In the area of engines, for example, Toyota and Nissan have pursued the replacement of metal with ceramics, and all carmakers are forging new links with electronics suppliers to provide the new components, gadgets, and back-up systems, such as digital indicators for speed control or fuel gauges, that may account for as much as one fifth of a car's total cost in the 1990s.

Robotics

A visitor to Japan can recognize robots in the most unlikely places, from traffic direction on highways and in construction zones, to beautifully decorated and lifelike animals on display in department stores. Robots are widely used in hospitals, old people's homes, and nursing wards to perform basic tasks for sick and disabled patients. Even children are at home with robots, through a bewildering range of robotic toys and the extensive literature of comic books (*manga*). Television programs depict the use of robots in a wide variety of non-manufacturing fields: spraying insecticides and fertilizers on farms; inspecting and packing eggs; cutting trees in forests; working underwater in fish farms – such are the everyday examples educating Japan about the human side of robots. Surveys of robot manufacturers claim that there will soon be robots to look after handicapped patients, sweep streets, guide blind people, and feed bedridden patients. In other words, even though 98% of robots are used in manufacturing, the population is being educated for their general use in society at large. Japan today is the undisputed world leader in robotics.

Kawasaki Heavy Industries produced Japan's first industrial robot in 1968, under license from a US firm, Unimation Inc. As many as 150 companies are now engaged in robot production. Machine tool makers have become major players in robotics, because of the close technological link with numerically controlled machine tools. Originally designed in the United States, the first robots in Japan were imported in small numbers. Production engineers studied their uses and structures, and applied them to domestic production in specific tasks, such as welding and paint-spraying.

After a decade of study, improvement, testing and innovation, robotics in Japan became a growth industry, not only in the domestic market but around the world. The term "flexible manufacturing systems" became a generally accepted description of the transformation of traditional assembly lines, using electronics to replace mechanical systems and robots to take over many human functions. This general trend to more robotics is usually analyzed in human terms, such as the number of displaced workers and the threat of unemployment. In point of fact, the behavioral issues are relevant, but mostly in the context of doing away with the most boring, unsafe and unhealthy jobs – involving, for instance, exposure to heat or noise – as well as the need for upgrading workers' skills and retraining. In this respect, robotics represents no greater threat than any other kind of technological change in industrial production.

The real impact of robotics, and the issue of most concern to the Japanese, is the revolutionary change in assembly line production. For large firms, robotics may bring the opportunity to introduce enormous flexibility in production scheduling for very diverse products on the same assembly line, even for small batches. Robots have entered the world of automated warehouses, linking the strengths of automated guided vehicles with various components of information technology, such as bar-coding, inventory control, logistics management, and checking systems. Downtimes for equipment during changeovers will be substantially reduced, since robots can be programmed to accept detailed instructions about changes on the line. Toyota, Nissan and other car-makers have already introduced elements of this through computerized scheduling and *kamban* inventory techniques, so that the same line can produce not only numerous variations on similar car models but also different models on the same shift. Robotics will speed up this process as their program software increases in sophistication. For large firms, the likelihood of an infinite range of extras and special features means that mass assembly can produce truly unique final outputs.

The impact of robotics on small firms is no less revolutionary. The small firm will have the opportunity of entering many market segments normally reserved for large firms because, within certain volume ranges, robotics will equalize costs for big and small firms alike. Programmable robotics can handle multiple functions – such as drilling, burring, or polishing – and provide superior productivity and quality. Small firms will not be faced with the cost disadvantages of the better work skills and staff support available to large firms, but will still be able to cope with the small lot production and flexibility characteristic of their sector of the economy. It is for this reason that the use of robots in small businesses in Japan is so widespread, and has received special government support, such as through low-cost leasing schemes.

An accurate interpretation of Japan's lead in robotics, both in production and in use, must take into account the traditional emphasis on production, and the potential to rationalize operations in an era of labor shortages, an aging workforce, and behavioral challenges in job design. Additionally, robots are a natural outgrowth of the country's obsession with best-practice manufacturing technology, because of the commitment to quality control and productivity. These domestic production issues have induced Japan's leading manufacturers to transform traditional assembly systems to more highly automated ones, and to aspire to flexible manufacturing systems and in some cases, totally automated factories.

At the level of work and work systems, robots can be viewed as an outgrowth first of mechanization of factories and the office, and then of automation. Mechanization refers not only to the work tools used by workers – hammers, typewriters, computers – but the change of energy sources from human (or animal) power to mechanical energy. Automation is a related but quite different aspect of the human/machine relationship, referring to the use of self-controlled machines to accomplish work. Modern factories in all societies engage in both processes, but the implications both for the analysis of job design and for the meaning of work are imperfectly understood (see Simon).

Historically, much of the introduction of mechanical equipment has served to eliminate the purely physical tasks performed by human effort. Increasingly, according to Herbert Simon, the comparative advantage of human work is in three areas: the use of the brain as a flexible, general-purpose problem-solving device; the flexible use of human sensory organs and hands; and the use of legs, on rough terrain as well as smooth, to make this general-purpose sensing, thinking, and manipulation system available whenever it is needed. The human capacity for flexibility is not an absolute advantage against machines. In some areas, production engineering can change tasks based on principles of flexibility to relative inflexibility, so that machines gain a cost advantage. There are also inherent costs of using human beings – the costs associated with motivation, physical weaknesses (fatigue, boredom, bathroom relief), and sensory limitations (hearing low noise, eyesight precision, touch sensitivity). Such issues are in the background of job design, of course, but they are also now germane to having robots match human sensory and brain skills. Far more attention has been paid to the first area, namely, designing work flow to let machines substitute for human physical and sensory movement. The growing literature on the quality of working life, which raises such issues as "blue collar blues", reflects the social costs of this traditional engineering approach to job design. The basic aim is to remove workers' discretion and to make the machine "idiot proof" or "sabotage proof".

A number of forces are converging in Japan to support more widespread use of robots in production. There is a double trend in the labor force – namely the aging of workers in manufacturing, especially of employees aged between 30 and 49, and the recruitment of more educated new employees aged 29 years or younger. Younger workers are increasingly unwilling to accept the jobs which tend to be boring and repetitive (traditional car assembly), or which are dangerous, dirty or unpleasant (painting or welding). Like their older counterparts, younger workers aspire to white-collar jobs, which involve greater variety of tasks, intellectual challenge, and continuous learning. Both age and higher skills add to wage costs which, in the absence of offsetting gains in productivity, can threaten a firm's competitive position. Robots are an accepted response.

Labor market conditions reinforce this acceptance, since permanent "lifetime" employment in the larger companies makes job layoffs less likely. The shortage of up to 800,000 workers in heavy-duty skilled jobs means that the broader demographic issues are recognized by workers and managers alike, particularly for the most tedious or dangerous jobs. Japan's automobile industry and its celebrated success against Detroit have done much to educate workers in other industries and in firms of all sizes about how productivity is central to industrial performance internationally. Seen in this light, robots are simply an extension of broader technological and production trends, such as the shift of mechanical systems to electronics ("mechatronics"), or the applications of computers and on-line information systems to everyday fields as diverse as banking, shopping, television games, and offices.

The more sophisticated line of robots involves advances in the range of movements and skills built into the robot machines. The skill factor represents the more basic aspects of task design, such as dexterity, precision, and flexible speeds. Numerically controlled robots have revolutionary potential in these areas, and are best exemplified in industrial assembly such as automobiles. Numerically controlled robots are capable of driving screws, tightening bolts and nuts, welding seams, inspecting joints, and measuring lengths.

The more advanced robots are sensor-aided machines capable of visual, touch or audio feedback. Japan's strengths in cameras and electronics are contributing to the development of electronic sensing systems in robots, which have made some headway in truly smart machines. Matsushita, for example, has a sensor robot capable of recognizing different-colored parts, such as red, white, and blue balls in a container. However, robots which can interpret complex visual data, respond to many different surfaces, or listen to a variety of sounds are in the very early stages of development, and offer only long-term possibilities of improvement. Sensory robots are expensive – costing more than US$100,000 each – and become cost-effective only in particular conditions. Sensory robots already in use can remove foreign matter in bulk drug powders, group fruit by grade, and check the proper positioning of labels on bottles. Yamanouchi Pharmaceuticals, for instance, has introduced a robot costing more than US$200,000 which uses laser beams and video cameras to inspect liquid medicine in ampoules. Working four to five hours a day, the robot can inspect 10 times the volume of human inspectors, who require a minimum of six months training.

Conclusions

The Japanese emphasis on advanced manufacturing went largely unnoticed by the outside world until the 1990s. The dramatic productivity and quality of Japanese automobile producers, and the global dominance of Japanese consumer electronics firms, have changed that myopic view. Basic production principles and the organization of work have received only scant attention in western countries. The overwhelming attention of academic studies of Japanese productivity has been given to issues of behavioral processes, such as slow promotion and consensual decision-making, and organizational issues, such as human resource policies and long-term strategy horizons. These "software" management elements are important, but they too readily ignore the underlying emphasis on hardware technology, at both the societal and corporate levels.

In Japan, engineers have a much more prominent role in management thinking than in most western countries, although no more so than in, for example, the aerospace industry of the United States or Europe. What is striking about the Japanese approach – in contrast, for instance, to France, where engineering is a high-status profession – is the extent to which industrial engineering practices have been linked to human resource policies and marketing needs. In automobiles, steel and consumer electronics, Japanese manufacturers have introduced innovative engineering design policies, and turned their productivity standard into the world's best, including in the state-of-the-art Japanese factories located abroad.

The three areas stressed in this chapter – experience curves, just-in-time inventory management, and robotics – all focus on techniques widely accepted and practiced in Japan manufacturing, and all three are increasingly being copied by western companies. The wider issue is that, in their application and diffusion, Japanese managers have given the lie to the claim that they are copiers rather than innovators. These engineering issues related to hardware technology also demonstrate that many of the productivity issues in Japan are not culturally determined, nor established by some mystical concern for national values.

In this regard, it is no accident that QC circles, so widely touted as a panacea for many productivity problems, grew out of an emphasis on statistical analysis of production, not on human resource policies. Shopfloor work in Japan, as well as operations management, reflect many of the philosophical concerns expounded by Taylorism, including scientific decision-making, and cooperation between management and labor. Stereotypes of Japan's human relations practices have for the most part ignored the underlying production emphasis on simplifying work, standardizing products, designing to cost, reverse engineering, and eliminating waste.

Fortunately, detailed comparisons with Japanese management practices and productivity, especially in cars, steel, machine tools, consumer electronics and other specific industries, have started to highlight the weaknesses in business practices in North America and Europe. Curiously, the alleged management triumph of the United States in the 1960s, so readily copied in Canada, as well as in Britain and elsewhere in Europe – namely the academic business school – has been found seriously deficient in the very area of Japanese strength: production management. Many prominent graduate schools of business have eliminated production courses from their curriculums because of the courses' unpopularity among students; enrollments in remaining programs have been small and declining; and departments concerned with production and operations management have been either disbanded or merged.

Japan, meanwhile, is making a national commitment to high-technology industries, and has plans to spend up to 4% of GNP on research and development. Changes in electronics, machine tooling, and other areas of technology are making fundamental – indeed revolutionary – advances in production techniques. These advances have been brought about as part of the universal and continuous Japanese practice of *kaizen* ("improvement"), in which nothing is sacred, and everything can be changed, or, even better, perfected.

Why, then, is Japan different? Why the constant effort to adapt, learn, challenge, succeed? The answer begins with the country's extremely literate and educated population, now perhaps the world's best workforce. The practice of *kaizen* has been embraced by the large corporations, precisely because of their emphasis on guaranteeing "lifetime employment" (see Chapter 13), fostering the deep involvement of the workforce in productivity enhancement, and relentlessly pursuing decreases in unit costs. As Peter Drucker has noted, in a perceptive analysis of the changing theory of manufacturing:

> Every manufacturing manager 10 years hence will have to learn and practice a discipline that integrates engineering, management of people, and business economics into the manufacturing process. Quite a few manufacturing people are doing this already, of course – though usually unaware that they are doing something new and different. Yet such a discipline has not been systematized and is still not taught in engineering schools or business schools (quoted from Drucker p. 102).

Robotics is the best known case, but other state-of-the-art new technologies, such as computer-aided design, computer-aided manufacturing, flexible manufacturing systems and automated service bays, are also at the forefront of this evolving industrial change. The education system, the supply of highly qualified professional personnel, and the system of continuous training all reinforce this broad trend. They are also being extended to areas where Japanese productivity has been relatively poor – government departments, trading firms, distribution, and other areas of the services sector. As the next chapter indicates, not all of Japan's management practices have been a total success.

Further Reading

Argyris, Chris, "Single-Loop and Double-Loop Models in Research on Decision-Making", in *Administrative Science Quarterly*, number 21, 1976

A profile of alternative modes of decision-making and feedback.

Drucker, Peter F., "The Emerging Theory of Manufacturing", in *Harvard Business Review*, May-June 1990

An insightful perspective on the multidisciplinary nature of modern manufacturing.

Ferguson, Charles, "Computers and the Coming of the US *Keiretsu*", in *Harvard Business Review*, July-August 1990

A perspective on Japan's strengths in electronics technology and methods of organization.

McMillan, Charles J., *The Japanese Industrial System*, second edition, Berlin and New York: Walter de Gruyter, 1996

A modern review of Japan's economy and industrial organization, its strengths and weaknesses.

Nakase, T., "The Introduction of Scientific Management in Japan and its Characteristics: Case Studies in the Sumitomo *Zaibatsu*", in K. Nagawa, editor, *Labour and Management*, Tokyo: University of Tokyo Press, 1979

An excellent profile of the transfer of scientific management to Japan.

Patchell, G., "Shinchin Taisha: Japanese Small Business Revitalization", in *Business and the Contemporary World*, number 4, 1992

An insightful look at the development of small businesses in Japan.

Simon, Herbert A., *The Science of the Artificial*, Cambridge, MA: MIT Press, 1979

Wheelwright, S. C., "Japan: Where Operations Are Really Strategic", in *Harvard Business Review*, July-August 1981

An early work on Japanese management's attention to detail in manufacturing.

Woodward, Joan, *Industrial Organization: Theory and Practice*, London and New York: Oxford University Press, 1970

A classic study of technology and industrial production systems.

Yelle, L. E., "The Learning Curve: Historical Review and Comprehensive Survey", in *Decision Sciences*, number 10, 1979

An excellent overview of the history and application of experience curves in manufacturing systems.

Charles McMillan is Professor of International Business, and Tom Wesson is Assistant Professor of Strategic Management, at York University, Toronto.

Table 6.1 Japanese and Western Manufacturing Processes Compared

	Traditional western practice	*Latest Japanese practice*
Craft production	small lots sales to design to manufacturing high level of skills interactive technology high quality, high costs	large lots design to sales to manufacturing high level of automation sequential technology high quality, low costs
Mass production	large batch lots standardization quality/cost trade-off production complexity vertical integration	large lots, continuous flows customization quality/cost production simplicity lean production
Process production	large batches, continuous flows capital-intensive production small labor component high supervision ratios control of import costs	large continuous flows total automation tight supplier relations high work skills international flow of inputs (location on deep water ports)
Reinforcing practices		*kaizen* (improvement) zero defects training process improvements (cycle times, product churning)

Chapter Seven

Marketing, Distribution, and Other Service Industries

Charles McMillan and Yoshikawa Toru

Of the many enigmas concerning Japan, the services sector stands out. Japanese products flood the foreign markets of the world, often at prices which seem lower than those in Tokyo or Osaka. In foreign countries, Japanese business prowess appears the model of success; at home, Japanese services and distribution appear as a model of byzantine inefficiency. Aside from foreign suspicions of pervasive non-tariff barriers, the Japanese market is bewilderingly complex, as when soy sauces come in several brands from the same producer, and foreign manufacturers find consumer demands more exacting than in virtually any other country. The services sector now faces an organizational and management revolution – the removal of entry barriers and import quotas, deregulation and decline of bureaucratic restrictions on foreign investors and licensed retailers, and a new wave of innovative marketing tools – private labels, mail order shopping, discounting, and new convenience stores. In short, Japan's traditional complex, multi-level service and distribution system is on a one-way track to systemic change.

The public understanding of Japanese marketing abroad is often linked to the general trading firms (*sogo shosha*). In one sense, this perception is accurate. These general trading firms control such a significant portion of total imports and exports – around 50% – that they influence almost every facet of economic life, from the sourcing of raw fish to the supply of equipment for commuter trains. Internationally, the *sogo shosha* are among the world's largest commercial enterprises. Firms such as Mitsui, Mitsubishi, and Sumitomo are at the apex of Japanese industrial groupings, the *keiretsu* controlling not only huge foreign subsidiaries but shaping the domestic marketing and distribution network, from large department stores to new shopping centers (see Chapter 10).

For the Japanese, an important influence on marketing strategies has been exerted by foreign retailers. Japanese manufacturers have arranged to sell product lines to foreign retail channels, often using the retailer's brand name. For example, in television and stereos, giant retailers such as Montgomery Ward and Sears Roebuck, which have traditionally had links with US manufacturers, pioneered the sale of Japanese products and brands. By developing these different distribution channels, the Japanese manufacturers took advantage of local consumer tastes and buying habits. Similar trends are now found in Europe. The huge presence of foreign franchise outfits, especially in food, has also altered the behavior of retailers and consumers alike – a trend now underway in retailing with investments by US chains such as Toys-Я-Us and Seven-Eleven.

Entering the next century, Japan's marketing prowess abroad is severely hindered at home by the antiquated service and distribution system. Japan's services sector is huge, made equally significant by the staggering pool of savings held by Japanese citizens, an estimated US$10 trillion in 1998. It is the paradoxical combination of modern organizational efficiency matched with corner shopkeepers which at once defines and shapes services in Japan.

Where the one is a model of technical prowess, the other is a model of village welfare and social cohesion.

Traditional Service Institutions

Of all the major influences on Japan's traditional services sector, the rise of the *sogo shosha* and other trading firms, and their contemporary role as the engines for worldwide communications and information scanning in the global shopping center, stand out.

The trading firms trace their modern structures and policies to the early years of the Meiji period (1868–1912). In Europe, trading firms by this time had two centuries of experience: they included the Dutch East India Company, the Hudson's Bay Company, and mercantile houses in Liverpool, Rotterdam, and Hamburg. Such firms had relatively simple management structures. The organizational linkages were personal, between leading personnel in the home country and carefully trained administrators in the colonial territories. Aside from military or political staff, the main personnel in the subsidiaries were men of finance, and communications were very much a problem of geographical distance. In such circumstances, personal relations formed the basis of trusted bonds and shared goals in the organizational network.

This framework lasted for centuries in the European case, and competition for scarce resources was relatively weak. The descriptive term "mother/daughter relationship" is particularly apt – it connotes a comfortable and relaxed commercial relationship which lasted until decades after World War II (see Franko). Today, many of the European firms with colonial histories have survived, such as the East Asiatic Company, the Compagnie Français de l'Afrique Occidentale, and the Société Commerciale de l'Ouest African, or the United Africa Company. The last named, for instance, has joint ventures in Africa, owns department stores and shipping companies, and sells products for more than 200 foreign firms.

When Japan first opened its doors to modern international commerce, companies such as Mitsui and Mitsubishi established trading houses with government support to cope with the specialized needs of both exporting and importing (see Morikawa). The government attempted to develop Japanese commercial practices, including policies not uncommon today. For example, Japan participated in major international expositions, such as at Vienna in 1873 or Philadelphia in 1876, and government representatives overseas were used as sources of commercial intelligence. There were subsidies for shipping firms, assistance for marine insurance companies, and facilities for converting yen-denominated bank checks. Despite such measures, the existence of foreign merchant houses (*shokan*), with their superior skills in international trade compared to the evident inexperience of the Japanese, promoted a new approach. Led by Inoue Kaoru, a leading political and commercial figure, the government mobilized domestic organizational strengths. The vehicle was the house of Mitsui (see Roberts).

From an organizational viewpoint, there are two general perspectives on the linkage between the manufacturer and the customer. In the first case, the one prevailing in North America, the linkage should be direct with no, or at least minimal, intermediaries. Manufacturers thus desire direct access to customers and their feedback, avoiding any loss of information control. This argument for integration usually involves the range of incentives for scale economies of distribution, inventory management, and customer servicing. In short, there are high transaction costs for these functions, and the typical response is to internalize them through forward integration (see Williamson).

In the second case, the manufacturing organization specializes in the production function and "contracts out" the marketing function to an external organization, the trading firm. Conceptually, the main argument for this approach is the gaining of economies of scale in the total marketing function of many products and specialized services. For the trading firm, the major financial obstacles are high start-up costs in personnel, market research, and distribution channels. Once developed, however, the incremental costs of volume expansion are quite low, since the main transaction costs are informational. Properly organized for bridging information linkages between many manufacturers and many consumers, the trading firm can afford to work on high volumes and low margins, even for standardized products with low value added. Japan historically has provided the best case studies of this approach.

The powerful information base of the *sogo shosha* has to be recognized in the context of their historical evolution and their contemporary relationships to the *keiretsu* structure of Japanese business groups, as well as in relation to their marketing strategies. In the first instance, the *sogo shosha*, while historically concentrating on bulk commodities, minerals and other highly standardized imports, as well as undifferentiated exports, have had close relationships to manufacturing companies in the various major groups – the prewar *zaibatsu* and the postwar *keiretsu*. The *sogo shosha* have operated as information screening houses for these groups and, indeed, it is not too farfetched to interpret their role in this informational sense rather than in their pure trading role.

This perspective is reinforced by the strategic evolution of the *sogo shosha* and other trading firms in recent years. They have been in the vanguard of Japanese direct investment overseas, in contrast to the situation in western Europe or the United States, where such outward investment has been led by manufacturing firms. Both immediately after the war and in recent years, few Japanese manufacturing firms could afford the substantial investments in informational infrastructure required for gathering and interpreting data on foreign market developments, least of all to match that of the European or US multinationals. The fact that Japan had several *sogo shosha* to develop this information system, and an efficient means of digesting and using it in the domestic market via the large corporate groupings, added to the efficiency and speed with which corporate strategies could develop. That is why, for instance, the trading firms have become the key organizational linkage in export consortiums where "turnkey" projects and sales of entire projects, from subway systems to petrochemical complexes, require a major firm to act as prime organizer for many separate manufacturers, suppliers and subcontractors.

This huge informational base and communication system are somewhat akin to what Norman MacRae has called the organizational confederation of entrepreneurs (see MacRae), since, in practice, *sogo shosha* operate with a very high element of decentralization. This provides unparalleled early warning systems for new products, technologies, and markets, making the *sogo shosha* key players in the corporate strategies of the *keiretsu* business groups. To function as such, however, *sogo shosha* need not only informational resources, but financial asset bases to provide credit, absorb foreign exchange rate risks and costs, and act as financial agents for producer firms, particularly small businesses. The sheer size of the largest *sogo shosha*, and their pivotal role in the major business groupings, make their financial dealings part and parcel of the total banking system. The *sogo shosha* stand as the main intermediaries between separate manufacturing companies and small business groups and the major banks. In some cases, the financial role of the *sogo shosha* may go beyond these functions to include equity investment, direct loans, or guarantees for loans, although this approach has developed only recently.

Actually, there are three kinds of trading firms in Japan, and each serves quite different needs. The first group are the best known, namely the huge *sogo shosha*, nine in number, which dominate Japan's export and import trade. The *sogo shosha* are not only giants in themselves, with total sales turnover well in excess of US$1 trillion, they are huge in any comparison of international corporations. The trading firms stand at the center of the dominant *keiretsu* with the leading banks, and indeed often serve as the financial arm of export sales, technology transfer, and market feasibility studies. The *sogo shosha* have been the central institutions in Japan's total international trade strategies, and account in sales for around one third of Japan's US$3 trillion economy. By combining the economies of information flows with their enormous bargaining leverage in commodity and raw materials markets (where there are usually few sellers), these firms provide the Japanese economy with a steady flow of competitively priced imports for downstream processing. Further, through their logistical systems for imports and exports, they have reduced the transportation costs of global distribution in markets relatively isolated from Japan. Their activities are not just the buying and selling of goods, services and commodities; they serve as bankers in lending credit and finance, and act as principals and coordinators in exporting turnkey projects and plants, establishing overseas joint ventures, and serving as linking agencies for third-country trade. Their key strength lies in their enormous global information systems. These firms are enormously diverse, and operate on huge sales volume and low profit margins. Mitsui, for example, had total sales in 1997 of US$162.9 billion, less than half of which originated in Japan.

The extreme diversity and large size of the *sogo shosha* contrast with the second type of Japanese trading firm, the specialized traders (*senmon shosha*). These firms number in the thousands, and limit their operations to quite specialized and focused strategies. Some firms may handle only particular product lines in a large number of markets. Some specialize by product and market, or even a region of Japan. For example, one small trading firm concentrates only on hospital supplies manufactured in southern Japan for Asian markets. Another medium-sized firm concentrates on motors and pumps, and focuses on the markets of the Middle East and Francophone Africa, mainly from an office in Tokyo and an agency in Paris. Almost all these specialized trading houses handle the full range of trading functions, from transport and freight to customs and credit (see Tsurumi).

The third type of trading firm in Japan, and among the fastest growing, comprises the captive sales arms of manufacturing firms, such as those in electronics, automobiles, office equipment or precision machinery. In this category are the sales arms of the automobile firms (such as Toyota Motor Sales) and the electronic firms (such as Matsushita Electric). The main reason for the development of these trading firms is the need of the manufacturer to establish direct contact with customers, to handle after-sales servicing, and to establish brand names and dealer networks. Most of the largest consumer manufacturers – Matsushita, Sanyo, Toshiba, Sharp – have extensive domestic networks of stores, many of which deal only with their own products. Matsushita, for example, has 60,000 stores, of which about 25,000 are exclusive brand outlets. It is for this reason that the wholesale distribution system in Japan acts as an import impediment for foreign products. As it turns out, many of these same firms are branching out into other products, as well as participating in foreign ventures through direct investment.

Despite their variety of sizes, types, and functions, the Japanese trading firms are extremely competitive, even within product groups. For example, while the *sogo shosha* are spread throughout Japan and around the world, in a bewildering array of products, markets, services, and third-nation trade flows, competition for trade is fierce within specific market segments in Japan. As of 1990, according to the Japan External Trade Organization (JETRO), there were 1,000 small trading firms importing food products, 900 handling textiles, 500 dealing in logs and lumber, amd another 1,000 in machinery. In the same year there were more than 50,000 Japanese employed in the overseas operations of Japan's trading houses. The nine *sogo shosha* alone had more than 1,000 overseas offices, from Moscow to Mexico City.

Obviously, Japan's position as a resource-poor, "throughput" economy contributes to the concentration of the import functions of trading firms. Such imports tend to be unprocessed, standardized commodities with selling prices set in world markets. Dealing in high import volumes tends to cushion the financial risks of export trade, since costs of global information and distribution systems, currency fluctuations, and buffer inventories can be spread across both imports and exports. It follows that in recent years the *sogo shosha* have led the way in Japanese direct investment overseas, often with domestic manufacturers participating in equity alongside host country firms.

The trading firms play the linchpin role within the *keiretsu*, coordinating trading functions across the group companies, with the financial and manufacturing strategies to provide a competitive edge over stand-alone multinational firms. This stems from the sheer diversity and scope of the trading firms' operations, and their capacity to link products to many markets through national subsidiaries. This coordination role of Japanese trading firms is improved only in part by the interlocking ownership patterns of organizations in the *keiretsu* groups. A more direct linkage, promoting shared goals and policies, is the system of presidents' clubs or councils (*shacho-kai*), involving the presidents of each company in the group voting as individual and equal partners (see Chapter 10).

Unlike the situation in US conglomerates, there is a strong incentive in the Japanese *keiretsu* to share and exchange information, since what may have very limited value to one company may be invaluable to another member of the group. It is the effect for the total group which is paramount, and the trading firm is the main nexus for coordinating information. That partly explains the enormous diversity of products and services of Japanese trading firms, since even small volumes and a narrow range of some products can still be handled economically: the overheads are spread over everything from *ramen* (Japanese noodles) to missiles. There is thus enormous concentration of economic power, and potential for harmful corporate behavior from the perspective of consumers.

The influential role of all trading firms in imports, exports, and the domestic market is central to modernizing the country's inefficient distribution system. Foreigners often argue that Japan's distribution system forms an effective set of non-tariff barriers, along with government regulations, technical barriers (for example, emission standards for automobiles), and Japanese consumers' reluctance to buy foreign goods. Unfortunately, foreigners not only place unwarranted credence in such shibboleths, they often attempt to enter the Japanese market with domestic trading firms as local partners. In so doing, they fail to recognize the basic differences between industrial products and consumer products, where distribution channels and, often, sale service requirements are quite different. In point of fact, such problems are largely misunderstood, and show the need to recognize the cultural and institutional factors explaining the evolution of Japan's services sector, which now exceeds that of Germany and equals that of Italy as a proportion of GNP (see Table 7.1).

The Path to Change

Japan's vast distribution system is culturally rooted in its agricultural past and its small communities. Even Japan's major cities can be seen as no more than assemblages of villages and neighborhoods. What is so striking is the extent to which this village structure has persisted in distribution, despite the vast modernization in manufacturing (see Chapter 6). The number of "mama papa" stores has actually increased in the past decade. The net result, as shown in Table 7.2, is a diverse and complicated distribution network, which differs for each product, and varies by the size of the producer and the end-retailer or user.

Even today, small retail outlets account for about three quarters of all consumer shopping. While international comparisons are flawed by differences in the definitions used and the dates of statistical gathering, the clear picture is that Japan has larger numbers of retailers and wholesalers than any other major country on a per-capita basis. The comparative statistics on retailing across countries tell only part of the story about this aspect of the service economy, as anyone who has lived in Japan can readily observe. Britain, traditionally "a nation of shopkeepers" (to use Napoleon I's apt description), actually has more shops per 100,000 people than Japan. Both countries are in a different league from North America when it comes to the large, efficient store chains, shopping centers, franchises and retail networks.

Japan's complex retail distribution system – the flow from manufacturer to ultimate retail outlet – arises from a mixture of government edicts, company cartels, political lobbying and disguised social policy (jobs for senior citizens). To illustrate, in 1996 Japan's wholesale sales were 4.2 times retail sales, compared to a multiple of only 1.7 in the United States, or 1.4 in France. The result is a constant history, industry by industry, of difficult marketing channels and government-imposed regulations, including zoning of real estate and consequent pricing policies, which have diminished competition, raised prices to consumers, and stifled innovation. These issues assume international importance when elevated to the level of trade disputes about rice imports, public sector infrastructure, income support for marginal farmers, senior citizens as shopkeepers, and cartels among big businesses.

In reality, Japan's retail system is changing quickly. Paradoxically, the country's manufacturing and trading companies were among the first to exploit North American retail efficiencies, by introducing their products through major US merchandise firms such as Sears and J. C. Penney, and immediately gaining market entry to a broad spectrum of consumers throughout the United States. Today, equivalent trends are evident in Japan, as large retail groups begin to provide easy entry to Japanese consumers. In 1976, the retail system had 1.61 million outlets, which had not really changed at 1.60 million in 1988, although the number then fell to 1.4 million by 1996. However, between 1976 and 1988 the volume of sales doubled from ¥56.0 trillion to ¥114.8 trillion. The number of small retail stores decreased by 54% between 1982 and 1988, and more than 2,500 large-scale stores opened since then.

The recent evolution of Japan's services sector – discount retailing, foreign competition, gradual deregulation – and the continued crisis of the banking industry reflect a range of economic influences. At the macroeconomic level, the strengthening of the yen against the US dollar and Japan's continued current account surplus (amounting to around US$60 billion in 1997) strongly influence the basic restructuring taking place in the domestic economy. Japan's manufacturers, not only the largest companies but also small and medium-sized companies, have made huge shifts into Asia to gain product advantages (see Chapters 9 and 18). At the same time, Japanese manufacturers have invested heavily in new plant and equipment, and in process technology. This trend has accelerated the shift of emphasis to service functions, service jobs, and highly sophisticated information networks based on advanced technology. The notable examples of this trend have been in financial services, although Japan's large construction industry displays many parallels.

The net effect of these macroeconomic trends reveals a set of microeconomic corporate influences as well. For one, there is the growing internationalization of Japan's smaller companies, particularly towards Asia. Second, Japanese managers are trying to apply their successful work methods and labor restructuring to the services sector, in such areas as hotels, department stores, banks and construction, where Japanese firms are expanding offshore, notably in North America, where US service productivity leads the industrialized world. The crisis in Japan's financial services, rooted in the overlending practices of the "bubble economy" in the late 1980s (see Chapter 1), reflects the lack of openness and deep political paralysis of the regulatory regime of the services sector, and illustrates the depth of the competitive pressures facing the services sector at large.

Services shape the competitive performance of entire economies and the competitive success of global industries. Accordingly, services are now politically controversial – between advancing industries and those in retreat, and between countries squeezed by the demands of new technologies. There are widely held views that services are the trump card of the rich, industrialized West against the poor, less developed countries. Services are the ultimate embodiment of the advanced countries' comparative strengths: information, knowledge, technology and finance.

The establishment of the World Trade Organization (as from January 1, 1995) has brought services into the world's trading regime, especially on such principles as transparency of rules, "national treatment" (non-discrimination against foreign entrants), and reduction of trade barriers. The failure to deal with services in previous international and regional trade agreements meant that few precedents or models of trade administration existed, despite the fact that services account for so much of world trade. Further, as the services sector itself changes so dramatically, there is a new equation. As services grow in importance, the role of the mode of dispute settlement, and the stakes involved, also increase (see Segal-Horn).

Japan's New Service Economy

Japan's services sector is important for several reasons. For one thing, its very size is too big to ignore, with so many spillovers into the export sector and a critical impact on imports. The sheer size of Japan's services sector offers enormous opportunities to deal-hungry foreign companies wanting to overcome the inanities of Japan's national and local government regulations, which add to inefficiencies and thus promise entrepreneurial rewards if they can be reformed. Further, the retail opportunities are staggering, especially if barriers on land control, retail distribution and certain forms of transport are reduced or removed. The size of Japan's middle class illustrates the incentive for customer-driven reforms – a point that severely threatens high-cost retailers of all descriptions – department stores and general merchandisers alike.

Japan's development and expansion as a service economy stems from a different macroeconomic force: its position as a net creditor to the world economy. Japan's trade surpluses, the growth in domestic productivity and the appreciation of the yen interact to reinforce this creditor position, meaning that the earnings on the stock of overseas capital will add to Japan's importance in such critical areas as foreign aid, direct investment overseas, corporate alliances, and technology partnerships. Indeed, on present trends, Japan stands poised to eliminate the traditional deficit on the service account and create a very significant trade surplus. Two factors which produce negative effects on the balance of payments – transportation costs for Japan's merchandise trade, and payments for patents and royalties in technology – have either declined or turned to surplus. Japan's structural deficit in energy, mainly on oil from the Middle East, is real, but held in check in part by the country's large and growing nuclear power industry and sophisticated policies on energy conservation (see Chapter 4).

Japan's services sector is also in the vanguard of the information revolution, exemplified by credit cards, debit cards and other forms of new software. The information revolution is at its most dramatic in microelectronics, transforming at once entirely separate industries and national boundaries, and making nonsense of the old belief that "software" and "hardware" are comfortably separate. Charles Ferguson, a prescient student of these trends, has observed that:

> We are witnessing the digitalization of everything. Previously unrelated industries – cameras, computers, stereos, photocopiers, typewriters – are converging to form a huge, unified information technology sector, itself based on common digital components and standard interfaces. Increasingly, competition in all kinds of hardware is driven by the same new logic governing competition in computers – growing commoditization of product markets and growing advantage for companies with superior components technologies, manufacturing systems and strategic leverage.

This technology is particularly important for its direct impact on economic performance and international competitiveness. The services sector well illustrates the changes now under way, from technology to new organizational forms, from the impact of government regulation to the entire question of international trade in financial services.

As a consequence, Japan's "new" services sector now faces a challenge in overseas markets. Contemporary trends make information in all its forms, its composition and transmission, fundamentally important to competitive success. Japanese corporations have singled out telecommunications as an area of explosive demand, growth and technological development. Domestic trends to digitize, wire and apply new technology to factories – robotics, lasers, computers, smart machines, and flexible manufacturing – are no less important for offices – fax equipment, imaging technology, supercomputers and expert systems. The same technologies are also being made increasingly available for other applications in the Japanese market, from automated homes and housing estates to "smart" buildings. The domestic and international developments of these trends involve long-term perspectives, patient capital and emphasis on commercial applications, precisely the areas of Japanese corporate strength. Japan's unique organizational groups, and the financial strength of the corporate sector in an era of high interest rates and expensive technology, pose special challenges to North American and European competitors. New strategic alliances and international consortiums are forming to meet this information challenge.

Behind the statistics lie a variety of social, economic and political factors which explain the persistence of Japan's distribution system. The three types of factors are interrelated. For example, there is the price of land which, at a cost of about ¥35,000 per square meter, was 10 times the equivalent price in the United States at the height of the bubble economy, and which deregulation is only starting to correct. For small "mama papa" retailers and wholesalers (*ton'ya*), which typically combine a business and a home in the same premises, there is a clear incentive to have one member of the family earn a salary from government or a large manufacturer and run the business with the family. As a political lobby, such merchants have been effective in requiring department stores and supermarkets to obtain regulatory approval from local authorities. Since 1979, stores larger than 500 square meters have required approval from local chambers of commerce, which usually consist of small retailers and wholesalers. The key legislation is the Large-scale Retailing Establishment Law (*Daiten Ho*) of 1973, which stipulates that any large store can be created only when neighboring small merchants approve its size, opening days, and hours of business. High land costs and legislative restrictions provide cost advantages to small retailers, which in turn favor neighborhood shopping – even for highly priced items such as electronic appliances or automobiles. Since, as of 1996, supermarkets in Japan averaged 2,050 square meters of floor space and department stores 17,100 square meters, small stores still accounted for 83.3% of the country's retail market in that year. In some of the newer suburbs, supermarkets have been in the vanguard of urban development, by buying huge tracts of land and building stores linked to housing and new shopping districts.

Foreign pressure for changes to Japan's regulatory regime has been fierce, notably by way of the Structural Impediments Initiative with the United States. In all, 17 laws and 45 separate administrative regulations apply to retailing, explaining in part not only the vast number of layers in the retail chain but the entry barriers to large stores, as well as the convoluted flow of products from manufacturers to customers. Even large Japanese retail groups, such as Daiei, have become fierce critics of the regulatory system, in part because of rising competition from goods and companies from Asia, and have welcomed the reforms implemented in 1993, including fewer controls on new stores, easing of imports of construction materials, an end to minimum production quotas for breweries, and a reduction in dairy subsidies.

Foreign products are often subject to fixed prices and fixed margins. The irony is that many Japanese goods sold abroad are actually cheaper than in Japan, especially in those countries, such as the United States, Canada or Britain, which have relatively efficient distribution sectors. To foreigners, the alleged issue of "dumping" is really one of distribution, and Japanese consumer groups and many businessmen are no happier about it than foreigners are.

In this context, trade in services has become of vital concern to less developed countries. Information technologies have fundamentally transformed entire manufacturing and services industries, including traditionally labor-intensive industries such as textiles or automobiles, as well as new growth areas such as construction or medical services. Less developed countries know the impact of trade barriers in services, especially the high entry costs of technology and financial sunk cost, and the cycle of dependence on western firms that this creates. They know as well that their fields of comparative advantage, such as commodities and raw materials, depend on access to foreign markets in service areas such as transportation, finance and distribution. These are, of course, the natural strengths of the developed world, as information technology advances and brings new forms of international organization. Yet today, it is not just the richer countries that have a stake in trade in services. Tourism is an obvious case in point for less developed countries, but so are construction and engineering (South Korea and Singapore), shipping (Hong Kong, Panama, South Korea), and finance, software, and data processing (Jamaica, Singapore and India), to name some prominent examples.

As a matter of government policy, Japan has fostered overseas tourism, inducing Japanese citizens to travel abroad by the millions (almost 17 million a year, up from a few hundred thousand two decades ago). The second leg of this change has been an aggressive overseas aid policy, turning Japan into the largest foreign aid donor in the world – yet another social force opening up Japan's economy to the world (see McMillan).

Franchising is the most telling example of firms shifting the growth strategy from a domestic to an international orientation. In 1996, the number of US franchise outlets in overseas markets increased by 10.6% to 35,045, according to the International Franchise Association Educational Foundation. Franchising is an area of rapid growth for foreign companies in Japan, which is the second largest market for franchisees after Canada.

"Just in Time" Marketing: The Retailing Revolution

In the 1980s, Japanese manufacturers revolutionized manufacturing with the new concepts of lean production – shortening the production cycle, reducing the quantity and flow of inventory, improving coordination with fewer suppliers and subcontractors, and simplifying of product design and subcomponents. The 1990s have shown a similar trend, but this time the revolution is with the customer. The new approach is just-in-time marketing – a radical restructuring of the marketing chain by reducing the distribution links across the board.

American and Japanese companies alike are racing ahead to implement just-in-time retailing, but the changes ahead are not difficult to outline. Like just-in-time production (see Chapter 6), just-in-time marketing will shorten cycle times, reduce the role of middlemen, and turn the process of customer purchasing into an on-line, computer-based system. Huge retail establishments such as Wal-Mart are in the vanguard of this approach, supported by enormous investments in information technology. Data from customers and cash registers provide on-line feedback to Wal-Mart's suppliers, which deliver products on a just-in-time basis.

This change in retailing has the potential to revolutionize the entire approach to buying decisions. In the traditional approach, aspects of retailing, such as educating the buyer, viewing the product, serving the customer, and making the sale were integrated processes in the major retail establishments. Innovations such as credit cards and home delivery now add to the convenience of one-stop shopping.

Just-in-time retailing turns the traditional approach on its head. It relies on a flow of customers to a centralized inventory warehouse, called a department store or supermarket. The presence of on-time information delivery vehicles, such as cable systems, television, mail delivery and express companies, coincides with the proliferation of product information sources – catalogues, consumer product guides, buyer services, specialty magazines, and books. The result is that customers have ready access to products without ever entering a store, since the products are available by telephone, computer, mail or television.

Various pressures have been developing for changes in the distribution system, some of them external to Japan, such as foreign exporters. Limited change has also come from marketing innovations by manufacturers on the one side, and by modern retailers on the other. There is no question about the complexity of distribution, but it applies equally to Japanese and foreigners alike. In recent years more rapid penetration of products from Southeast Asia, compared to the United States, lies in their greater understanding and effective use of the Japanese system (see Gregory, and McMillan). Some Japanese companies have attempted to assist western importers to cope with the Japanese market. Sony, for example, imports refrigerators from Whirlpool, cleaners from Hoover, and food equipment from Oster. However, the major threat to the cumbersome distribution system is not coming from the trading firms but rather from the department stores, the supermarket chains, and the manufacturers. As is usual for Japan, each sector also presents a competitive threat to the other (see Ohbara et al.).

The department stores have lost market share, mainly to chain stores, franchise outlets, and supermarkets. As a result, they have tended to develop links with foreign companies to handle exclusive rights for particular goods, in exchange for agreed levels of purchases, regardless of sales. The department stores operate only in the large cities. The largest, Takashimaya, has only 19 stores throughout the entire country. They cater to consumers for more expensive items, especially high fashion apparel, which accounts for 47% of sales. The department stores also handle the vast market for presents during the gift and bonus seasons (in July and December). The large retailers have also established private brand manufacturers at home and abroad to circumvent the wholesaler channels. Wholesalers have responded in kind by equity investments in their own outlets, and some have even established private brand manufacturing – a trend on the rise in Japan as well as in the United States (see Shill et al.).

However, it is the supermarket chains which have grown the fastest, averaging 5–6% in the 1990s to capture about 25% of total retail sales. In many ways, the supermarket chains have succeeded by introducing retail practices that are well-developed in the United States and Europe. Daiei, which has been Japan's largest retailer since 1991, and which increased its sales to around US$26.5 billion in 1996 (as compared to just US$75,000 in 1957), has been the most aggressive innovator. It has developed about one fifth of its sales through in-house brands and acquired foreign expertise from Lawson's of the United States. In an attempt to get into the department stores segment, it tried to acquire one of the major groups, Takashimaya. Daiei has developed links with Au Printemps of France, and markets goods from Marks and Spencer of Britain, and J.C. Penney of the United States. Similarly, Seibu retails merchandise supplied by Sears Roebuck of the United States.

Perhaps not surprisingly, it is the chain stores which are introducing the fast food outlets of Japan. This is a development from the practice of licensing goods for sales (*shohin teikei*). In retailing, well-known brands such as Cardin or Yves St Laurent are prominent examples. In the same way, the chains have sought out licensing arrangements for fast foods. Ito Yokodo operates Denny's Inc. of California under license; Daiei is the franchisee for Wendy's hamburgers and the Victoria Station restaurant chain; Nichii operates Arby's roast beef sandwiches. By contrast, two of the best-known US fast food companies, Kentucky Fried Chicken and McDonalds, have joint venture arrangements. Mainly as a result of US influences (31% of all US retail sales come from franchisees), Japanese marketing is adapting to franchise outlets in areas as diverse as hairstyling, sports centers, travel and convenience stores. However, while there were almost 7,000 franchise food outlets in Japan by 1997, they have largely been in new areas of retailing, rather than changing traditional distribution outlets to any significant degree. Shopping habits remain deeply ingrained despite more cars, larger malls, and more automated service bays, characteristic of US and European cities. Japan's changing lifestyle habits – better standards of living, shorter working weeks, diversification of eating habits – have shifted buyers' behavior in basic ways, so that the permanence of the 850,000 restaurants is open to modernization.

External pressures for change have come from foreign retailers, which have increased in importance as a result of growing consumer awareness and high disposable incomes. The term *hakurai-hin* – literally meaning "goods that arrive by ship" – refers to foreign-made products that often command premium prices in the domestic market. The key challenge for foreign importers is to gain access to existing distribution channels. In some cases, new channels may have to be created, but the very high cost of land, and the enormous expense of media advertising, inhibit this approach. The United States, among other trading partners, has argued that Japan's traditional business customs and regulations, including those governing large-scale stores, are really non-tariff barriers, a point underscored in the lengthy bilateral trade talks known as the Japan/US Structural Impediments Initiative.

Conclusions

The opening up of Japan's market to import penetration goes hand in hand with the country's evolving industrial structure and emergence as a mass consumption society. The protracted arguments about modernizing the distribution system to make Japan more open for foreign imports will probably continue, especially in North America, where distribution is considerably less complex. Just how much of a practical impact this debate will have is rather questionable. The trading houses will still control a very significant proportion of all imports and exports, because of their global scale and specialized information systems. The department stores will concentrate on higher-priced items with lower sales turnover, while the supermarket chains will gain greater control over wholesale channels, along North American lines. The trend for manufacturers to encompass the marketing function will continue, but mostly for products requiring after-sales service or exclusive outlets. The examples of cars, consumer electronics, cameras and cosmetics stand out. Tokyo and Osaka dominate large store sales, with 60% of the total, but this still leaves a vast market of retail sales in smaller cities and communities. For foreign importers, these trends mean that the Japanese market will remain a great challenge, requiring long-term horizons and investment.

Japanese consumers are also changing, as Japan itself has internationalized, foreign goods have become better-known, and the rising yen has made imports competitive. The trading firms have played a significant role in bringing the world's best products to Japan, so today the typical Japanese consumer is more demanding than probably any in the world. Salary levels and spending patterns have played a part, but so have education, travel, and voracious Japanese reading. Since housing is so expensive in Japan, there are differences in household budgets. The Japanese spend more on food, clothing, and household fixtures, but less on housing as such, compared to people in most western countries. The aging of the population is also another factor, as are the trend to increasing numbers of working mothers, and an education pattern where 40% of the population between 18 and 24 attends university; by the year 2000, almost 20% of the population will have a university or college degree.

Japan's service economy provides tangible evidence that Japan's future is largely on the same path as western economies are, away from manufacturing of traditional products, especially heavy industry, and oriented to knowledge-intensive industries providing high value added. Second, Japan's adjustment to global competitiveness and international trade pressure has added to the momentum of the domestic structure to place increasing emphasis on a host of new industries and corporate strategies, many of which will have an impact on services throughout the world economy. Examples include medical services, education, leisure and sports, engineering and construction, travel and tourism, financial services, communications and transportation.

Third, the Japanese increasingly recognize that the management demands of "smoke-stack" industries in heavy manufacturing, exemplified by iron ore, steel, shipbuilding, automobiles, machinery and petrochemicals, no longer represent the future job opportunities of the international dimension of the global economy. Their place has been taken by airlines, financial services, communications, fashion and clothing, leisure, health, and automation in all its forms – areas where Japan clearly trails US productivity by most measures.

Japan's wealth-creating strengths lead to all kinds of social impacts, few of which are understood in the West. Japan's leading manufacturing companies plan and manage their "service" functions – distribution, advertising and financing – more than their western counterparts do. This explains why service-related job content is higher in the US economy than in Japan, by a factor of about 5:3, although this itself is changing. The Japanese government's encouragement of holidays and shorter working weeks has promoted a domestic leisure boom since 1994, when the yen began strengthening against most other currencies, and predictable indicators show the results. The number of overseas travelers doubled in each of four consecutive years. Spending on such areas as horse racing, bicycle races and non-commercial sports is at record levels. At least 17 million Japanese now travel overseas each year, forcing comparisons of living standards and price points across the board, and thereby drawing Japanese consumers to attractive pricing bargains, regardless of whether the seller is Japanese or foreign. Customers' loyalty to Japanese sellers is thus on the wane.

What all of this means, for Japanese firms and foreign competitors alike, is that both consumer and industrial marketing will require new levels of sophistication and adaptability. As Japan's domestic market evolves, it may well set a standard for quality, price and reliability such that foreign competitors and Japanese companies alike will face new ground rules and new challenges. An indication of just how significant this change will be is the emergence of Japanese multinational service firms and increased penetration of foreign-based service firms in the Japanese market, already well under way in Japan's troubled banking industry. Japan's service revolution is now well under way.

Further Reading

Ferguson, Charles, "Computers and the Coming of the US *Keiretsu*", in *Harvard Business Review*, July-August, 1990

A provocative view of how US corporations should adopt Japanese organizational practices.

Franko, L. S., *The European Multinationals: A Renewed Challenge To American and British Big Business*, Stamford, CT: Grewlock, 1976

An assessment of multinational corporations based in continental Europe, and their spread to Britain and North America.

Gregory, G., *Japanese Electronics Technology: Enterprise and Innovation*, Tokyo: Japan Times, 1986

An insightful view of Japanese technology practices applied to new industries, including services.

McMillan, Charles J., *The Japanese Industrial System*, second edition, Berlin and New York: Walter de Gruyter, 1996

A comprehensive overview of Japanese management practices at the level of government, industry, and corporations.

MacRae, Norman, "The Coming Entrepreneurial Revolution", in *The Economist*, December 25, 1976

A radical approach to the use of small scale in big business.

Morikawa, H., "The Organizational Structure of the Mitsubishi and Mitsui *Zaibatsu*, 1868–1922: A Comparative Study", in *Business History Review*, number 44, 1970

One of the best historical descriptions of the origins of Japanese big businesses.

Ohbara, Tatsuo, et al., "The Emperor's New Stores", in *The McKinsey Quarterly*, Volume 2, 1994

A study of the competitiveness of the new retail industry against traditional retailers, such as department stores.

Robarts, J. G., *Mitsui*, New York and Tokyo: Weatherhill, 1973

A definitive study of one of Japan's most powerful business groups.

Segal-Horn, S., "The Internationalization of Services Firms", in *Advances in Strategic Management*, number 9, 1993

An overview of the strategies and structures of service enterprises.

Shill, Walter, et al., "Cracking Japanese Markets", in *The McKinsey Quarterly*, Volume 3, 1995

A consultant's assessment of the opportunities for foreign retailers in Japan.

Tsurumi Yoshio, *The Sogo Shosha*, Rutland, VT, and Tokyo: Charles E. Tuttle, and Montreal, PQ: IRPP, 1980

An insight into Japan's general trading houses as a model for practice in other countries.

Williamson, O. E., *Markets and Hierarchies: Analysis and Anti-Trust Implications*, New York: Free Press, 1975

A brilliant discussion of organizational dynamics and why managers prefer hierarchies or bureaucracies to markets.

Charles McMillan is Professor of International Business at York University, Toronto. *Yoshikawa Toru* is Assistant Professor of International Management at Nihon University, Tokyo.

Table 7.1 Composition of Group of Seven Economies, by Sector, 1997 (%)

	Primary	*Manufacturing*	*Services*
Japan	5.8	34.0	60.2
Canada	4.1	22.6	73.3
France	5.1	27.8	67.3
Germany	3.3	37.6	59.1
Italy	7.7	32.1	60.2
United Kingdom	2.1	27.7	70.2
United States	2.9	24.0	73.1

Source: OECD

Table 7.2 Japanese Retailing, by Size of Business, 1991

Number of employees	*Number of stores (thousands)*	*Total employees (thousands)*	*Annual sales (¥ billions)*	*Sales area (sq. meters)*	*Share of sales (%)*	*Sales per employee (¥ thousands)*
1–2	847	1,381	15,224	26,996	10.8	11,023
3–4	417	1,404	23,006	22,345	16.4	16,386
5–9	214	1,336	28,877	18,657	20.5	21,615
10–19	72	948	21,408	10,564	15.2	22,582
20–29	20	477	10,673	4,713	17.6	22,375
30–49	13	479	10,478	5,117	7.4	21,874
50–99	5.8	384	9,216	6,195	6.6	24,000
100+	2.3	525	21,754	15,255	15.5	41,436

Source: Ministry of International Trade and Industry, *Census of Commerce*

Chapter Eight

The Financial System

James D. Malcolm

Thanks to its apparently endless capacity for crisis, scandal and intrigue, the Japanese financial system has provided huge quantities of morbidly interesting material for its observers in the 1990s. At the same time, Tokyo remains one of the world's leading financial centers, alongside London and New York, and therefore occupies a central position within the global economy. However, more than simply serving to contextualize these facts, an understanding of the Japanese financial system is crucial to an understanding of contemporary Japan itself, for several reasons. First, the financial system is widely recognized as having played a central role in underwriting the country's postwar economic miracle. Second, developments in this one area of the economy alone have continually overshadowed all others in driving the country's economic performance since the mid-1980s, to a degree not seen elsewhere. Third, as a microcosm of the nebulous debate about the convergence of socioeconomic systems, the financial system provides a tangible and highly representative object for empirical analysis.

The Development of the Postwar Financial System

Despite extensive attempts by the Allied Occupation authorities to restructure the Japanese financial system, largely on the US model, what emerged and became institutionalized during the 1950s and 1960s, while superficially new, reflected the prewar system in its underlying structure and regulatory ideology.

In seeking to establish a modern financial system after the Meiji Restoration of 1868, Japan's financial authorities borrowed heavily but not indiscriminately from the West. The banking system was inspired by the US model, stock exchanges in Tokyo and Osaka drew on the UK model, and the central bank copied the Belgian model. However, the practices of regulation remained intrinsically Japanese. Initially espousing a *laissez-faire* attitude towards the development of a financial system centered on private banks, the government sought only oblique influence over the flow of industrial funding, both through selective patronage and by direct control of a number of "special financial institutions" which were used to supplement private capital flows in trade financing and industrial development. Even after lax supervision of private banks led to systemic instability, and eventually precipitated bank runs and expensive bail-outs in the mid-1920s, the government did not implement fully the "modern" (western) regulatory norms of the day. The comprehensive Banking Law of 1927, promulgated in response to the crisis, included only a bare minimum of newly codified regulations. Ostensibly, the legislation served to maintain the benefits of the existing system – its operational and administrative flexibility – while responding to a new need for greater stability. Nevertheless, it also worked to increase substantially the discretionary powers of the already powerful Ministry of Finance (MoF).

The 1930s saw Japan's embarkation on the path to an increasingly centralized wartime economic structure, and over the next decade and a half MoF guided consolidation among private banks, the number of which fell from

1,031 in 1928 to 61 in 1945, and extended ever tighter control over public and private financial flows for the sake of "national efficiency". The scale of existing special financial institutions was increased dramatically and new organs were added to their number. Wartime legislation, in 1937 and 1942, further strengthened the state's control over the financial system, but the strains of the time eventually led to industrial collapse in 1943, after which crippling inflation engulfed the financial system.

Following the defeat in 1945, a primary goal of Occupation policy was to democratize Japan by implementing sweeping economic, political and social reforms. With respect to the financial system, the Occupation authorities secured a number of changes. Under their direction, Japan's bank-centered *zaibatsu* (industrial conglomerates) were broken up, in pursuit of an ideal legitimized by both anti-trust ideology and the implication of the *zaibatsu* in the state's recent militarist project. In addition, the banking and securities industries were segregated; the stock exchanges were transformed from private companies into institutions governed by their members; and the Bank of Japan (BoJ) was made as independent from MoF as the US Federal Reserve is from the Treasury. Finally, the Occupation authorities saw to the establishment of a Securities Exchanges Commission (SEC), on US lines, again in order to diminish the power of MoF.

However, with the exception of Article 65 of Japan's Securities and Exchange Law, which separated the securities and banking industries, attempts to rebuild the Japanese financial system from the bottom up fell short of initial expectations. Not only did MoF escape the militarist purges, but there were numerous policy amendments following the end of the Occupation in 1952. The SEC was absorbed into MoF and replaced by an independent advisory committee with no legal powers, the BoJ reform initiative was abandoned, and the *zaibatsu* banks were allowed to re-emerge at the center of new *keiretsu* (industrial groups). Such retrenchment was permitted by inconsistencies and inherent flaws in the Occupation's agenda, and fed by the "reverse course" in the Japan policy of the United States following the onset of the Cold War in 1948 (see Tsutsui). As a result, the structure of Japan's financial system remained largely intact in spite of the Occupation reforms.

For the Japanese, the financial system itself had not been discredited by the war – quite the opposite in fact – and little incentive existed for adopting potentially unsuitable reforms. Returning to a tried and tested method, pragmatic intervention in the financial system was re-embraced by the newly democratized state as a means of coordinating scarce resources to meet the needs of postwar reconstruction. The authorities were, nevertheless, compelled to substitute new indirect means of influence over the allocation of capital for direct wartime controls. In the Temporary Interest Rate Adjustment Law of 1947, which mandated that the BoJ set ceilings on deposit rates, MoF obtained a basis for substantially suppressing lending rates to foster economic growth. Due to the BoJ's subordinate position, MoF was thus in a position to coordinate all other interest rates by administrative guidance (*gyosei shido*). Large public financial institutions were established to channel funds, most of which came from the postal savings (*yucho*) system, into favored industrial or public infrastructure development projects and underfunded areas of the economy. Even after the volume of such flows began to decline, they continued to play an important role until the early 1960s, in that the approval of a loan by a public financial institution was often used as a signal to spur private sector loans because it could be interpreted as an implicit government guarantee. Further, when in the late 1950s private banks were unable to satisfy the demand for industrial funding, the shortage was often alleviated by the BoJ extending credit lines to city banks, a practice known as "window guidance" (*madoguchi shido*) which lasted into the 1970s. Other steps were also taken to centralize bureaucratic control over the financial system. For example, balance sheet, branching, and entry requirements were all reformulated, and guidelines were established to regulate the income and expenditure of individual firms, right down to the level of specifying ceilings on employee wage-levels and share dividend payouts.

The Main Characteristics of the System

Within this framework, several definitive characteristics became institutionalized in the financial system during the 1950s and 1960s. These included a structural bias towards indirect financing; rigid functional segregation of financial institutions; administrative guidance; "convoy" regulation; and international isolation.

First, the authorities sought to maintain the system's prewar bias towards indirect, bank-intermediated financing. While making bank loans available at below competitive market rates, MoF continued to restrict the free development of the securities market by using an established organ, the Bond Issue Committee (*Kiseikai*), to establish stringent requirements for potential issuers, and to manage the pricing, size, and timing of all public offerings (see Hamada and Horiuchi). Further, its strict interpretation of the 1948 Securities and Exchange Law gave financial institutions little room to introduce new types of products, because securities were defined according to prespecified categories, rather than by their attributes, while the stock market listing of riskier or complex securities was prohibited outright for the sake of protecting supposedly naive investors.

Second, in line with the principle of "division of labor", MoF reinforced the prewar system's structure of compartmentalization. Not only were the roles of financial institutions defined according to the notion of separating the banking, insurance, and securities sectors, but subsectoral segmentation was invoked to contain specializations within each sector (see Table 8.1). This fostered clearly defined patterns of funding with, for example, city banks serving big business, *shinkin* (credit) banks serving small and medium-sized firms, and long-term credit banks meeting long-term financing needs. It also had the effects of ensuring that funds flowed into areas of the economy which would otherwise have been underfinanced, and of preventing any industry-wide front against MoF being constructed. As a result, intrasectoral competition among financial institutions was at times intense, but intersectoral competition was almost non-existent.

Third, administrative guidance (*gyosei shido*) became more central as a means of conducting financial regulation. The Securities and Exchange Law of 1948 (amended in 1992), and the Foreign Exchange and Foreign Trade Control Law of 1949 (amended in 1979 and 1997), were added to Japan's existing framework of statutory financial regulations – principally, the Banking Law of 1927 (amended in 1982 and 1992) – and, as with earlier legislation, their skeletal nature enhanced MoF's scope for applying its own dynamic interpretations.

The formal tools of administrative guidance available to Japanese bureaucrats consist of ministerial ordinances (*shorei*), regulations which are drafted by individual ministries on the basis of existing legislation and implemented without the authority of Parliament; and notifications (*tsutatsu*), a means of discretionary bureaucratic action without a legal basis except that a ministry's role is to oversee activity within a particular jurisdiction. Informally, these are backed up by an extensive array of institutionalized channels of contact between the ministry in question and individual firms and organs within its jurisdiction. In the case of MoF, these include personal relationships with specifically designated bank employees, so-called "MoF-watchers" (*MoF-tan*), whose job it is to attend to their firm's relations with the Ministry; former MoF officials who have taken positions in the private sector or in government bodies through *amakudari* (see Appendix 2); and periodic meetings with sectoral and subsectoral umbrella organizations, such as the Federation of Bankers Associations of Japan or the Association of Regional Banks.

MoF's mandate encompasses an unusually high concentration of regulatory functions – financial supervision as well as budgetary, tax, and customs policies – which, in other OECD countries, are shared at least with the central bank, and often with other branches of national or federal government. Nevertheless, care should be taken not to overestimate MoF's discretionary power. More specifically, sectionalism makes talking about MoF as a unitary actor potentially misleading, in that its seven bureaus often represent different interests in accordance with their own functional mandates. In the face of considerable personnel movements among bureaus, considerable jurisdictional competition continues. The Banking and Securities Bureaus broadly defend their respective industries; the International Finance Bureau defends the interests of larger, internationally active financial institutions; the Finance Bureau has close relations with Japan's public financial institutions; and so on. Competition among ministries also curtails MoF's power in areas where various ministries which share responsibility for financial regulation compete. When dealing with interest rates MoF must contend with the Ministry of Posts and Telecommunications (MPT), which has primary responsibility for the massive postal savings and postal insurance (*kampo*) schemes; and when dealing with supervision issues it must vie with the Ministry of International Trade and Industry (MITI), which has primary responsibility for certain institutions (see Table 3.1). Finally, the prospect of conflict with politicians exists in several areas of financial regulation which touch on the interests of powerful political lobby groups, such as post offices and regional financial institutions, both of which are important sources of local support for many elected politicians (see Rosenbluth).

Fourth, the "convoy system" (*gososendan hoshiki*), by which MoF sought to ensure that the evolution of the financial system progressed with all firms in step, developed significant immobilist tendencies. Since the 1920s, MoF had used the granting of branch and product licenses, as well as the threat of forced mergers, to hurry the pace of slower firms and slow the pace of faster firms. However, after a spate of mergers between small financial institutions following financial instability in the late 1960s, MoF was forced by political pressures to tone down its active support of all but non-hostile mergers. By default, this paved the way for greater direct intervention in the affairs of weak financial institutions, as MoF was left with little other choice but to try to improve the management of inefficient firms by other means. When such a firm encountered difficulties, MoF would use its influence with other financial institutions to have them extend support, while insisting that the troubled firm accept the introduction of new skilled senior managers in the form of retiring MoF officials. For MoF, this also provided a useful conduit for retiring staff, but paradoxically the practice also worked to increase the new immobilist tendencies in the convoy system brought about by political intervention. Most MoF *amakudari* who entered private financial institutions "descended" into small firms which were highly sensitive to deregulation, and this served to make such firms adept at protecting their own interests against periodic attempts by MoF to push ahead with reforms. Moreover, other small firms, seeing the potential benefits of representation, began to solicit MoF *amakudari* (see Horne).

Fifth, most of these characteristics depended on the international isolation of Japan's financial system which, as with almost every other state under the Bretton Woods system of fixed exchange rates and strict exchange controls, was effectively separated from the rest of the world. Ostensibly providing a basis for controlling the flow of funds into and out of the country in order to maintain the value of the yen, Japan's Foreign Exchange and Foreign Trade Control Law did much more than this, securing for the government absolute control over both private and public spheres of financial activity. This power was used to serve the country's needs for economic rehabilitation and industrial development. Isolation protected an embryonic and unique financial system which could not have become as deeply rooted under terms of international engagement.

Changes in the System

When examining change in any financial system, a useful analytical strategy is to separate pressures which are brought to bear on the system into three categories: those emanating from developments in the domestic arena; those emanating from developments in the international environment; and those emanating from developments in the financial services industry itself, designated here as "global structural developments". All three types of pressure have brought about changes in the Japanese financial system during the last three decades of the 20th century (as summarized in Table 8.2).

Economic Transformation in the 1970s

Two important events coincided to put pressure for change on the Japanese financial system during the 1970s. One was the country's transition from a developing to a developed economy around the start of the decade. The other was the externally induced oil price hikes of 1973 and 1979, which accentuated the slowdown in growth that was already occurring. The rate of real economic growth fell from the double-digit levels of the 1950s and 1960s, to 8.3% in 1970 and 5.3% in 1973, and the economy actually contracted in 1974. This resulted in major changes in the flow of funds within the domestic economy, which created pressure for reform in the financial system.

As the corporate sector's demand for funds shrank rapidly and the personal sector's surplus increased, financial institutions began to press MoF for the right to market new financial instruments and diversify their customer bases beyond the bounds of their original mandates. As government borrowing increased dramatically, MoF was left with little choice after 1975 but to scrap the artificially low interest rates on government bonds and begin deregulating the bond market. While inherent immobilist tendencies, particularly MoF's sectionalism and political resistance associated with vested interest groups, slowed the progress of liberalization considerably, the financial system's rigid functional segmentation and its fixed interest rate structure were undermined perceptibly by the end of the decade.

In the international arena, despite the breakdown of the Bretton Woods system in 1971, it was not until the very end of the decade that Japanese leaders began to concede the damage which their narrow mercantilism was inflicting upon the international political economy. Thus, their formal participation in several highly visible rounds of international economic diplomacy seeking to resolve trade friction in the 1970s had no immediate ramifications for the Japanese financial system (see Volcker and Gyohten).

However, the enormous expansion of the Eurodollar market in the 1970s did have some effects on the Japanese financial system. First, Japanese banks which were licensed to conduct foreign exchange business saw a rapid drop in demand, as their Japanese multinational clients gained direct access to cheap funding through the Euromarkets, and the small number of foreign banks in Japan found themselves with a lucrative monopoly as demand rose for medium-term "impact loans", denominated in foreign currencies and sourced through the Euromarkets. Japanese banks pressed MoF for access to the impact loan business and their demands were eventually granted. Second, pressure then began to build after a credit squeeze associated with the first "oil shock" of 1973–74 (see Chapter 1) compelled MoF to grant more foreign financial institutions licenses to operate in Japan. MoF took this opportunity to support greater overseas expansion by Japanese banks on the grounds of reciprocity, a move which assuaged these firms' immediate calls for domestic deregulation. However, foreign firms in Japan soon found that the domestic slowdown, coupled with increased competition in "their" market sector, eroded any profitable opportunities which might have existed. They began to press for permission to compete in areas of business open only to domestic banks, and to introduce into Japan some of the more sophisticated products which they used overseas. This latter demand was echoed by large Japanese banks which had become familiar with such products in the course of their own overseas activities. Limited concessions were gained on both of these fronts (see Pauly).

Thus, the way in which Japan's financial authorities responded to domestic, international and systemic developments in the 1970s was characteristically reactive and piecemeal. Of the three, domestic developments had by far the most significant and direct influence on regulatory changes. By the end of the decade, however, the Japanese financial system could only be said to have undergone peripheral changes to its fixed interest rate structure, its rigid functional segmentation, and its international isolation. Administrative guidance and convoy regulation had been reasserted strongly throughout the decade as methods for shoring up the system.

Internationalization in the 1980s

As the financial regulatory changes of the 1970s had paid only minimal attention to Japan's rising external tensions, it was inevitable that MoF would be faced soon with the necessity of comprehensively restructuring the financial system in light of international developments. In the context of soaring bilateral trade imbalances, the new administration of US President Ronald Reagan approached the Japanese government in 1981 with a set of formal demands for changes to be implemented in its financial system. In response, a "Yen-Dollar Committee" was set up in late 1983 for negotiations to take place between the US Treasury and MoF. Six months later a comprehensive agreement was announced under which the Japanese started to deregulate specific sectors of the domestic financial system, according to an agreed timetable which stretched over the rest of the decade. The Japanese authorities were thus committed to liberalizing interest rates, establishing domestic markets in new financial products, granting domestic firms greater access to overseas financial markets, granting foreign banks access to Japan's trust market, and opening membership of the Tokyo Stock Exchange to foreign financial institutions. Commitments were upheld on almost all of these fronts. In retrospect, it appears that MoF used external pressure (*gaiatsu*) surrounding the highly politicized issue of bilateral trade imbalances to circumvent domestic resistance in furthering its own agenda for progressive reform of the financial system. In this case, the initial US demands had been so extensive that MoF was able, to a great extent, to choose which items it acquiesced in, and on what time scale.

Global structural developments during the 1980s were dwarfed by the domestic "bubble economy" which defined the second half of the decade. Essentially, the structural effects of the deregulation set off by the Yen–Dollar Committee combined with the macroeconomic effects of two major international currency agreements, the Plaza Accord of 1985 and the Louvre Accord of 1987, to produce a massive expansion of credit within the Japanese economy. As financial and property markets soared, businesses and individuals were drawn into a vortex of speculative investment. Large manufacturers, for example, exploited newly licensed financial derivatives to compensate declining overseas revenues caused by the high yen with profits from "financial engineering" (*zaitekku*), using financial derivatives and other instruments to hedge against cross-border currency and interest rate risks. As manufacturers reduced their dependence on indirect finance, banks solicited new borrowers with easy credit, making large new loans to the luxury end of the market, including condominiums, hotels, and golf courses. Because MoF had recently lifted many regulations governing international transactions, the "bubble" also engendered global reverberations. External developments, such as the "Big Bang" in the United Kingdom in 1986, which deregulated the stock market there, opened up significant regulatory gaps between Tokyo and other international financial centers, with the result that large amounts of Japanese business were drawn overseas to cheaper and more advantageous locations. Newly cash-rich Japanese financial institutions ascended rapidly to highly prominent positions in the international financial community by pursuing aggressive strategies on low-margin lending and commissions (see Arora). Nevertheless, the impact of the bubble on the structure of the Japanese financial system at the time was negligible. In other words, pressures for change from all quarters were actually assuaged by the tremendous boom conditions.

In comparison with the reforms of the financial system achieved in the 1970s, the changes in the 1980s were of a much greater magnitude. International pressure proved to be the critical variable in bringing about change. By the end of the decade, deregulation had diminished the system's bias towards indirect financing, all but demolished its rigid interest rate structure, and significantly undermined its functional segmentation. However, MoF's administrative practices of financial regulation had shown little sign of change. Ironically, also, because much more drastic financial deregulation measures had been implemented elsewhere, the formal regulatory gap between Japan and its main competitors was actually greater at the end of the decade than it had been at the outset.

Globalization in the 1990s

Global structural developments have been the main agent for change in the Japanese financial system in the 1990s. They exerted pressure through domestic developments, many of which were related to the aftermath of the bubble, and through market forces, which threatened to marginalize the Japanese economy in international competition if deregulation was not enacted decisively.

The bubble economy and its aftermath gave rise to a seemingly endless stream of financial scandals which dominated the media for much of the 1990s. Their effects on the structure of the financial system can be seen in the greater codification of regulations, the narrowing of MoF's scope for administrative guidance, and the undermining of the convoy system.

First, legislation banning insider trading was passed in 1988, following two high-profile cases involving Tateho Chemical and Recruit Cosmos. Compensation for losses on shares was then made illegal in 1991, following discoveries that many brokers, including all the "Big Four" (Nikko, Nomura, Daiwa and Yamaichi), had solicited clients by guaranteeing minimum returns on stock investments. The new legislation, which also banned discretionary trading, was uncontroversial in so far as it was necessary to reassure existing and potential investors that Japan's markets were fair. However, subsequent revelations in 1997, relating to massive loss-compensation pay-offs by Nomura and others to a corporate racketeer (*sokaiya*), did tremendous damage to the reputation of the Japanese financial system in general, and that of MoF in particular. Penalties were increased substantially in 1998, but remain more lenient than, for example, those in the United States.

Second, a series of bureaucratic scandals showed that the untransparent nature of MoF's main regulatory tool, administrative guidance, had made it highly susceptible to conflicts of interest. Public outrage emerged as a result of the half hearted nature of its original investigation into the loss-compensation scandal, which had to be reopened after the Fair Trade Commission chose to involve itself. In response, the government launched an initiative to re-establish an independent securities regulator modeled on the SEC in the United States. However, by the time that the proposal passed into law in 1992, it had been watered down so much that the new Securities and Exchange Surveillance Commission became a *de facto* organ of MoF. Domestic "MoF bashing" soon resurfaced with the emergence of one after another scandal which MoF might have prevented with more rigorous preemptive checks. Its initial attempt to conceal from the US authorities a case of fraud, involving billions of US dollars, at Daiwa Bank's US subsidiary also caused considerable international hostility. In addition, the startling political insensitivity with which MoF proposed a liquidation scheme involving the injection of unspecified amounts of public money for the failed housing loan corporations (*jusen*) led to calls in the domestic media for it to be broken up. Political debate on the issue was again initiated by the government in 1996, and legislation was passed in 1997 to create an independent Financial Supervision Agency, and to revise the 1942 BoJ Law to give the central bank greater independence. How much substance the changes will have remains to be seen, but initial indications are reasonably positive. Equally importantly, however, MoF has taken steps to make its regulation more transparent, and in 1998 it will begin to apply a new system of "prompt corrective action" in bank supervision, based on the capital adequacy guidelines of the Bank for International Settlements.

Finally, it became clear that MoF was fighting a losing battle to paper over the extent of the bad loan crisis when a string of small financial failures began to undermine the convoy system in 1995. Since then, the situation has continued to escalate. The first bank to fail in the postwar period went under in 1995 (Hyogo Bank), while the first postwar failures of an insurance company (Nissan Life), a securities company (Ogawa), a city bank (Hokkaido Takushoku), and a Big Four brokerage (Yamaichi) all made domestic and international headline news during 1997.

There were also structural developments which were not directly related to scandals or the excesses of the bubble economy. Two important sets of financial reform legislation in the 1990s showed that the country's financial authorities were cognizant of Japan's relative position in reference to wider changes which were occurring in the global financial services industry. The Financial System Reform Law, passed in 1992 and effective from April 1993, was the product of several years of discussions within MoF, which had ended in agreement on a conservative compromise to allow banks and securities companies to enter each other's business areas on a limited basis. Through the establishment of separate subsidiaries, banks would be allowed to underwrite bonds, but not sell stocks; ordinary banks and securities companies would be permitted to enter the trust business, but not the most profitable loan and pension segments of that industry. Slowness would be offered as a concession to the inevitable losers, with weaker firms being given priority in the queues for new licenses (see Vogel). It was clear that, in coming to terms with financial globalization, MoF was conceding that it was impractical to maintain its policy of rigid functional separation, yet also making every effort to continue to support the convoy system.

However, in response to the continuing and serious decline in the international competitiveness of Japanese financial institutions and domestic financial markets, Prime Minister Hashimoto (a former Minister of Finance) announced in late 1996 that Japan would move to implement a radical and comprehensive "Big Bang" program of deregulation. He promised that, by making the country's financial system "free, fair, and global", the plan would bring about convergence with international norms by the end of the century. This stunning announcement appeared to signal a fundamental departure from the traditions of financial reform in Japan.

The Big Bang

Undoubtedly, the Big Bang plan is the most important proposal for reforming the Japanese financial system in the postwar period. Not only does it aim explicitly to abolish all of the historical characteristics of the Japanese financial system outlined above, but the way in which it emerged as a political initiative of the Prime Minister appeared to indicate the displacement of MoF's hand in guiding financial reform. Nevertheless, as the plan was fleshed out during 1997 in the advisory councils (*shingikai*) attached to the ministries concerned, the familiar logic of administrative guidance and convoy regulation was still evident, albeit in a less obtrusive form.

Hashimoto's proposal, launched as a headline policy of his new government following the general election of October 1996, was put forward in the context of pressing domestic, international, and systemic developments. These included a rapidly rising government deficit; a widening gap in competitiveness between Japan and its main competitors; the prospect of European Monetary Union proceeding on schedule; the declining international competitiveness of the Japanese financial services industry; and the hollowing out of the Tokyo stock market. Against this background, the Prime Minister and his staff constructed the Big Bang proposal from a number of sources. The umbrella concept came from the Economic Council, an advisory group attached to the Economic Planning Agency. Its core ideas drew on proposals which MoF's Securities Bureau and International Finance Bureau had been refining quietly for a couple of years. Finally, other items were added, such as proposals drawn from political initiatives to restructure MoF, and ideas from the Prime Minister's separate agenda of general administrative reform.

The integrated concept was then made public with great ceremony and handed to the bureaucracy to translate into concrete policy through discussions held in various *shingikai* attached to MoF. If this in itself was not enough to raise alarm bells, the progress of discussions soon showed that Hashimoto's original proposal was being undermined by the *shingikai* system's inherent proclivity for compromise. For example, in discussing the notion of ensuring "fairness" of opportunity for market participants, an unwieldy distinction was adopted to differentiate between "the weak, who need protection, and those who lose in fair competition" (according to a report in the magazine *Shukan Toyo Keizai*, April 23, 1997). With regard to external competition, support for a managed "fairness" was promoted overtly by MoF, which introduced a neat rhetorical analogy to stress the negative attributes of the Anglo-American model against the positive ones of the traditional Japanese model (according to a report in the newspaper *Nihon Keizai Shimbun*, March 12, 1997). Whereas British financial deregulation had created a world-class competition, in which most of the players and all of the winners are foreigners, as in the Wimbledon tennis tournament, it was claimed that a uniquely Japanese deregulation policy could establish a successful national competition based on international rules and energized, but not dominated, by "star" foreign players, as in the J-League soccer tournament.

By the time that final *shingikai* reports were submitted, in June 1997, it was clear that the Big Bang would not live up to the high expectations which Hashimoto's initial presentation had generated. Opinions were still split on a number of sensitive issues where deadlines, details, and even broad policies had been omitted – such as tax and accounting issues, and the reform of public financial institutions – while on others deadlines had been set back, for example, on the deregulation of the life insurance industry. Some of these matters were settled in subsequent political discussions, but the fact that the Big Bang had been drawn up in a fragmented way by many different panels, most of which had ties to interest groups, meant that the end-result displayed many of the signs of that balancing of interests which has been historically characteristic of the Japanese financial system.

On top of this, the concentration in November 1997 of a string of unprecedentedly large financial failures unnerved the markets by threatening to undermine Japan's already faltering economy. In response, ruling party politicians initiated moves to postpone and otherwise amend one or two important items of the Big Bang agenda, such as the implementation of the new framework for bank regulation (described above). Table 8.3 provides an outline of the Big Bang plan as it stood at the beginning of 1998.

Nevertheless, such developments do not suggest that monumental change is not afoot. Together, the liberalization of commissions on securities and foreign exchange transactions, the licensing of many new and complex financial products, the further lowering of functional barriers separating different financial sectors, and the legalization of financial holding companies are transforming the competitive terrain of Japan's financial markets. Faced with the prospect of competing for a much larger share of the country's estimated ¥1,200 trillion in personal financial assets, technologically superior foreign financial institutions have boosted their presence in Japan. They have already secured significant market shares in recently liberalized fields such as bond underwriting and sales of investment trusts. For their part, domestic firms have been compelled to step up their post-bubble restructuring efforts. In unprecedented ways, they are abandoning long-cherished management practices, such as seniority pay and lifetime employment, and shedding large numbers of "excess" staff and assets. However, reports from international credit rating agencies, such as Moody's Investor Services, suggest that for all but a handful of firms the immediate outlook is bleak. While some have already gone out of business (as noted above), many others have sought to buttress their positions by retreating from foreign markets, pursuing mergers with affiliates, and establishing joint ventures with foreign firms.

In sum, the Big Bang is effecting a dual process of polarization among domestic financial institutions. On the one hand, the differences between strong and weak firms are being exaggerated; on the other, firms are either expanding their capital base and acquiring the expertise necessary for them to become full-line financial service providers, or they are retreating to take up niche positions. To the extent that these trends reflect the dominant currents in finance more generally, they offer strong evidence that the globalization of the Japanese market is now well under way. Nevertheless, both explicitly and implicitly,

Japan's regulatory authorities remain as intent on creating a "third way" in restructuring their financial system at the end of the 20th century as they were a hundred years ago. As the eminent Japanese political economist Yamamura Kozo concluded recently, in a broader context, "*Plus ça change . . .*"

Conclusion

Several lessons could perhaps be drawn from this overview of the Japanese financial system. The first is not to be taken in too easily by the political rhetoric of deregulation. The case of the Big Bang shows that radical proposals are made public in order to test the political waters in the hope of soliciting public support, but this in itself is not a foolproof barometer of their chances of being implemented. The second is to focus on the substance of debate in the policy-making arenas that count. In the financial sphere, non-Japanese observers have consistently underrated the importance of the *shingikai*. However, as recent developments have also shown, political dynamics at the intra- and interparty levels can also affect outcomes under certain circumstances, for example, coalition politics or economic crisis. The third is to be skeptical about changes that appear to be completely out of character. In the case of Japan, as elsewhere, new pressures associated with globalization are certainly significant, but institutionalized historical characteristics do still matter.

Further Reading

Arora, Dayanand, *Japanese Financial Institutions in Europe*, Amsterdam and New York: Elsevier, 1995

A study of the internationalization of Japanese banks and securities companies in the 1980s.

Hamada Koichi and Horiuchi Akiyoshi, "The Political Economy of the Financial Market", in Yamamura Kozo and Yasuba Yasukichi, editors, *The Political Economy of Japan*, Volume 1, *The Domestic Transformation*, Stanford, CA: Stanford University Press, 1987

The classic English-language introduction to the Japanese financial system.

Hayakawa Shigenobu, editor, *Japanese Financial Markets*, Cambridge: Gresham Books, 1996

A recent practitioner's guide to Japan's financial markets.

Horne, James, "Politics and the Japanese Financial System", in J. A. A. Stockwin, Alan Rix, Aurelia George, Daiichi Ito and Martin Collick, editors, *Dynamic and Immobilist Politics in Japan*, London: Macmillan, and Honolulu: University of Hawaii Press, 1988

A study of domestic financial policy-making and deregulation.

Nihon Keizai Shimbunsha, editors, *Do Naru Kinyu Biggu Ban?* [What Will Happen with the Big Bang?], Tokyo: Nihon Keizai Shimbunsha, 1997

A comprehensive guide to the Big Bang (nothing comparable has yet been published in English).

Pauly, Louis W., *Opening Financial Markets: Banking Politics on the Pacific Rim*, Ithaca, NY: Cornell University Press, 1988

A comparative study of financial deregulation in the United States, Japan and Australia.

Rosenbluth, Frances McCall, *Financial Politics in Contemporary Japan*, Ithaca, NY: Cornell University Press, 1989

A widely cited study of domestic politics and financial policy-making.

Schwartz, Frank J., *Advice and Consent: The Politics of Consultation in Japan*, Cambridge, MA: Harvard University Press, 1997

A recent study of Japan's *shingikai* system.

Tsutsui, William M., *Banking Policy in Japan: American Efforts at Reform during the Occupation*, London and New York: Routledge, 1988

A re-evaluation of the success of attempts to "Americanize" the Japanese financial system.

Vogel, Steven K., *Freer Markets, More Rules: Regulatory Reform in Advanced Industrial Countries*, Ithaca, NY: Cornell University Press, 1996

A comparative study of regulatory reform in the fields of finance and telecommunications in Japan and the United Kingdom.

Volcker, Paul, and Gyohten Toyoo, *Changing Fortunes: The World's Money and the Threat to American Leadership*, New York: Times Books, 1992

A first-hand account of international financial diplomacy during the 1980s.

Yamamura Kozo, "The Japanese Economy After the Bubble: *Plus Ça Change?*", in *Journal of Japanese Studies*, Volume 23, number 2, 1997

A review article on the current state of the debate about economic convergence.

James D. Malcolm is a doctoral candidate at the University of Sheffield, and a Visiting Researcher at the University of Tokyo.

Table 8.1 Segmentation of the Postwar Financial System

Deposit-taking institutions
- Ordinary commercial banks
  - City banks
  - Regional banks
  - Second-tier regional banks
  - Foreign banks
- Long-term credit institutions
  - Trust banks
  - Long-term credit banks
- Small business finance institutions
  - Mutual (*sogo*) banks
  - Credit (*shinkin*) banks
  - Labor association banks
  - Credit cooperatives
- Finance institutions for agriculture, forestry and fisheries
  - Agricultural cooperatives
  - Fisheries cooperatives

Consumer credit institutions
- Housing finance institutions
- Consumer finance institutions

Insurance companies
- Life insurance companies
- Non-life insurance companies

Securities companies (brokers)

Note: Public sector financial institutions are included in Table 3.1.

Table 8.2 Causes of Change in the Postwar Japanese Financial System

Source of pressure for change	*1960s*	*1970s*	*1980s*	*1990s.*
Domestic developments	rising	high	low	medium-high
International developments	low	rising	high	low
Systemic developments	very low	rising	medium	high

Table 8.3 Outline of the Big Bang Proposal

	To come into effect
Supervision	
Revised Bank of Japan Law (passed by Parliament 1997)	April 1998
Revised Foreign Exchange and Foreign Trade Control Law (proposed by Committee on Foreign Exchange)	April 1998
New Financial Supervision Agency (established by Parliament 1997)	July 1998
Banking (proposed by the Financial System Research Council)	
Removal of barriers among city banks, long-term credit banks and trust banks	fiscal 1997–98
Revision of rules on banking, securities and trust subsidiaries	from April 1997
Implementation of system of "prompt corrective action"	from April 1998
Permitting the formation of financial holding companies	from January 1998
Reform of public sector financial institutions by Parliament	being considered
Reform of postal savings system by Parliament	not yet set
Securities (proposed by the Securities and Exchange Council)	
Liberalization of brokers' commissions	from April 1998
Removal of restrictions on investment trust products	by 2001
Registration system for investment trust business	by 2001
Banks selling investment trust products over the counter	about 1999
Pricing according to public demand	fiscal 1998–99
Introduction of stock-lending in the over-the-counter market	fiscal 1998–99
Removal of prohibition on private investment trusts	fiscal 1998–99
Accounting (proposed by the Business Accounting Council)	
Introduction of marking to market	from April 1997
Pensions (proposed by MoF and the Ministry of Health and Welfare)	
Revision of rules on public and private pension contributions	from April 1997
Introduction of investment advisory business for private pensions	fiscal 1997–98
Relaxation of fixed interest rates	from April 1997
Introduction of fixed contribution system	fiscal 1999–2000
Insurance (proposed by the Insurance Council, except as shown)	
Liberalization of premiums for non-life insurance	fiscal 1997–98
Banks selling life insurance products over the counter	about April 1999
Expansion of interpenetration between life and non-life businesses (proposed by MoF)	fiscal 1999–2000

Chapter Nine

The Role of Trade in Japan's Re-emergence

Carin Holroyd

It is international trade that forces the world to pay attention to Japan. Since the end of World War II the country has been a comparatively quiet member of the community of nations. One rarely hears of the international efforts of Japanese diplomats, of major social movements spawned from Japanese soil, or of significant cultural phenomena that catch on across the globe. However, one does hear, fairly routinely, about Sony or Toyota, about Japanese robots, *tamagochi*, or other electronic and mechanical innovations, and about the continuing efforts of western businesses to break into the Japanese domestic market. Japan is the world's second largest trading nation, an exporter of some of the world's highest-quality and most technologically sophisticated manufactures, and an importer of considerable importance to countries from Sweden to New Zealand.

That Japan stands at the end of the 20th century in a position of economic importance exceeded only by the United States is the outcome of one of the most remarkable national stories of recent decades. The country is relatively small, no larger than California, and its 125 million people place it only eighth on the world's list of most populous nations. This small archipelago in the North Pacific is also not well-endowed with natural resources. Japan is no Brunei or Kuwait and is not blessed with the readily available raw materials necessary for a modern manufacturing nation. As a consequence Japan imports a great deal of what it needs to survive, from oil to foodstuffs, and much of what it requires to operate its industrial plants. In turn the country has positioned itself at the forefront of the world's trading nations, with a commercial reach that extends to all corners of the globe and that places Japan near the center of the globalizing economy. The path has been difficult and uneven – only 25 years ago the phrase "made in Japan" was synonymous with cheap, low-quality products – but the determined and coordinated effort to carve a place for Japan in the global economy has achieved results that few would have forecast only decades ago.

Thus Japan's rise from the ashes of World War II has been one of the most significant economic developments during the second half of the 20th century. Starting as a war-ravaged economy unable to provide for the basic subsistence of its citizens, Japan re-emerged on the global trading scene, firmly entrenched as one of the world's great trading nations. Japan's complex network of trading relationships, including reverse trading structures with many nations, continued reliance on the importation of raw materials and a vital role in the development of high-technology international trade, has transformed this once marginal nation into a global economic powerhouse.

More than anything else, trade has meant national and individual prosperity. Japan's GNP reached an all-time high of US$5 trillion in 1996. This represented 70% of the GNP of the rest of Asia combined, and fully 80% of the GNP of the United States, a nation more than twice as populous as Japan. Further, income per capita is now around 36% higher in Japan than in the United States. The vast majority of Japanese citizens consider themselves to be "middle class" and only slightly more than 5% feel that they are members of the "lower class" (see also Chapter 11). *Fortune* Magazine's Global 500 list of the top corporations included 140 Japanese firms in 1996, and Japanese savings, the cornerstone of the national economy, account for more than 40% of the world's total savings (see JETRO).

Japan at the End of World War II

The emergence of Japan as a major trading nation, and as one of the most innovative and complex international economies in the world, can be traced back to the late 19th century. After significant economic expansion in the early decades of the 20th century, offset in part by the impact of the depression in the 1930s and all but destroyed during the war, the Japanese government and its business partners set out in 1945 to rebuild the industrial infrastructure and re-establish trading relationships in a country which was physically and psychologically flattened. The population was exhausted and starving; most of Japan's cities and almost all of its industries had been destroyed; agricultural production had been drastically reduced. The country was in desperate need of food for its hungry citizens and raw materials for its crippled industries, but it had little to sell in return, and postwar hostility meant that few countries were eager to buy what Japan did have.

The two main objectives of the Allied Occupation were to demilitarize and to democratize Japan, both politically and economically. Economic reforms included anti-monopoly measures to break up the prewar *zaibatsu* (business conglomerates), land reform policies, and the legalization of labor unions (see Ito). While the Occupation forces concentrated on creating a strong economic foundation for Japan, the government immediately started to try to build up the productive capacity of its main industries, beginning with the coal and steel industries. Recovery took time. The late 1940s were a time of economic and social uncertainty. Poverty was still widespread, inflation was high and US anti-Communism had led to a retrenchment of political freedoms. Just when the economy seemed to be heading for another recession, relief appeared in the form of the Korean War (1950–53). The United States and its allies decided to procure many of the war supplies they needed from Japan, providing exactly the stimulus that the Japanese economy needed.

Japanese Trade During the Period of Industrialization, 1955–70

During the Korean War demand for Japanese goods soared, leading to rapid industrial expansion. The San Francisco Peace Treaty came into effect in April 1952, restoring Japan's sovereignty and allowing the country to create its own economic policies. In 1955 the government's Economic Planning Agency announced the first of its five-year plans – The Five-Year Economic Independence Plan – with a target of 5% growth. Up to the late 1980s the government announced 11 different plans or forecasts. Their purpose was to indicate the government's commitment to growth by setting targets for GNP growth (see Ito). As with many of these plans, the first was scrapped before its term was over because the economy was growing so quickly that the plan had become obsolete.

Economic growth was the nation's main focus and the population threw themselves into work with single-minded purpose. Environmental and labor considerations came second. Iron and steel, and petrochemicals, both targeted by the Ministry of International Trade and Industry (MITI) for special assistance, remained important export items throughout the 1960s and 1970s. The steel industry made rapid improvements in its costs of production, its mill facilities and its technological know-how. By 1976 Japan boasted seven of the 10 most efficient blast furnaces in the world (see Shinohara). The petrochemical companies also expanded their capacity and upgraded their facilities. Both of these industries spawned the development of additional industries as diverse as automobiles, plastics and synthetic fibers, and rubber (see Shinohara).

Japan's trade structure shifted dramatically in these 15 years (see Table 9.1). In 1955 Japan's main export was textiles, which represented 37.3% of all exports. Textiles represented only 12.5% of exports by 1970 and a mere 4.8% by 1979. Machinery and equipment, on the other hand, represented 13.7% of exports in 1955 but leapt to 46.3% by 1970 and 61.3% by 1979. The value of Japanese exports absolutely skyrocketed, from less than US$3 billion in 1958 to US$19 billion in 1970, and an astounding US$103 billion in 1979. This was an increase of 4,300% in just over 20 years. Other major trading nations such as the United States, West Germany, the United Kingdom, France or Canada saw their international trade grow at much less dramatic rates in this period (see IMF, various years).

The 1960s saw the Japanese enter many new areas of manufacturing. Japanese consumer durables such as refrigerators, television sets, cameras and watches flooded the North American and European markets. These products at first were low-priced, inexpensively made and of mediocre quality, but as each year passed their quality improved. Japan's manufacturing industries were the power behind the country's extraordinary growth rates, which averaged above 10% in the years between 1956 and 1972.

Automobile-making was probably the most prominent of these new Japanese industries. In the same way that the petrochemical industry had fueled expansion of the automobile industry, automobiles in turn had a positive impact on the growth of other industries. When the Japanese first began exporting cars to North America the initial reaction was one of surprise and good humor. Initially Japanese cars were small, unattractive and poorly equipped for the rigors of the North American climate. The American "Big Three" (General Motors, Ford and Chrysler) did not see Japanese producers as offering much competition. They did not notice as the Japanese producers upgraded and modified their cars, or respond when urban drivers slowly began to appreciate these smaller, less expensive and more fuel-efficient alternatives.

The United States has remained Japan's most important trading partner since the end of World War II: it is both the largest recipient of Japanese goods and the biggest supplier (see Tables 9.2 and 9.3). Asia has also become a consistently important destination for Japanese exports. Between 1958 and 1978 six or seven of Japan's top 10 export destinations were in Asia. The most important of these were South Korea, Hong Kong and Taiwan, while the Philippines and Thailand dropped away in the 1970s, to be replaced by Singapore and Indonesia. The United Kingdom was an important market for Japan in the late 1950s and early 1960s, but gradually faded out of the top 10. Large purchases by Liberia, probably ships, propelled that nation sporadically to the top of Japan's export destination list between the late 1950s and the mid-1960s. Australia was an important market throughout the 1960s and 1970s as, to a lesser extent, was Canada, while West Germany became a very important market in the 1970s.

The countries that sold the most to Japan between the 1950s and the 1970s were predominantly suppliers of energy and raw materials. The presence of five Middle Eastern oil exporters in the list of Japan's top 10 importers for 1974 and 1978 is clear evidence of Japan's dependence on oil imports, and of the high prices that followed the first "oil shock" in the early 1970s (see Table 9.3). Australia and Canada were both important providers of minerals, coal, timber, and other resources to Japan throughout the 1960s and 1970s.

Japan's Emergence as a Trading Powerhouse, 1970–85

The 1970s brought with them a number of changes, both positive and negative, for Japan. One of the first was the end of the Bretton Woods regime of fixed exchange rates. (The yen had been fixed at 360 to the US dollar in 1949.) Adjustment to the floating exchange rate, which caused the yen to appreciate quickly and made Japanese exports more expensive, took some time. However, adjustment to the new exchange rate paled in comparison with adjustment to the 1973 and 1979 oil shocks, the massive increases in oil prices and the embargos on oil supplies imposed by the major petroleum-exporting countries. No other major industrial nation in the world was as dependent on imported oil as Japan. Much of the previous decade's rapid industrialization had depended on an inexpensive supply of oil, which at times in the mid-1970s made up almost 50% of Japan's total imports (see Dore).

Growth ground to a halt in 1974 but the nation and its industries quickly worked on becoming more energy-efficient. The economy picked up gradually but the average growth rate in the 1970s dropped from 10% to 5%. By the time that the second oil shock hit in 1979, Japan was much better prepared. Japanese companies had begun looking for alternate kinds and sources of energy: steel companies searched the world, particularly Canada and Australia, for sources of coal. While Japan coped well and quickly with the rise in oil prices, the country was unlikely ever to forget the sense of vulnerability that the oil crises had triggered. The issue of energy had lasting implications for Japan's industrial structure. Many companies in industries which were heavy users of energy diversified into other activities, while those that remained dependent on energy worked hard to make their industries exceptionally energy-efficient.

On the positive side the 1970s saw a transformation in consumers' attitudes to Japanese products. "Made in Japan", which had once been synonymous with poor quality and low-end trinkets, became a hallmark of quality and reliability. This shift, imperceptible at first but gathering strength so quickly that it had all but faded from memory by the 1980s, seemed to originate with the rapid expansion of the Japanese electronics industry. Although transistors were initially developed in the United States, Japanese firms were the first to capitalize on the commercial opportunities they presented. Led by Sony, Japanese companies flooded the market with an array of miniature radios and other products. Similarly, Japanese car producers learned from their earlier mistakes, and the quality and appearance of their cars dramatically improved, resulting in soaring sales.

By the early 1980s, Japan was moving higher up the value-added ladder. Exports of products such as communication/electronics equipment, automobiles, trucks, motorcycles, electrical machinery and precision instruments increased at a rapid rate. Japanese firms established reputations for providing some of the most innovative, efficient and high-quality items on the market. Whereas Japanese goods had once been relegated to discount aisles, they now stood at the forefront of the burgeoning consumer marketplace. Japanese companies continued to move into new product lines, and to improve on ideas and products originating elsewhere.

While IBM, Apple and other US companies provided the lead in the development of the personal computer market, Japanese firms moved quickly into the field. The Japanese government was determined to have a domestic computer industry, and fiscally and organizationally encouraged Japanese companies in their research (see Anchordoguy). By the middle of the 1980s companies such as NEC, Toshiba, Hitachi and Fujitsu were front-line competitors in personal computing. Perhaps most significantly, several Japanese firms were critical in the development of certain key components, such as the technology associated with the screens in portable computers or retail scanning equipment. This has given them a virtual global monopoly in these high-technology, high-value-added products.

Japan's imports throughout this period continued to be composed mainly of raw materials and barely processed goods. As the years passed the nation's expanding industries needed increasing quantities of resources such as oil, coal, lumber, and aluminum. Despite the increased sales, many of Japan's trading partners were not so pleased. While Japan was happy to buy those raw materials that it did not possess, it did not make it easy for companies who wished to sell manufactured products in Japan. Japan's trade surpluses kept growing at a phenomenal rate and other nations, particularly the United States, were becoming upset.

Trade friction between the United States and Japan increased over the years. Textile negotiations had taken place between the two countries in the early 1960s, followed by steel negotiations in the late 1960s. By the 1970s, television sets, machine tools and, most seriously of all, automobiles, were all focuses of trade tensions (see Ito). Japanese cars had flooded the US market, and their price and reliability had made them a hit with consumers. US cars, on the other hand, had barely made a dent in the Japanese market: dentists and *yakuza* (gangsters) are known to be among the few customers. By 1980 the US car-makers were in dire straits and Chrysler was actually close to bankruptcy. The United States and Japan negotiated a voluntary export restraint agreement which saw Japan agree to limit the number of cars it could sell in the United States. This arrangement remained in place for more than a decade, although the ceiling was raised over the years.

As Japan's trade surplus continued to increase throughout the 1980s (see Table 9.4), international frustration began to mount. The United States, in particular, complained that one of the reasons was the heavily undervalued yen and pushed for exchange rate corrections to be made. In September 1985, at the Plaza Hotel in Washington, DC, the leaders and finance ministers of what was then the Group of Five nations (the United States, Japan, West Germany, the United Kingdom and France) met and agreed to put in place mechanisms that would see the yen and the Deutschmark increase in value relative to the US dollar.

The effects of the Plaza Accord were immediate. (see Table 9.5). Just before the meeting the exchange rate was ¥240 to the US dollar, but only nine months later it was ¥150, a 60% shift. As expected, this dramatic change initially sent the Japanese economy reeling and there was a brief recession, which lasted until late 1986. Soon afterwards, however, manufacturers adjusted to the appreciation of the yen. While it made their exports more expensive, the raw materials that their industries required were now less expensive. The economy bounced back and began to expand once again.

This expansion was primarily due to the expansion of domestic demand – a growth in capital investments and in personal spending. The boom, often referred to as the "bubble economy", lasted for four and a half years until the middle of 1990, when its foundation of unrealistically high real estate values collapsed. Even after the collapse Japanese companies continued their export onslaught. In 1993 Japan's trade surplus with the United States again reached record heights (see Table 9.2) and the United States again began pressuring Japan to liberalize its markets. As talks stalled the value of the yen took off once more, even falling below ¥100 to the US dollar.

The Transformation of Japanese Trade

The Plaza Accord and the consequent rapid rise in the value of the yen became the catalyst for a number of dramatic changes in Japan's trading structure. One of the more obvious was the immediate increase in direct overseas investment (see Table 9.6). Japanese manufacturers wished to take advantage of lower labor costs abroad and to use local production as a way to avoid increasing protectionism in Europe and the United States. The bulk of overseas investment initially, therefore, was in manufacturing. Many of the automobile companies and their affiliated parts-makers set up production lines in the United States and other countries. For example, production in Canada was mainly designed to obtain access to the US market under the Canada/US Free Trade Agreement of 1989 (since expanded, via the North American Free Trade Agreement, to include Mexico).

Direct overseas investment has had an impact on Japan's total trade surplus and its surpluses with individual countries, although the total surplus has remained at around the levels of 1985 (see Table 9.4). The anticipated declines in Japan's trade, however, did not occur. Overseas production of automobiles and electrical equipment by Japanese companies has also resulted in "reverse importing", whereby manufacturers ship their products back to Japan for assembly or sale. In 1989 10% of Japan's imports of finished products were reverse imports (see *Japan Economic Almanac* 1990).

Additional production shifted offshore after 1993 when the value of the yen again began to climb. For many manufacturers an exchange rate of ¥100 to the US dollar meant that they would simply not be viable if they did not find new ways to survive. For most of the larger companies, from Sony to Matsushita, Nikon to Toyota, this has meant a further shift to overseas production (see *Japan Economic Almanac* 1994).

Overseas production by Japanese companies has instigated a regional division of manufacturing in Asia in what the Japanese economist Akamatsu Kaname has called the "flying geese" pattern of development. As incomes rose in the newly industrializing countries, production shifted out of Japan. Products in the middle range of value added, such as cellular phones, standard fax machines, computer displays, and large color televisions, are being shifted to Singapore, Malaysia, and other countries, while lower-priced, labor-intensive products, such as small color televisions, lower-priced audio-visual products, computer assembly, and electrical parts, which were once produced in Singapore and Malaysia, are being transferred to Indonesia and the Philippines (see Ohta et al.; see also Chapter 18).

The implications for Japan's trade structure are enormous. As Japan moved much of its production overseas, it had to ensure that its industrial base did not "hollow out" by losing too much of its industrial infrastructure and, thus, too many of the attendant high-paying jobs. It therefore needed to produce more value-added products domestically and develop new industrial technologies (see *Japan Economic Almanac* 1995). Thus Japan is now producing high-value-added products, such as high-performance home appliances, laser disk players, high definition and digital televisions, CD-ROM multi-media products, digital compact cassette recorders, and certain computer-related parts, including customized memory chips and high-performance work stations (see Ohta et al.). While consumers in western industrialized countries will remain interested in these products, Japanese producers see the large and expanding markets of Asia as being even more important. As these countries become more prosperous, they represent enormous opportunities for the sale of consumer goods.

Japan is also working hard to develop new technologies. The country has taken the lead in the production of industrial robots, and more than 60% of the industrial robots operating in the world are in Japan. Industrial robots have been traditionally targeted at automobile and electronics manufacturers, but slow sales have encouraged their makers to develop robots which can be used in warehousing and distribution, as well as personal robots which do jobs such as cleaning offices or retrieving objects. Japanese companies are trying to develop cheaper robots and robots with improved visual techniques and the ability to handle fragile objects (see *Japan Economic Almanac* 1995). While neither industrial nor personal robots have yet attracted much attention, the industry holds considerable potential. Japan is also in a good position to lead many energy-related industries. It is at the forefront of the development of energy-saving applications for superconductivity, the design of energy-efficient modes of transportation, such as the magnetic levitation ("maglev") train, and the commercialization of solar power for use in pocket calculators, street lights and air conditioners (see Fingleton).

The prospects for growth in the electronics industry are also strong, particularly as Japanese businesses are continually searching for new applications of existing technologies. High-definition television, for example, is being used to teach medical students and has potential for use in higher-definition radar screens for aviation, and in visual databases for architecture and engineering (see Fingleton). Car navigation systems, video CD systems, flat-screen and wide-screen televisions, and videophones all have considerable commercial potential as the world embraces the digital revolution (see *Japan Economic Almanac* 1995).

Another noticeable change that occurred in the late 1980s was in the composition of Japan's import profile. Resource products, which in 1985 made up two thirds of Japanese imports, comprised only 50% in 1991 (see Lambert). Japan began to import more manufactured products, particularly relatively easily manufactured products such as textiles. This change stems from a number of factors. First, as mentioned, Japan's investment in overseas manufacturing increased greatly and imports from these overseas affiliates have been quite steady. Second, consumer attitudes toward foreign products have been changing. In 1988 a survey by the Manufactured Imports Promotion Organization revealed that 75% of the Japanese population would not discriminate between imports and domestic products. This is a major change from previous years, when nationalism meant that most consumers would buy Japanese products if at all possible. Third, Japan's trade surpluses with numerous countries, particularly the United States, have led to pressure on the Japanese to purchase from abroad.

In 1993 Ron Wickes and Aldith Graves of Australia's Department of Foreign Affairs conducted a study of the changes in the Japanese market for a wide range of manufactured imports. They concluded that in more than half of the sectors that they studied imports of manufactured goods had tripled between 1985 and 1990. In particular, imports of road vehicles had increased 900%, and there had also been high growth in imports of works of art, clothing, electrical machinery and non-ferrous metals (see Wickes and Graves). Japanese consumers with their new-found wealth enjoy imported luxury products, such as French perfume, Italian clothing and expensive art.

Japan's export destinations during the 1980s were not radically different from those of the previous decade. The United States remained the largest purchaser of Japanese products, followed by West Germany and South Korea. Other East Asian markets, particularly Hong Kong, Singapore and Taiwan, and the United Kingdom increased in importance, while Australia and Canada remained steady customers.

On the import front in the 1980s, the oil-exporting nations of the Middle East lost some of their prominence in Japan's import profile. Only Saudi Arabia and the United Arab Emirates remained among the top 10, and as the decade progressed they declined in importance. Meanwhile the nations of East Asia were beginning to sell more to Japan. Indonesia had always been important, primarily due to its oil and other resources, but Taiwan, South Korea and China were beginning to sell more too. West Germany also gradually increased sales. This change in countries exporting to Japan was a reflection of many of the changes described earlier, including Japan's overseas investment, reverse importing and the change in the types of products that Japan imports.

Japanese Trade Towards the End of the 1990s

In 1998 Japan is well-positioned as to the diversity of its exports, and the distribution of those exports among the world's trading nations. While the United States remains Japan's largest market for both exports (27%) and imports (22%), the trading profile is diversified enough not to be unbalanced. The EU has remained an important market: its member states now purchase about 15% of Japan's exports and are the sources of 14% of its imports. East Asia has been an important export market for Japan for decades. South Korea (7%) is now Japan's second most important export market, followed by Taiwan (6%), Hong Kong (6%), China (5%) and Singapore (4%) (see Table 9.8). The members of the Association of Southeast Asian Nations (ASEAN) form the destination for 17% of Japan's exports. When the ASEAN total is combined with those for South Korea, Taiwan, Hong Kong and China, it becomes clear that more than 40% of Japan's total exports now go to East Asia. While the shift towards these markets presents some challenges during times of economic distress in the region, as during the financial crisis of 1997–98, long-term expectations of continued regional expansion suggest that Japan will profit substantially from this reorientation of its trading activity.

As was discussed above, East Asia increased in importance as a source of Japanese imports in the 1980s, and by 1996 four of the top five sources of Japan's imports were in that region (see Table 9.9). This is at least partly a reflection of the increase in Japan's imports of manufactured goods, and in the number of manufacturing companies that have moved their operations to other countries in East Asia and now export products from those factories into Japan.

Japan has made marked changes in its export mix. Thirty years ago the country's exports consisted largely of low-end, low-cost manufactured products. With the recent and rapid expansion of Japan's high-technology industries, an initiative that rested on extensive relations between government and industry, it is hardly surprising that the composition of these manufactured exports has shifted dramatically. Japan's major exports are high-value-added manufactured products, particularly machinery, electrical machines and vehicles (see Table 9.10). In fact, these three categories make up 65% of all Japan's exports. Included in these rather broad categories are goods ranging from industrial robots to fax machines, and from computer parts to television equipment. Japanese goods have found impressive markets because of the reputation of Japanese corporations for leading-edge technological innovation and high quality. Given the continued growth of high-technology industries, Japan has established an important foundation for this vital area of international trade.

Japan's leading imports continue to be mineral fuels and oil, which make up slightly less than 20% of the total. Other major imports include machinery and vehicles, electrical machinery, wood, and seafood and fish products. Manufactured imports have become an increasingly important part of Japan's import profile (see Table 9.11): six of the top 10 imports are in the manufactured category. A sizable proportion of the manufactured products imported into Japan are made by Japanese corporations overseas. Because Japan has also moved away from producing certain low-end manufactured products to focus on the higher end, there is more room for other countries to sell it products that it no longer produces.

Although the changes are not yet readily apparent, the shifting Japanese import mix is resulting in substantial alterations in international trading patterns. Those countries that are closely attuned to the Japanese market, realizing that the Japanese are willing, as never before, to open their borders to selected manufactured items, have excellent opportunities to expand their shares of Japan's import trade. However, where there are winners there are also losers. Several countries, most notably Canada and New Zealand, that rely on Japan as a ready market for raw or partially processed materials, such as lumber, fish or food products, are already finding that they face increasing international competition to hold on to existing Japanese sales. By not paying sufficient attention to Japan's growing demand for manufactured goods, a highly specialized demand that must meet the longstanding and exacting quality expectations of Japanese consumers, these countries stand to lose out on important trading opportunities. After several decades of seemingly unchanged trade relationships, based on the importation of raw materials and the export of manufactured goods, the shifting Japanese trading scene presents some formidable challenges and opportunities for other trading nations.

Future Prospects

Western observers have mixed opinions on the future of the Japanese economy. While there appears to be some agreement that Japan is likely to remain a major global force, how important the country will be, and how long it will remain at the forefront of international trade, are hotly contested questions. The extremes of the debate can be summarized by reference to two recent forecasts of where Japan will be by the year 2000. In 1991 Marvin Cetron and Owen Davies declared that Japan would "not even be among the top 10" economies in the world; in 1995 Eamonn Fingleton argued strongly that, in the absence of a major disturbance in global trade patterns, Japan's economy will be even larger than that of the United States. The majority of writers on the future of Japan make more moderate and cautious predictions than these, and the rapid and diverse changes in the country's situation make predicting the future a difficult task in any case. Most believe that the economy will remain reasonably strong, despite a number of serious challenges that the nation faces in the short and medium term. The main problems include its rapidly aging population, concerns about the education system, disarray in financial services, public disillusionment with politicians and a new distrust of bureaucrats, tensions in Japan's relations with the United States, and the potential hollowing-out of the economy as manufacturing goes overseas.

While these are real problems, Japan also possesses an abundance of strengths. The most prominent of these is the emphasis on manufacturing, with its particular orientation towards high technology. Fingleton supports his view that Japan will become the world's largest economy in the very near future by pointing to the broad range of industries and technologies that Japan dominates, either directly – robotics, cameras, automobile industry manufacturing equipment, copiers, musical instruments, superconductivity, energy-efficient transportation technologies, and more – or through its domination of key components – notebook computers, printers, compact disk players, semiconductor materials and equipment, supercomputers, cellular phones, fax machines, optical scanning equipment, and others. These manufacturing strengths, and in many cases monopolies, underpin the economy (see Fingleton). Combined with a culture that is accustomed to sacrificing short-term personal pleasure for long-term national gain, and a government bureaucracy that ensures that industries and consumers put the nation first, Japan is likely to remain an important global trader for the foreseeable future.

Japan's trading profile is likely to continue in the patterns that have been established. As Japan shifts more and more of its low-end manufacturing abroad, its demand for resources will continue to decline, with the exception of those for the production of high-technology products and for basic subsistence items, such as foodstuffs. It should be noted that Japanese companies do not move production of high-technology items or of cutting-edge technology out of the country. Japan will continue to focus on the domination of key industries and technologies, moving to the value-added end of the spectrum. Japan's evolving division of labor sees it producing the most expensive and most technologically demanding products while other countries, further down the line, produce the less technically sophisticated items. Japan's recent direct investment in manufacturing in East Asia (discussed above) is evidence of this division. Leon Hollerman made this point a decade ago in a discussion of Japan's economic strategy in Brazil, referring to the "calculated disaggregation (or 'unbundling') of the production process" as between Japan and host countries, through which Japanese companies keep control of operations yielding higher value added and therefore higher returns (see Hollerman).

Japan's international trading prominence brings both power and challenges, for its trading activities invariably carry the potential for conflict. Other countries worry about the rapid influx of Japanese products and, in particular, about Japan's ability to dominate key industrial markets. They worry about the entrenched Japanese surpluses that seem to have put the United States, Canada and Australia, among other countries, perennially in Japan's economic debt, and that lead to periodic demands for protectionist legislation. Japan now possesses tremendous international clout, primarily through its import and export activities, and not, as is more usual, military or political activities. Many countries are wary of Japan's strength and its companies' growing control over key high-technology industries, but few have the power or the will to risk losing access either to the Japanese market or to Japanese products.

International trade is largely responsible for the development of any country's international economic reputation, whether it be Switzerland's tradition of producing precision machinery, or Canada's longstanding image as a hewer of wood and drawer of water. In a remarkably short period of time Japan has transformed itself from a producer of trinkets and low-quality products into a manufacturer of high-quality consumer goods. It is now in the process of reinventing itself once again and establishing a dominant international trading position in the broad area of high technology. Through the transformation of its economy, and the restructuring of its trade relationships with nations around the world, Japan has recast its image on the international scene. Long gone is the isolation of earlier centuries and gradually fading, too, is the image of Japanese imperialism left over from World War II. In their place is the image of the new Japan: a country of corporate innovation and technological inventiveness, and one of the most important participants in international trade. The image fits well with the prevailing self-image of most

Japanese people, and with their desire to assure their country a prominent place on the world scene.

Further Reading

Anchordoguy, Marie, *Computers Inc.: Japan's Challenge to IBM*, Cambridge, MA: Harvard University Press, 1989

This is a fascinating in-depth study of Japan's efforts to develop a national computer industry. It illustrates, perhaps better than any other single study, the singlemindedness of Japanese governments in their pursuit of economic advantage, and the capacity of government and industry to cooperate in the national interest.

Argy, Victor, and Leslie Stein, *The Japanese Economy*, New York: New York University Press, 1997

This recent overview of the development and contemporary state of the Japanese economy provides an accessible, well-documented analysis of current issues and accomplishments. Three chapters are devoted to aspects of Japan's trade and trading relationships.

Cetron, Marvin, and Owen Davies, *Crystal Globe: the haves and have-nots of the new world order*, New York: St Martin's Press, 1991

One of the many books published in recent years that offer projections of international economic and political developments, *Crystal Globe* stands out from the pack for its comparatively gloomy forecasts for Japan's performance in the coming century.

Dore, Ronald, *Flexible Rigidities: Industrial Policy and Structural Adjustment in the Japanese Economy, 1970–80*, Stanford, CA: Stanford University Press, 1986

A detailed study, with an emphasis on the textile industry, of Japanese economic transition in a key decade, when Japan wrestled with the costs and consequences of the oil shocks, and with the upgrading of its industrial capabilities.

Fingleton, Eamonn, *Blindside: Why Japan is Still On Track to Overtake the United States by the Year 2000*, London: Simon and Schuster, and Boston: Houghton Mifflin, 1995

Fingleton provides a provocative and informative analysis of Japan's current and future economic performance. Rather than focus on macroeconomic analysis of Japan's trading situation, Fingleton examines Japanese attempts to dominate specific key industries, and argues that the coordinated national effort, involving government and business, and designed to enhance the country's economic performance, remains very active.

Hollerman, Leon, *Japan's Economic Strategy in Brazil: Challenge for the United States*, Lexington, MA: Lexington Books, 1988

A detailed analysis of Japan's trading and investment activities in Brazil, this book provides a valuable study of the expanding global involvement of Japanese corporations.

Huber, Thomas M., *Strategic Economy in Japan*, Boulder, CO: Westview Press, 1994

Huber, like Fingleton, argues that the Japanese government remains actively involved in managing and protecting the national economy. He contrasts Japan's economic management with national military strategies and highlights the concerted, coordinated and well-planned nature of Japan's trading, investment and development activities.

IMF, *Direction of Trade Statistics*, annual publication, various years.

One of the most useful and systematic sources for trade statistics and economic data.

Ito Takatoshi, *The Japanese Economy*, Cambridge, MA: MIT Press, 1992

This introduction to, and overview of, the Japanese economy provides a useful description of economic development, emphasizing quantitative and theoretical analyses of economic developments in Japan. Only one chapter is devoted to the study of Japan's trading relationships.

Japan Economic Almanac, 1990, 1994 and 1995 editions of an annual publication

The specific articles referred to in this chapter included Shimizu Isaya's 1990 piece, "Overseas investment"; Kato Hidenaka's 1994 piece, "Strong yen weakens manufacturers, may hollow industrial base"; and three pieces from 1995: Hirose Toru, "Growth seen in industrial items, but consumer goods still failing"; Shirae Hideshi, "Production flees abroad as dollar crashes through ¥100 floor"; and an unsigned article, "Makers seek to add machine value as market shrinks for second year". This is a journalistic publication concerned with topical issues, which are generally addressed with a high level of competence; other editions are also worth consulting.

JETRO (Japan External Trade Organization), *Nippon 1997 Business Facts and Figures*, 1997

JETRO analyzes numerous aspects of the Japanese economy, and provides researchers and the general public with accessible and detailed information. This document is the 1997 edition of a popular and useful small "fact book".

Lambert, James, "Japan's Changing Marketplace: Japan's Major Trading Partners", in *Doing Business in Japan*, The Canadian Chamber of Commerce in Japan / Key Porter Books Ltd. Canada, 1994

This article provides a succinct and useful overview of recent internal changes in the Japanese economy and considers the implications of these changes for Japan's trading partners.

Nakamura Masao and Ilan Vertinsky, *Japanese Economic Policies and Growth*, Edmonton, AB: University of Alberta Press, 1994

This volume provides a statistical analysis of economic growth in Japan, focusing largely on the period after 1985, and relates these developments to Japanese government policies and directives.

Ohta Hideaki, Tokuno Akihiro and Takeuchi Ritsuko, "Evolving Foreign Investment Strategies of Japanese Firms in Asia", in Nomura Research Institute and Institute of Southeast Asian Studies, editors, *The New Wave of Foreign Direct Investment in Asia*, Nomura Research Institute and Institute of Southeast Asian Studies, 1995

This important article offers a detailed analysis of the changing nature of Japanese investment in Asia, and illustrates how profoundly Japan's international investments and trading relationships have shifted in the past decade.

Shinohara Miyohei, *Industrial Growth, Trade, and Dynamic Patterns in the Japanese Economy*, Tokyo: University of Tokyo Press, 1982

This volume, produced before the "bubble economy" and as Japan was emerging from the oil crises of the 1970s, provides a very useful analysis of trading activities in the 1960s and 1970s. This was a key period in Japan's economic transition, for it saw the rapid expansion of the country's international trade, and the shift from the production of low-value to higher-value manufactured goods for export.

Wickes, Ron, and Aldith Graves, *The Japanese Market for Manufactured Imports: the Door Opens Wider*, East Asia Analytical Unit Working Paper No. 1, Canberra, ACT: Department of Foreign Affairs and Trade, Australia, 1993

This short article offers an introduction to a still little-noted development in the Japanese economy: the lowering of barriers to the importation of foreign manufactured goods. This change sits at the heart of the recent and major changes in Japan's import profile.

Carin Holroyd is a doctoral candidate in the Faculty of Business at the University of New Brunswick at Saint John, and was coauthor of the book *Pacific Partners: The Japanese Presence in Canadian Business, Society and Culture*, published by Lorimer in 1996.

Table 9.1 Composition of Japan's Exports, by Product Category, selected years 1955–79 (%)

	1955	*1960*	*1965*	*1970*	*1975*	*1979*
Metals	19.2	13.8	20.3	19.7	22.5	17.8
Iron and steel	12.9	9.6	15.3	14.7	18.3	13.7
Machinery	13.7	25.3	35.2	46.3	53.8	61.3
Industrial	–	–	–	10.4	12.1	14.4
Electrical	–	–	–	14.8	12.4	16.9
Transport	–	–	–	17.8	26.1	25.0
Precision	–	–	–	3.3	3.3	5.0
Chemicals	5.1	4.5	6.5	6.4	7.0	5.9
Textiles	37.3	30.0	18.7	12.5	6.7	4.8
Non-metallic mineral ores	4.9	4.4	3.1	1.9	1.3	1.5
Food	3.3	6.3	4.1	3.4	1.4	1.2
Total value (US$ billions)	2.011	4.055	8.452	19.318	55.753	103.032

Source: Shinohara Miyohei, *Industrial Growth, Trade, and Dynamic Patterns in the Japanese Economy*, Tokyo: University of Tokyo Press, 1982, p. 88

Table 9.2 Japan's Top 10 Export Destinations, by Value in US Dollars, selected years 1958–78

1958	*1962*	*1966*	*1970*	*1974*	*1978*
United States	United States	United States	United States	United States	United States
Liberia	Hong Kong	Hong Kong	South Korea	South Korea	South Korea
United Kingdom	Philippines	South Korea	Taiwan	Australia	West Germany
Hong Kong	Australia	Liberia	Hong Kong	China	Saudi Arabia
Taiwan	United Kingdom	China	Australia	Canada	Hong Kong
Philippines	Thailand	Thailand	China	West Germany	China
India	Indonesia	Australia	Canada	Indonesia	Iran
Thailand	India	Philippines	West Germany	Brazil	Australia
Singapore	Taiwan	Taiwan	Philippines	Singapore	Singapore
Canada	South Korea	Canada	Thailand	Hong Kong	Indonesia

Source: IMF, *Direction of Trade*, various editions. It should be noted that for the years 1974–84 the IMF did not list figures for Taiwan.

Table 9.3 Japan's Top 10 Sources of Imports, by Value in US Dollars, selected years 1958–78

1958	*1962*	*1966*	*1970*	*1974*	*1978*
United States	United States	United States	United States	United States	United States
Australia	Australia	Australia	Australia	Saudi Arabia	Saudi Arabia
Saudi Arabia	Kuwait	Canada	Iran	Iran	Indonesia
Canada	Canada	Saudi Arabia	Canada	Indonesia	Australia
Malaya	West Germany	Iran	Saudi Arabia	Australia	Iran
Philippines	Malaya	Philippines	Indonesia	Canada	Canada
Mexico	Philippines	China	West Germany	Kuwait	United Arab Emirates
West Germany	Saudi Arabia	Soviet Union	Philippines	United Arab Emirates	South Korea
Kuwait	Soviet Union	Malaysia	Soviet Union	South Korea	Kuwait
India	United Kingdom	Kuwait	Malaysia	West Germany	China

Source: IMF, *Direction of Trade*, various editions. It should be noted that for the years 1974–84 the IMF did not list figures for Taiwan.

Table 9.4 The Balance of Trade: Merchandise Imports, Exports, and Trade Surplus, selected years 1965–94 (US$ billions)

	Exports	*Imports*	*Surplus*
1965	8.33	6.43	1.90
1970	18.96	15.00	3.96
1975	54.65	49.71	4.94
1980	126.74	124.61	2.13
1985	174.02	118.03	55.99
1986	205.59	112.77	92.82
1987	224.62	128.20	96.42
1988	259.77	164.77	95.00
1989	269.55	192.66	76.89
1990	280.35	216.77	63.58
1991	306.58	203.49	103.09
1992	330.87	198.47	132.40
1993	351.31	209.74	141.57
1994	384.18	238.25	145.93

Source: IMF, *International Financial Statistics Yearbook*, 1995, pp. 470–1

Table 9.5 The Average Market Exchange Rate of the Yen Against the US Dollar, selected years 1965–94

	Yen to US$1.00
1965	360
1970	360
1975	297
1980	227
1985	239
1986	169
1987	145
1988	128
1989	138
1990	145
1991	135
1992	127
1993	111
1994	102

Source: IMF, *International Financial Statistics Yearbook*, 1995, pp. 468–9

Table 9.6 Japanese Direct Investment Abroad, selected years 1965–94 (US$ billions)

1965	0.03
1970	0.26
1975	1.53
1980	2.39
1985	6.45
1986	14.48
1987	19.52
1988	34.21
1989	44.16
1990	48.05
1991	30.74
1992	17.24
1993	13.74
1994	17.97

Source: IMF, *International Financial Statistics Yearbook*, 1994, pp. 450–1, and *International Financial Statistics Yearbook*, 1995, pp. 470–1

Table 9.7 Japan's Top 10 Export Destinations and Sources of Imports, by Value in US Dollars, selected years 1984–90

Export destinations			*Sources of imports*		
1984	*1988*	*1990*	*1984*	*1988*	*1990*
United States	United States	United States	United States	United States	United States
South Korea	West Germany	West Germany	Saudi Arabia	South Korea	Indonesia
China	South Korea	South Korea	United Arab Emirates	Australia	Australia
West Germany	Taiwan	Taiwan	Australia	China	China
Hong Kong	Hong Kong	Hong Kong	China	Indonesia	South Korea
Saudi Arabia	United Kingdom	United Kingdom	Canada	Taiwan	West Germany
Australia	China	Singapore	Malaysia	Canada	Saudi Arabia
United Kingdom	Singapore	Thailand	South Korea	West Germany	United Arab Emirates
Singapore	Australia	Australia	Iran	Saudi Arabia	Canada
Canada	Canada	Canada	West Germany	United Arab Emirates	Taiwan

Source: IMF, *Direction of Trade*, various editions. It should be noted that for the years 1974–84 the IMF did not list figures for Taiwan.

Table 9.8 Top 10 Destinations of Japanese Exports, by Value in Millions of US Dollars (and as Proportions of Total Exports), 1994–96

Country	*1994*	*(%)*	*1995*	*(%)*	*1996*	*(%)*
United States	118,071	29.75	120,944	27.29	111,867	27.22
South Korea	24,453	6.16	31,302	7.06	29,341	7.14
Taiwan	23,885	6.02	28,971	6.54	25,956	6.32
Hong Kong	25,833	6.51	27,783	6.27	25,340	6.17
China	18,765	4.73	21,923	4.95	21,809	5.31
Singapore	19,681	4.96	22,993	5.19	20,778	5.06
Thailand	14,761	3.72	19,707	4.45	18,287	4.45
Germany	17,858	4.50	20,327	4.59	18,214	4.43
Malaysia	12,399	3.12	16,786	3.79	15,328	3.73
United Kingdom	12,789	3.22	14,147	3.19	12,478	3.04

Source: *The World Trade Atlas*, Global Trade Information Services, 1993 and 1997

Table 9.9 Sources of Imports to Japan, by Value in Millions of US Dollars (and as Proportions of Total Imports), 1994–96

Country	*1994*	*(%)*	*1995*	*(%)*	*1996*	*(%)*
United States	62,839	22.82	75,414	22.44	79,385	22.74
China	27,666	10.05	35,888	10.68	40,374	11.56
South Korea	13,571	5.14	17,259	5.14	15,957	4.57
Indonesia	12,930	4.69	14,204	4.23	15,188	4.35
Taiwan	10,759	3.91	14,356	4.27	14,972	4.29
Australia	13,626	4.95	14,561	4.33	14,234	4.08
Germany	11,166	4.05	13,710	4.08	14,170	4.06
Malaysia	8,233	2.99	10,543	3.14	11,747	3.36
United Arab Emirates	9,176	3.33	10,188	3.03	11,490	3.29
Saudi Arabia	8,405	3.05	9,732	2.90	10,654	3.05

Source: *The World Trade Atlas*, Global Trade Information Services, 1993 and 1997

Table 9.10 Japan's Top 10 Exports, by Product Category, 1996 (%)

Machinery	24.55
Electrical machinery; sound and television equipment	23.17
Vehicles (excluding railroad)	18.01
Optical and medical instruments	6.05
Iron and steel	3.00
Organic chemicals	2.71
"Special" (reimported) goods	2.40
Ships and boats	2.35
Plastic products	2.21
Iron and steel products	1.53

Source: The World Trade Atlas, Global Trade Information Services Inc. 1993 and 1997

Table 9.11 Japan's Top 10 Imports, by Product Category, 1996 (%)

Mineral fuel	17.41
Electrical machinery	10.43
Other machinery	9.40
Wood	4.57
Fish and seafood	4.08
Vehicles (excluding railroad)	3.80
Woven apparel	3.12
Optical and medical instruments	3.10
Meat	2.50
Knitted apparel	2.21

Source: *The World Trade Atlas*, Global Trade Information Services Inc., 1993 and 1997

Chapter Ten

Corporate Structures and the Dual Economy

Graham Field

The corporate structures of Japan are as prone to being caricatured as just about any other aspect of the country's economic and social life. The rhetoric of "Japan-bashing" has characterized the country's dominant corporations as scheming, ruthless, methodical behemoths, presiding over a sweatshop hinterland of myriad subcontractors. US commentators, in particular, have argued that close intercompany ties constitute a major impediment to market access. Undoubtedly Japan does have a number of very large companies. For example, Toyota, and Nippon Telegraph and Telephone (NTT) both ranked among the world's top 10 companies by market capitalization at the start of 1998; both also figured among the 12 Japanese companies on the list of the world's top 20 by sales. At the same time, as D. H. Whitaker has pointed out: "Japan ranks alongside Italy as having the highest proportion of small firms – and employees in them – amongst OECD countries" (Whitaker p. 1). Japan's corporate structures are certainly varied and distinctive, but none is necessarily unique or, indeed, inexplicable, given the overall pattern of economic and political development. As elsewhere in the world, corporate structures in Japan have evolved in response to a mixture of economic, political, regulatory and technological pressures and changes.

The *Zaibatsu*, 1868–1948

The present pattern of corporate structures owes much to Japan's rapid and "late" industrialization (as compared to that of the United States and some western European nations). The absence of a thriving indigenous bourgeoisie in the 1870s meant that the capital accumulation required for industrial development was initially undertaken by the state, with the help of what have been called "political merchants" in the development of specific industries. The early fortunes of the Mitsui and Yasuda *zaibatsu* - family-owned conglomerates – were founded on the role that they played in providing fiscal and financial services to the government. Again, the Mitsubishi *zaibatsu* was rewarded with government help after its founder, Iwasaki Yataro, assisted an attack on Taiwan in 1874, by providing troop transports, and then gave material support to the suppression of the Satsuma rising in 1877. At a later stage, Iwasaki acquired the Nagasaki shipbuilding yard when it was sold by the government in the 1880s. This sale was one of many that the state made as it divested itself of non-strategic industrial enterprises in cement, glass, textiles and mining. Not all the companies that made acquisitions at this time went on to become *zaibatsu*; nor did all the emerging *zaibatsu* make such purchases, Sumitomo and Yasuda being notably absent from the list of purchasers. However, the foundations of the growth of the *zaibatsu* unquestionably lay in their relationship with the government. As E. H. Norman put it: "In Japan, the concentration of capital, as distinct from its accumulation, was accelerated by the government's policy of subsidy and artificial encouragement" (Norman p. 219). Although there were some industries, such as cotton-spinning, in which large capital requirements led to development through joint stock companies, the pattern of a relatively small number of large, closely held companies (or groups of companies) occupying the commanding heights of the economy was settled early in Japan's industrial history. By 1930, the banks controlled by the four largest *zaibatsu* – Mitsui, Mitsubishi, Sumitomo and Yasuda – controlled 20% of all deposits and 14% of loans in the banking system.

Family ownership continued to characterize the *zaibatsu* until their dissolution after the end of World War II (and it is interesting that this pattern of ownership, which was never uniquely Japanese, has persisted elsewhere to this day, notably in Sweden). The *zaibatsu* formalized their corporate structures from the 1890s onwards, but avoided the joint stock form with its disclosure requirements. Between 1894 and 1913, all seven major *zaibatsu*, with the exception of Furukawa, pursued policies of diversification, establishing a pattern of broadly based groups which continue in a different guise today (as will be described below). Increasingly, the *zaibatsu* were run by professional managers, and the strategy of diversification was reinforced when World War I created a fresh demand for the output of heavy industries. The wartime boom also gave an opportunity for the emergence of new *zaibatsu*, such as the Nomura group, which expanded into plantations and manufacturing as well as financial services. In the 1920s, the *zaibatsu* were responsible for an innovation in corporate structures, with the adoption of the multi-subsidiary system. Incorporation of subsidiaries allowed these offshoots to raise their own funds; reduced the risk of the collapse of one part of the group endangering its other, healthy parts; and gave management the incentive of being able to obtain extra directorships.

In the 1930s, the *zaibatsu* came under attack from both the Left and the Right, as Japan became increasingly militarized and the owning families sought to reduce their active involvement. The years up to and during World War II saw the rise of a number of other new industrial groupings, sometimes also referred to as "new *zaibatsu*", which, however, were not family-owned. These were concentrated in the heavy industries needed to support Japan's growing war effort after 1937.

The *zaibatsu* were a prime target of the Occupation authorities, as they sought to demilitarize and democratize defeated Japan. A total of 83 concerns had their stockholdings turned over to a Holding Company Liquidation Commission, and 56 "designated persons" had their assets frozen and their stockholdings transferred. In April 1947, the Anti-Monopoly Law rendered holding companies illegal, while the Law for the Termination of *Zaibatsu* Family Control, passed in 1948, banned family members from taking management positions (see also Chapter 1). In the event, however, SCAP did not have a far-reaching effect on Japanese corporate structures. The ban on holding companies did not lead to the dissolution of the ties among the companies which had made up the *zaibatsu*. Senior managers were largely left in place and, predictably, preferred to continue cooperating with former associates. The defeat of the Nationalists in China in 1949 also meant that the United States, by far the dominant player in the Occupation, changed its policy towards Japan, in order to build up its economic strength. The *zaibatsu* were soon being reassembled as *keiretsu*.

The *Keiretsu*

Following the break-up of the *zaibatsu*, it did not take long before corporate groups began to coalesce again, being referred to by a newly coined term, *keiretsu*. From 1952, old company names came back into use: Mitsubishi Corporation, for instance, had been re-formed by 1954, and Mitsubishi Heavy Industries was reborn 10 years later. Although holding companies remained illegal until 1997, the main banks which had been one of the other core ingredients of the *zaibatsu* had been left largely untouched by SCAP, and it was around these institutions that the industrial companies gathered.

In connection with the rise of the *keiretsu*, some commentators have stressed the contrast between SCAP's determination to have the Japanese government set up a Fair Trade Commission (FTC) to enforce anti-monopoly measures, and the apparent lack of determination of governments in subsequent years to make the FTC into a strong force in the administrative framework. This claim has been summed up by Miyashita Ken'ichi and David W. Russell in strong terms: "Had the FTC been allowed to do its job, the Big Six *keiretsu* would not exist today" (Miyashita and Russell p. 35). In fact, this is wide of the mark. The FTC was meant to control cartels, price-fixing and collusion, which, of course, exist in virtually all economies in some shape or form, rather than to regulate corporate structures. Whatever the FTC has been allowed to do, or failed to do, the existence of *keiretsu* has been compatible both with price-fixing and with stiff competition.

Keiretsu have been created in two forms, horizontal (*yoko*) and vertical (*tate*). A horizontal *keiretsu* is a group of companies, sometimes but not always linked by a common name, but almost invariably centered on a "main bank" and a general trading company (*sogo shosha* – see Chapter 7). The relationships within the group are underpinned by cross-holdings of one another's shares, by lending from the main bank, and by the mutual exchange of information. A vertical *keiretsu*, in contrast, is a group subdivided into production and distribution networks. The strength of the ties within a vertical *keiretsu* varies in accordance with the degree of dependence which subcontractors or distributors have on what is usually termed the "parent" company at the apex of the pyramid.

Whether in a vertical structure or a horizontal one, *keiretsu* companies and their employees are assumed to have a preference for using one another's products and services. It is this belief which has given rise to descriptions, in the United States in particular, of the "*keiretsu* problem". The fact is, however, that the persistence of *keiretsu* structures in Japan is, in many cases, an alternative to forms of vertical or horizontal integration in other economies, and the same critics who attack *keiretsu* can often be found applauding integration elsewhere as a rational way of reducing costs and uncertainties (or, depending on fashions in management theory, counterposing it to "contracting out"). *Keiretsu* relationships can convey the same advantages but have more mystique because they are not governed purely by formal ownership relations.

The "Big Six" *keiretsu* referred to above are all horizontal *keiretsu*, known in Japanese as the *roku dai kigyo shudan* ("six big industrial groups"). They are identified with three "city" banks – Dai-Ichi Kangyo Bank (DKB), Fuji Bank, and Sanwa Bank – and with three sets of industrial companies – Mitsubishi, Sumitomo, and Mitsui. Since Fuji Bank was the Yasuda Bank until it was renamed in 1946, it is evident that four of the six can be directly traced to their origins as *zaibatsu*. Between them, they may account for as much as one third of the Japanese economy, at least according to one questionable "guesstimate" (Miyashita and Russell's). There are also two other, significantly smaller horizontal *keiretsu*, one centered on the Tokai Bank and the other on the Industrial Bank of Japan.

Calculations of the size and scope of the horizontal *keiretsu* depend, of course, on how tightly the membership of each group is defined: more elastic definitions of membership naturally produce larger estimates of their overall significance to the economy. Thus, estimates by Jardine Fleming Securities (shown in Table 10.1) are based on relatively strict definitions, but those by the London *Financial Times* (November 30, 1994) suggest that the DKB *keiretsu*, for instance, has 48 member companies and a further 190 affiliates in which it has at least a 10% stake. The *Financial Times* also concluded, on the basis of a survey of 195 member companies of the Big Six, that together they accounted for 38% of market capitalization, for 4.5% of employment, 16% of sales and 17% of net profits in Japan.

Alongside the main bank at the center of each group are, typically, a life insurance company, a non-life insurance company, a trust bank, and two or three large manufacturing enterprises. Beyond this core, large horizontal *keiretsu* may have a company competing in each major industry. The bank acts as the principal lender to the non-financial corporations in the group; it is sometimes the largest single shareholder in its affiliated companies; and it acts effectively as credit monitor through its access to information from related corporations. The main bank has, until recently, also been expected to stand behind any *keiretsu* companies that find themselves in difficulties, and to take prime responsibility for bailing them out. The general trading company helps to coordinate commercial activities, and provides credit to small and medium-sized enterprises (SMEs), defined as companies which have fewer than 300 employees, or which are capitalized at less than ¥100 million.

A presidential council (*shacho-kai*) brings together the presidents of the inner group of *keiretsu* companies (perhaps 30 or so) for regular discussions. The content of these meetings remains the subject of speculation, particularly among critics of the *keiretsu*, who believe that they violate the spirit, if not the letter, of the Anti-Monopoly Law by acting as coordinators of group policy. Certainly the opportunity has existed at such meetings for exchanges of information and views about developments that may affect the *keiretsu* as a whole. The practice of assigning directors to affiliated companies is another way of binding them more closely to the *keiretsu*. It is a visible sign of commitment that also provides a conduit for information to flow back and forth. In all, the Big Six posted 4,000 board members to other companies in 1992, including 400 presidents and chief executive officers.

Cross-holdings among member companies of *keiretsu*, both vertical and horizontal, constitute the skeleton on which the sinews of actual trading relations are built. These holdings were assembled in the 1950s, when they were seen as a means of fending off possible takeover attempts, whether by foreign predators or by Japanese corporate raiders. They are by no means unique internationally. In South Africa, for instance, the isolation of the apartheid years meant that companies were unable to export capital, and ended up buying a multitude of stakes in one another. Nevertheless, Japanese cross-holdings have attracted the ire of foreign competitors, who argue that their existence makes *keiretsu* companies "all but impervious to hostile takeovers" (as the *Financial Times* put it on November 30, 1994). This language recalls the criticism leveled at the *zaibatsu* in the 1930s, which were judged "impregnable" by virtue of their preeminence in finance, commerce and industry. The extent of cross-holdings varies within the Big Six and among the *keiretsu* at large. Typically, no one affiliated company owns more than 5% of another, but, taken together, these holdings add up to 20–30%. In more extreme cases, as much as 90% of the shares of a company may be owned by affiliated companies. The DKB group has the loosest links among the Big Six, with cross-holdings among its main companies averaging 12%. The Fuyo group, led by Fuji Bank, has average cross-holdings of 15%, while those in the Sumitomo group average 27%, and those in the Mitsubishi group are the highest, averaging 35%.

Another complication is that each member of a horizontal *keiretsu* may have its own vertical *keiretsu* of suppliers or distributors. The largest of the vertical *keiretsu* tend to be in manufacturing, and the quintessential vertical supplier networks are in the automobile and electronics industries. While around 30% of automobile parts are produced by subcontractors in the United States, the figure is closer to 70% in Japan. Toyota is the largest corporate group in Japan after the Big Six, and 10 of its first-tier subsidiary suppliers are themselves listed on the first section of the Tokyo Stock Exchange. In fact, Toyota's development has meant that it has diversified beyond its well-known automobiles and trucks into other business areas, such as telecommunications, financial services and computer system development, and is thus on its way to becoming an important horizontal *keiretsu* as well as a vertical *keiretsu*. Another giant of Japanese industry, the electronics manufacturer NEC, may have as many as 6,000 companies in its *keiretsu*.

The degree of dependence entailed by *keiretsu* relationships of suppliers to manufacturers has provoked external criticism of the type made by Miyashita and Russell, for whom the uniqueness of Japan's corporate structure lies in "enormous formalized structures tying thousands of firms to a single manufacturer so rigidly" (p. 118). Yet, as we shall see below, the formality of these ties is changing, and is, in any case, open to question. The depth of *keiretsu* networks is, nevertheless, worthy of note. Relationships extend down, through as many as four or five tiers of subcontractors, to manufacturing plants employing fewer than 10 workers. This structure has frequently been linked to stratified bands of pay and working conditions (see Chapter 13). Similarly, distribution *keiretsu* extend down to neighborhood retail outlets, especially for electronics goods. These networks developed from the 1950s onwards and, by 1982, the five largest manufacturers controlled two thirds of Japan's 71,283 consumer electronics retailers. Matsushita alone controlled more than 25,000 stores. Around 30% of wholesalers are affiliated to a *keiretsu*, although, as one OECD survey of Japan has put it, encapsulating foreign dissatisfaction with Japanese corporate structures: "even without formal membership in a *keiretsu*, Japanese business practices, which favor continuity in transactions and discourage dealing with newcomers, tend to link wholesalers and manufacturers" (p. 189). Corporate structures and corporate culture can easily be conflated in this way.

Finally, there are a number of anomalous cases, in the form of companies which are nominally "independent" but which have ties to, or are "close" to, one of the major groups. Matsushita, for instance, is "close" to Sumitomo, while Honda is similarly linked to Mitsubishi Bank. Nippon Steel, Sony and Bridgestone are all outside the Big Six but necessarily have relations with the major banks. Hitachi is different again, in that it belongs to the DKB *keiretsu* and to two others.

The Dual Economy

In addition to its alleged "*keiretsu* problem", Japan has also been diagnosed as suffering from a "small firm problem", by which commentators mean the persistence of a large number of small companies, which they identify as a symptom of backwardness in the economy. This diagnosis was a feature of the writing on Japan during the years of rapid economic growth in the 1950s and 1960s. A Japanese government White Paper of 1956 saw SMEs as languishing in an economic backwater. As Seymour Broadbridge put it in 1966: "the small enterprise sector accounts for a very high proportion of employment, and contributes a substantial proportion of total output, but wage and productivity differentials are great" (p. 7). The frequently used term "dualism" refers to the gulf between the wage and productivity levels achieved in large companies and those pertaining among SMEs. Lower wages among subcontractors meant that subcontracting was "more profitable for the large concerns than producing goods themselves" (Nakamura 1995 p. 167). Exploitative relationships between first-tier companies at the top of vertical *keiretsu* pyramids and their legions of subcontractors have been seen as one of the main drawbacks of the dual economy. Wage cost differentials between large and small firms have shrunk from almost four times in 1953 to slightly more than double in recent years, but this clearly remains a significant gap. Further, when large exporting companies face difficult times, as they did when the rapid appreciation of the yen in 1985 made their products much less competitive internationally, they have forced their chains of subcontractors to accept lower prices. The relative powerlessness of these producers – many of which are tied exclusively to single customers – has meant that they must choose between accepting these demands (and reducing wages or increasing hours), and going out of business.

Almost 75% of Japanese employees work in SMEs (as defined above). Companies of this size have been an important part of the industrial landscape throughout the 20th century, initially relying on cheap labor to remain competitive before rising to greater prominence in the interwar period. The large corporations increased their use of subcontracting in these years (Whitaker p. 26), partly to avoid investing in capacity which they feared would prove to be surplus to requirements if there was a serious economic downturn. During World War II, military production took precedence, and most small companies in other industries regarded as non-essential were closed down, as labor and capital were diverted to the war effort. After 1945, there was a resurgence in the formation of smaller companies, and many of them survived in the major urban centers even after larger companies began to move out to new sites in the 1960s. Government support came in the form of the Delayed Payment Law of 1956 and was administered, through agencies attached to the Ministries of Finance, and of International Trade and Industry, from the 1960s onwards (see Table 3.1). The Large-scale

Retailing Establishment Law of 1973 afforded additional protection to smaller retailers. In recent years, financial support for SMEs has been at the level of around ¥200–300 billion a year, equivalent to around 0.3% of total government spending. Government loans have been available at slightly below market rates to provide long-term capital, and to help to redress the funding disadvantages which SMEs have suffered since the start of industrialization.

In recent years there has been a "revisionist" trend in the interpretation of the role of SMEs in Japan's corporate structure. By appealing to the tradition of economic analysis represented by Alfred Marshall and, more recently, Michel Piore and Charles Sabel, D. H. Whitaker, for one, has sought to demonstrate that SMEs can play a more positive role in the overall economy, and to counter the simplistic notion that all SMEs are tied into exclusive, tyrannical subcontracting relationships. For example, even in the 1960s, the idea of monolithic *keiretsu* pyramids was inaccurate. More than 30 large automobile parts suppliers belonged to the suppliers' associations for both Toyota and Nissan and, by 1990, the proportion of subcontractors dependent on single customers for 90% or more of their orders may have fallen as low as 16%. The proportion of manufacturing SMEs undertaking subcontracting work fell below 60% in the 1980s, although it remains the case that the smaller the company, the more reliant it is likely to be on subcontracting. Most commentators on the dual economy have tended to stress the drawbacks in the subcontracting relationship, including downward pressures on prices, additional costs associated with meeting "just in time" inventory requirements, and responsibility for design work, but Whitaker and other revisionists point to some of the benefits for smaller companies. They can obtain steady work – more than half of SMEs surveyed in the 1980s had been doing business with a parent for more than 15 years, and two thirds had never changed their parent – and their relationships with larger firms can help them to overcome their limitations in product development, marketing and design.

Nevertheless, even the revisionists only go so far as to argue that a transition is occurring: "Japan has lost its traditional small firm 'problem' – too many, too small – only to be confronted by another, namely how to nurture dynamic, new small firms" (Whitaker pp. 206–7). This clearly depends on the willingness of entrepreneurs to start up new firms, and it may be doubted whether there are as many individuals willing to go down this path as there were at the end of World War II. As in Italy, for example, the years immediately after the surrender saw an influx of workers from agriculture into manufacturing, and many of them brought with them rudimentary business skills acquired through peasant farming. This generation is now retiring and dying off, and its offspring appears, at least in some cases, not to relish the long hours and insecurity of small business life. Knowledge- and skill-intensive businesses are replacing manufacturing SMEs, and service businesses are growing in number. At the same time, the overall numbers of SMEs are falling, and the launching of new businesses within larger firms (sometimes labeled with the ugly and needlessly confusing neologism "intrapreneurship") has not been a conspicuous success.

A Changing Environment

Three main factors are reshaping Japanese corporate structures: the recession and prolonged economic sluggishness of the 1990s; the yen's periods of strength; and the search for better returns by both foreign and domestic investors in Japan. The government's response – its "Big Bang" reform of financial services (see Chapter 8) – is adding to the pressure for change as well as facilitating it. Corporate structures are more likely to be dictated by commercial criteria, especially the profit motive, in the future.

The depressed state of the Japanese economy in the 1990s has meant that companies have embarked on a number of cost-saving measures. The yen's periodic bouts of strength – particularly after the Plaza Accord of 1985 and in early 1995 – have made Japanese exports less competitive, and exerted similar pressures for cost reductions. This, in turn, has affected relations between large and small companies. In the past, subcontractors have been squeezed and have acquiesced in price-cutting, but in the 1990s the process is leading to a structural change in inter-company relationships. The traditional bonds which cemented subcontractors and manufacturers are cracking as producers look for the cheapest supplier, regardless of the affiliations embodied in *keiretsu* ties. For example, the automobile manufacturer Mazda, in which Ford of the United States increased its stake to 33.4% in 1996, announced in 1997 that its purchasing decisions would be allotted to the most cost-competitive suppliers. The company instructed 80 suppliers in Hiroshima Prefecture to cut parts prices by 30%, or risk losing contracts. This kind of action is the most drastic and, perhaps, reflects what Ford has called the "strengthening of its relationship" with Mazda. Other large companies are looking for ways to reduce costs without necessarily freezing out *keiretsu* suppliers. Cost savings in the region of 30% were achieved by automobile parts suppliers between 1994 and 1997, and both the cost and the length of time needed for the development of new models are being reduced through the participation of suppliers in earlier stages of the design process. More generally, there has been a broadening of purchasing patterns, with major manufacturers looking to source components from a variety of subcontractors, inside and outside their *keiretsu*. Between 1989 and 1992, the average proportion of *keiretsu* members' sales going to other *keiretsu* members slipped from 7.64% to 6.85%. Toyota, for instance, instructed its components affiliate Nippondenso to begin making parts for Nissan at the start of the 1990s. SMEs, too, have been looking for a wider range of customers, although their ability to do so hinges on the possession of relatively sophisticated positions in the division of labor.

Within the "upstream" distribution *keiretsu*, the recession and the high yen have also undermined the strength of distributors linked to major companies. In one case, Honda has responded by increasing the proportion of imported replacement parts available at its affiliated parts distributors and maintenance centers, and thereby reduced costs to help these affiliates remain competitive. This move is a reaction to the increasingly price-conscious behavior of Japanese consumers. Rising unemployment and consumption taxes have made them more sensitive to prices, while the high yen has also allowed new discount stores to introduce foreign goods at very much lower prices. Tied retail outlets for electronics goods, for instance, have been placed at a severe disadvantage. At the same time, manufacturers have been less able to invest freely in marketing and support for their affiliated retailers. New forms of selling – especially mail order – have made substantial inroads in retailing. In this context, the importance of *keiretsu* ties is being steadily eroded. The structure of retailing is also changing as a result of the relaxation of the Large-scale Retailing Establishment Law, which prompted record numbers of openings of new chain stores in the mid-1990s. Japan is likely to see the marginalization of the "mama papa" stores that have been a familiar feature of retailing, but also the rise of new discount retailing companies, and the expansion of such large retail chains as Seiyu, Ito-Yokodo or Daiei. New patterns of distribution may also adversely affect the trading houses at the heart of the *keiretsu*, which finance much of the flow of goods up and down the chains.

The difficulty in servicing a large debt burden in harsh economic conditions has also been a factor in change. A major restructuring announced by the Saison group early in 1998 was greeted by stockbroking analysts as a sign that restructuring was taking on a qualitatively different aspect. The group, which includes the Seibu department stores, embarked on a program of major divestitures as it retreated from an earlier strategy of diversification. This took it into the international hotel business, with over 200 hotels in 76 countries, and led it to expand its "non-bank" financial services subsidiary. The process left it with consolidated debts of ¥1.2 trillion. It agreed to sell its shares in Japan's third-largest convenience store operator, Family Mart, to the Itochu trading company, and to close 14 joint-venture stores operated with a Hong Kong company. It is expected to concentrate instead on its core supermarket and department store businesses.

Close behind debt come bankruptcies, of which there were 16,365 in 1997. Collapses are likely to continue, among large and small firms alike, as bad debts from the years of the "bubble economy" finally overwhelm them. The main impact on Japanese corporate structures is likely to be a reduction in the number of real estate and construction companies, of which more than 4,000 went bankrupt in 1997, as they are still the two industries worst affected by bad debts. Bankruptcies in these areas may in turn bring down more of the institutions which lent to these companies, particularly the "non-banks" (such as Saison's subsidiary, mentioned above), which are chiefly finance and leasing companies permitted to lend but not to take deposits. The impact of bankruptcies is probably more interesting in demonstrating the weakening of the "convoy system" (*gososendan hoshiki*, described in Chapter 8) and of the main banks' relationships with other *keiretsu* members (see below).

Currency movements have given further twists to the relations between large companies and their subcontractors, by exposing them to greater foreign competition, initially as a result of the sharp appreciation of the yen in 1985, and again in 1995. Mazda, for example, responded to the 1995 rise by announcing that it would increase the proportion of imported raw materials and components in its domestically assembled cars from 5% to 30%. Although the exchange rate fell from ¥80 to the US dollar in 1995 to below ¥120 in 1997, the sharp decline in other currencies in East Asia during 1997 meant that Japanese suppliers gained little advantage over their most direct competitors. In shipbuilding, for instance, a big increase in orders placed with Japanese yards in that year did not result in a parallel flow of increased business for Japanese subcontractors, because of intensified competition from suppliers in South Korea. These overseas subcontractors are more important where Japanese companies have themselves relocated production outside Japan. This has forced subcontractors to decide between following their customers into other markets, or staying at home and seeking new customers. In the former case, corporate relationships can be maintained; in the latter another blow is dealt to the *keiretsu* system. Only a limited number of subcontractors have been able to make the move into overseas production. In 1994, of 5,000 clients of the Japan Finance Corporation for Small Business, a government body, just over 6% had established overseas operations, while 20% were contemplating such a move or felt the need to do so. If the trend towards "hollowing out" continues in manufacturing, it will accentuate the growing divide between subcontractors which can invest in technological advances – moving from the production of metal parts, for instance, to integrated circuit components – and thus keep pace with the changing profile of large manufacturers, and those which cannot.

The search for better investment returns reflects the growing concern about poor returns achieved on investments by company pension schemes and insurance companies. There is a fear that these institutions will be unable to meet their obligations to Japan's pensioners unless there is a sharp improvement in returns. This fear has prompted the changes associated with the Big Bang. At the corporate level, the impact of this change is being felt in the dissolution of the structure of cross-holdings. Evidence indicates that the profitability of *keiretsu* groups has been below that of independent groups. Critics of the *keiretsu* system have argued that cross-holdings held by friendly affiliated companies protect managements from market pressures to increase profits and dividend payments, at the same time as they tie up the capital of those friendly companies in relatively unprofitable investments. As US and European advisers come to play a larger role in managing the assets of Japanese pension funds, following liberalization under the Big Bang plan, their preferences will tend to tilt investment, over the longer term, away from the less well-performing *keiretsu* groups and towards companies which achieve higher profits. Additionally, the growing presence of foreign investors, already the largest net buyers of Japanese shares between 1994 and 1996, may help to reinforce a bias – largely unknown in Japan until recently – towards companies perceived as placing an overriding priority on maximizing "shareholder value", narrowly defined by reference to short-term increases in dividends or share prices. It is at least open to question whether the resulting corporate structures will necessarily be more economically efficient than *keiretsu* firms, or SMEs, in the longer term – leaving aside, as many such investors do, considerations of the effects of short-termism and cost-cutting on workers, customers and the wider society.

However, it is not at all certain that Japanese companies in general are about to go down the same road taken by so many western companies in recent years, and the evidence for the actual dismantling of cross-holdings is decidedly patchy. Nearly 70% of all outstanding shares on the eight Japanese stock exchanges were held by other corporations in 1994, and as much as one half of these outstanding shares were held in order to help cement business relationships, either between banks and their clients or between suppliers and their customers. There is indeed some evidence that companies are selling off their stakes in banks. Between March 1992 and March 1996 the proportion of company holdings in all banks fell from over 58% of outstanding shares to just under 51%. There have also been sales of cross-holdings in traditional manufacturing industries, such as paper and pulp, cable, electrical machinery, and transport equipment. Yet such changes as have occurred do not represent an end to the *keiretsu* system, nor do they mean that *keiretsu* member companies are vulnerable to hostile takeovers. Despite reductions in cross-holdings, friendly companies still control enough shares to block takeover attempts, which are expensive undertakings even with low stock market valuations and a weak yen. It is still relatively easy to block takeovers in other ways, such as by the issue of a large number of shares to a third party, and, perhaps crucially in Japanese society, a hostile takeover would risk destroying the unwritten loyalties on which corporate culture still rests. These bind workers and management together within companies, as well as linking particular companies with their suppliers and other *keiretsu* members.

The New Corporate Landscape

Although hostile mergers and acquisitions (M&As) are still difficult to put into effect in Japan – because of the system of cross-holdings, rather than because of any legal or regulatory impediment – the number of agreed M&As is on the increase. The future is likely to see more consolidation in a variety of manufacturing and service industries. Estimates by Nomura Securities' Financial Research Center pointed to 520 M&A deals in 1997, 14% more than in the previous record year, 1993. Within this total, foreign acquisitions of Japanese companies rose to a record number of 50, helped by the strength of the US dollar which reduced the cost of acquisition.

M&As have been taking place mainly in response to tougher market conditions and in order to create stronger competitive platforms. The financial services industry has witnessed a number of mergers, as banks and stockbroking firms have started positioning themselves to cope with the increase in competition brought about by reforms up to and including the prospective Big Bang, scheduled mainly for April 1998. Cuts in spending on public works have already triggered similar moves among cement companies. In the field of telecommunications, two important mergers were under way in 1997, while NTT was negotiating a partnership with another company in the area of international services, following a decision by the Ministry of Posts and Telecommunications to split the giant company into three by the summer of 1999. The formerly fragmented bulk chemicals industry has seen consolidation and, in 1997, the FTC permitted the first major merger since 1969 in which the combined company will control more than a 50% market share, when it allowed Mitsui Petrochemical Industries and Mitsui Toatsu to merge. The FTC's decision was based on considerations of international rather than domestic competitive conditions, and was expected to lead to further restructuring and regrouping, on the basis of the same considerations, in the steel and chemicals industries. Such mergers are likely increasingly to involve companies from more than one *keiretsu*, thus contributing to a limited overhaul of corporate structures.

The extent to which the traditional *keiretsu* are being overtaken by market pressures is also apparent in the greater willingness of Japanese companies to enter partnerships and other collaborative arrangements with foreign companies. Mitsubishi, for instance, has forged ties with Daimler Benz of Germany, and banks such as the Swiss Bank Corporation and Bankers Trust have been providing assistance to Japanese institutions facing the Big Bang.

At a more formal level, the new holding company law which came into effect in December 1997 gives Japanese companies the chance to form overt ties and, it seems, to act together in a concerted fashion, reminiscent of the days of the *zaibatsu*. However, such interpretations should be treated with a dose of skepticism. Historically, the importance of holding companies is questionable, as Morikawa Hidemasa has stated: "the role of the holding company in almost all the *zaibatsu* was not one of directing from above but rather of providing informal cooperation" (p. 213). Nevertheless, the first holding company was set up by the country's largest supermarket group, Daiei, on the first day that it became legal to do so. The holding company, based in Kobe, will supervise 40 companies involved in services, restaurants and real estate. According to its President, Naka'uchi Isao (as quoted in the *Nikkei Weekly*, December 22, 1997), the reason for the change was "essentially, to divide capital and management. Given more autonomy, subsidiaries will do business independently and become enthusiastic about offering shares publicly." While stressing that Daiei and the *zaibatsu* were "too different in scale and history to compare", Naka'uchi added that: "owners of prewar *zaibatsu* corporate groups such as Mitsui and Mitsubishi had not become directly involved in management. We will follow the same direction." The intention ultimately is that all Daiei companies will be grouped in the holding company. Commercial motives seem to have been the most important driving force behind the change.

The legislation, however, retains restrictions on three types of holding company: large corporate groups that own leading companies in several different fields; groups that include both large financial and non-financial companies; and groups that include leading companies in related business fields. Revisions to the Anti-Monopoly Law also limit the scale of financial holding companies, which were expected to be allowed from March 1998. For instance, the Law now sets a limit of 15% on the stake that financial institutions under the same holding company can have in any single non-financial company. Before the revision came into effect, banks and securities houses were allowed a maximum stake of 5%, while insurance companies could hold up to 10%. Financial holding companies will be allowed to group together banks, securities houses and insurance companies, and this could be of considerable importance in the context of the Big Bang. If, as is expected, this process forces weaker institutions within these three segments of the financial services industry to seek alliances with larger and stronger banks or firms, then the framework of the holding company could facilitate acquisitions across what were previously strictly policed boundaries. Financial holding companies will not, however, be allowed to have non-financial subsidiaries. As the deadline approached for implementing the new financial holding company rules, there was speculation that the financial services units of, for instance, the Fuyo group, including Fuji Bank, Yasuda Mutual Life and Yasuda Trust, were about to form such a structure.

While the holding company represents a reincarnation of an earlier form of corporate structure, the "convoy system" of support from large banks or companies for smaller institutions is decaying. Main banks have shown a growing tendency to walk away from corporations or affiliates, such as finance companies, which find themselves in difficulties. Banks are also operating on the basis of more strictly commercial criteria, and are willing to accept that there may be some diminution in their standing, or loss of face, rather than saddle themselves with additional debts by rescuing insolvent companies or institutions. In one of the most prominent collapses of recent years, Fuji Bank decided in November 1997 not to come to the assistance of Yamaichi Securities, the fourth-largest stockbroker in Japan, which was allied to the Fuyo Group. This neglect of *keiretsu* obligations was another powerful reminder of the fact that corporate structures will continue to evolve, as Japan itself undergoes further economic, social and political changes.

Further Reading

Broadbridge, Seymour, *Industrial Dualism in Japan: A Problem of Economic Growth and Structural Change*, London: Frank Cass, and Chicago: Aldine, 1966

Written in the days when Japan was still an interesting developmental phenomenon, this book retains some interest, partly for the comparisons and contrasts that the passage of time has made possible as between the dual economy that Broadbridge lucidly describes, and the state of the country three decades later.

Field, Graham, *Japan's Financial System: Restoration and Reform*, London: Euromoney Publications, 1997

An overview of the elements of the financial services industry that impinge on corporate structures, written as the Big Bang was in its early stages.

Makino N. et al., *Nihon o Kaeru Shin Seicho Sangyo* [New Growth Industries that are Changing Japan], Tokyo: PHP, 1994

An example of the "paradigm shift" literature, looking at industrial transition.

Miyashita Ken'ichi and David W. Russell, *Keiretsu: Inside the Hidden Japanese Conglomerates*, New York: McGraw-Hill, 1994

This book takes as its starting point the authors' preoccupation with "exclusionist confederations" and is chiefly concerned with what they see as the baleful effects of the *keiretsu*.

Morikawa Hidemasa, *Zaibatsu: The Rise and Fall of Family Enterprise Groups in Japan*, Tokyo: University of Tokyo Press, 1992

A scholarly but readable account of the *zaibatsu*, concentrating on seven major groups.

Nakamura H. et al., *Gendai Chusho Kigyo Shi* [History of Modern Small and Medium-sized Enterprises], Tokyo: Nihon Keizai Shimbunsha, 1981

A multi-author work, broadly supporting the argument that SMEs have played a dynamic role in the Japanese economy.

Nakamura Takafusa, *The Postwar Japanese Economy: Its Development and Structure, 1937–1994*, second edition, Tokyo: University of Tokyo Press, 1995

A competent scholarly overview, in which Chapter Five has details of the dual structure of the economy.

Norman, E. H., *Japan's Emergence as a Modern State*, New York: Institute of Pacific Relations, 1940; reissued in *Origins of the Modern Japanese State: Selected Writings*, edited by John W. Dower, New York: Pantheon Books, 1975

A classic account of developments from the 1850s to the 1930s, which helped to underpin the fixation of US and other Allied policy-makers on the role of the *zaibatsu* in the Japanese economy and society.

Ohsono Tomokazu, *Charting Japanese Industry: A Graphical Guide to Corporate and Market Structures*, London and New York: Cassell, 1995

Handy diagrammatic summaries showing interconnections among major companies, including those in the Big Six horizontal *keiretsu* and a further 19 "giant companies".

OECD, *Economic Surveys, 1994–95, Japan*, Paris: OECD, 1995

Part IV includes a special analysis of the distribution system.

Piore, Michael, and Charles Sabel, *The Second Industrial Divide: Possibilities for Posterity*, New York: Basic Books, 1984

A groundbreaking work on new forms of industrial organization.

Rafferty, Kevin, *Inside Japan's Powerhouses: The Culture, Mystique and Future of Japan's Greatest Corporations*, London: Weidenfeld and Nicolson, 1995

The subtitle of this informative but somewhat superficial book, by a British newspaper journalist, largely speaks for itself.

Whitaker, D. H., *Small Firms in the Japanese Economy*, Cambridge and New York: Cambridge University Press, 1997

A partisan and consciously revisionist account, by the son of a small business owner, that seeks to compare Japanese machine industries, especially in the Ota ward of Tokyo, to small businesses elsewhere, such as in the British city of Birmingham.

Dr Graham Field, a former editor of the magazine *Asiamoney* and author of *Economic Growth and Political Change in Asia* (London: Macmillan, and New York: St Martin's Press, 1995), is the Director of the British Institute for Contemporary Economic and Political Studies.

Table 10.1 Composition of *Keiretsu*, by Type and Number of Member Companies, 1996

	Leading member companies	*Other member companies*	*Associated independent firms*	*Totals*
Horizontal (*yoko*)	117	578	151	846
DKB	17	79	15	111
Sanwa	12	38	17	67
Fuyo	17	62	19	98
Sumitomo	15	80	28	123
Mitsubishi	23	95	30	148
Mitsui	20	115	22	157
IBJ	11	106	17	134
Tokai	2	3	3	8
Vertical (*tate*)	89	165	n. a.	254

Source: Adapted from data researched by Jardine Fleming Securities, and kindly provided by Professor Charles McMillan

Society and Culture

Chapter Eleven

Diversity in Japanese Society

Mika Merviö

There is a long-established tradition of books, both in Japanese and in western languages, that emphasize the homogeneous and harmonious nature of Japanese society, and describe how traditions somehow keep it uniquely Japanese. In contemporary Japan this *Nihonjinron* literature (literature on "Japaneseness") fills the ideological needs of conservative politics, and some of these books provide explicit support for nationalism; while western *Nihonjinron* books more often try to give simple and holistic explanations of Japanese phenomena or "miracles". Their authors are happy to assert that the Japanese population is more homogeneous, in "race", language, religion or education, than the populations of many other countries, and this assumed homogeneity is often linked to the mechanisms in Japanese society which are said to resolve conflicts in harmonious ways, counting on the inbred harmony or conformity of the people. These perceptions of homogeneity are often based on contrasting idealized images of the Japanese with usually rather unrealistic images of foreign people. When there are references to diversity and heterogeneity in Japanese society the evidence is usually limited to references to obvious, relatively well-studied minorities – Koreans, *burakumin*, Ainu (see Appendix 2). Yet these may serve to divert attention away from less obvious forms of diversity and heterogeneity, which affect the whole society.

Nevertheless, there is an increasing amount of research being made available in western languages which challenges the accepted holistic models of Japanese society and tries to present it in all its diversity, variation and stratification (see, for example, Denoon et al., Mouer and Sugimoto pp. 234–373, Sugimoto, and Sugimoto and Mouer). In this regard, western scholars and commentators are to a large extent catching up with their Japanese colleagues, since social diversity is well-covered in the Japanese academic literature. It is also common knowledge among ordinary Japanese, despite the popularity of some of the *Nihonjinron* texts. As in other societies of human beings, this diversity reflects and expresses far more than "ethnic" differences, to embrace material differences of income and wealth, both among individuals and among localities, as well as less concrete but still very significant differences of class, generation, gender, and ascribed status. Japanese society, after all, comprises at least 125,570,000 disparate human individuals (as of 1995).

The "Middle Class"

Contrasting contemporary Japanese society with images of foreign societies and of its own past explains why about 90% of Japanese people identify themselves in surveys as members of the "middle class". After all, it can be argued, within the terms of the comparisons that most Japanese can reasonably be expected to make, that the postwar policies that have tried to correct social excesses and redistributed wealth have indeed created a greater degree of social and economic equality than at any previous time in Japanese history, or than in most comparable OECD member states. At the least, there is a widespread perception that such equality exists.

However, it should also be stressed that there have been some gradual and revealing fluctuations in the perceived proportions of "upper middle class", "middle middle class" and "lower middle class" within the total. From the early 1960s the majority of Japanese have perceived themselves as members of the middle middle class, the peak coming in the mid-1970s, at more than 60%, since when there has been a slow decline. Those locating themselves in the lower middle class fell from 32% of the surveyed population in 1958 to a low point of 22% in 1973, but there has since been a slow rise. Over the same period the proportion perceiving themselves to be among the upper middle class has remained stable, at around 8%. These changes apparently are related to general economic trends: during the years of high economic growth more people perceived that they had experienced upward social mobility (Kyogoku p. 125). For these people upper classes, or working/lower classes, are apparently phenomena that they associate with other societies but not with contemporary Japan.

It should also be noted, in passing, that the Japanese middle class is ill-represented by the standard western media stereotype of the white-collar corporate warrior, often known by the Japanese "English" term *sarariiman.* Most Japanese people do not work for large corporations with global reach, nor do they have particularly good job security (see also Chapter 13).

Nevertheless, it is clearly not easy to find people who are ready to reveal that they see themselves as different from the mainstream, and vague ideas about what that mainstream is do seem to be widely shared – albeit the same could be said of other advanced industrial societies, where what is seen as a "middle class" lifestyle, if not already achieved, is quite generally aspired to. In addition, however, the myth of the monoethnic state (*tan'itsu minzoku kokka*), first created to legitimize Japanese imperialism, is still powerful in Japan. This lingering belief in distinct biological and social origins, and the consequent marginalization of, and ignorance about, many forms of cultural diversity, have tended to reinforce narrow perceptions of what should be recognized as being part of Japan's "mainstream" (Macdonald and Maher pp. 3–23, Takita pp. 292–319, and Oguma).

Distribution of Income and Wealth

The remaining shortcomings in the realization of equality in Japanese society are common knowledge in Japan, and differences in income and wealth are still considerable (as is evident from such sources as the *Asahi Shimbun Japan Almanac 1998*, from which most of the figures in this section are taken).

According to the Ministry of Labor (*Rodosho*), in 1996 the national average monthly pay-packet for workers in major firms contained ¥413,100 (see also Chapter 13 for more details on wage structures). This figure, like most averages, can be misleading. First, by definition it takes no account of wages for workers in small and medium-sized enterprises, which, if sample surveys are reliable, average around 75% of this figure and, in many industries, are much lower. Second, about 25% of this figure represents bonuses and other special payments, which vary considerably from company to company. Third, this figure also fails to capture the impact of seniority, a major factor in the calculation of pay. Fourth, wages for women still tend to be substantially lower than for men, even for the same work. Women still tend to be forced to take lower-paying, often part-time jobs, although by 1996 they made up 40.5% of the labor force and were in the majority in some service industries. Finally, the people who have amassed the greatest fortunes in Japan are generally those few who have large holdings of urban land, or who own their businesses outright, and they are not surveyed by the Ministry of Labor. Interestingly, the proportion of the population who are in the top 2% of the income range (as measured by the tax authorities) has steadily declined, from 12.2% in 1965 to 5.8% in 1985. Some researchers have perhaps been too ready to conclude that this alone proves that income distribution is fairly equal in Japan.

Meanwhile, the welfare system continues to play a relatively restricted role, and government measures for the redistribution of wealth are still moderate when compared to the general pattern in OECD countries. Thus, in 1997 the combined burden of national taxes (14.9% of GNP), local taxes (9.5%) and social security payments (13.8%) was 38.2% of GNP, which was very close to the rate in the United States but was significantly lower than the average in member states of the EU. Those receiving assistance under the Daily Life Security Law of 1950 – mostly the elderly, single-parent households, the injured and the disabled – form a relatively low proportion of the population, which has actually fallen in recent years, from 1.22% in 1980 to 0.82% in 1990 (Kosaka pp. 35–46). Yet these figures do not necessarily reveal an accurate picture of Japanese poverty. Many disadvantaged people, such as the homeless, are too scared to contact the authorities, and others survive beyond the reach of the authorities with the help of relatives.

For the great majority of Japanese people, again as in most OECD countries, owning one's home remains the primary economic goal; yet in 1993 only 59.8% of households owned the home that they lived in. In Tokyo, Osaka and other major cities owning a house is an unrealistic dream for a majority of the population: thus, in 1993 the ownership ratios for these two cities were, respectively, 39.6% and 47.9%. (The average cost of residential land remains relatively high, ranging from ¥399,200 per square meter in Tokyo to ¥30,600 in Miyazaki Prefecture and even lower in Saga, Aomori, Akita and Shimane Prefectures.) Further, since such items as housing, food, education and transportation tend to be relatively expensive for Japanese consumers, few households have much left after paying for these necessities. Japan's notoriously high average proportion of household savings to income, which stood at 13.1% in 1995, can be largely explained by the need to save for old age, for housing, and for children's higher education.

Regional Inequalities and Government Spending

It is not only in the cost of residential land (cited above) that drastic differences continue among Japan's 47 prefectures, which vary widely in the level and composition of output, and therefore also in incomes. In 1994 (the latest figures available) just 1.8% of Japan's GNP of nearly ¥4,830 trillion was derived from the primary sector and 68.5% came from the tertiary (services) sector. However, within the output of Tokyo, the richest prefecture (producing about ¥826.7 billion), the respective proportions were 0.1% and 81.9%; while within Miyazaki Prefecture's output (nearly ¥31.24 billion) they were 7.0% and 65.6%. Similarly, in the same year, when the national annual income per capita was ¥3,080,000, the figure for Tokyo was ¥4,411,000, but that for Okinawa, the poorest prefecture, was ¥2,118,000. Recasting these average per-capita incomes as percentages of the figure for Tokyo would give a national average of 69.8%, a figure for the second-richest prefecture, Aichi, of 80.5%, and a figure for Okinawa of just 48.0%. The six prefectures placed just above Okinawa, and their 1994 per-capita incomes as percentages of Tokyo's, were Kagoshima (52.3%), Shimane (52.5%), Miyazaki (52.6%), Kochi (55.3%), Wakayama (55.3%) and Aomori (55.9%). What these seven low-income prefectures have in common is that they are geographically distant from the administrative and industrial centers of Japan and that – with the exception of Okinawa, which in many respects is a special case – the share of the primary sector within their local economies is at least twice the national average. (All figures in this paragraph are derived from Yano Tsuneta Kinenkai pp. 268–271.)

These differences are less stark than in the past, yet they persist, even though the Japanese state, which is relatively weak at administering redistribution of incomes on an individual basis, has consistently placed a strong emphasis on the redistribution of wealth among the regions (see also Chapter 12). This has long been one of the proclaimed goals of the system of public finances, which has ensured that more than half of the income of prefectural and other local governments comes from the national government, in the form of transfer payments and various subsidies. The corollary has been that the national government has exercised tight control over local and regional spending, and that many of the tasks performed by the prefectures and other municipalties are simply dictated to them by national laws, which give little scope for initiative or variation. National and local governments cooperate closely on implementing public spending programs, and there has been a strong political will to stimulate the economy in the less developed areas. As a result there has been considerable progress on alleviating some of the problems caused by uneven development. At the same time, however, public spending, especially during the 38 years of continuous government by the Liberal Democratic Party (LDP), from 1955 to 1993, has introduced other forms of inequality among localities. For example, those prefectures which are relatively more dependent on agriculture have been favored with generous support, and in turn have tended to favor the LDP politicians who have secured that support (see Chapter 5). Again, one may wonder whether, for instance, there is any real social need for the numerous public works programs which, in certain cities and prefectures, have literally layered cities, roadsides and riverbanks with concrete. It is all too obvious that there are strong political and business interests which favor technocratic, centralized, and politically convenient conceptions of "development", and are hostile both to greater local autonomy – perhaps especially for areas which tended to vote against the LDP – as well as to "softer" ways of spending the public revenues.

The lingering problems of uneven development may yet become more severe. As in other OECD countries in recent years, the social climate in Japan, as expressed through almost all the political parties (except the Communists), is now vehemently opposed to increasing taxes. The present government has responded to this new mood by reviving the goal of "administrative reform" (*gyosei kaikaku*) and planning for cuts in the national bureaucracy, a decrease in the budget deficit, deregulation of the economy, and the transfer of some powers (and costs) to local governments. However, the national civil service in Japan is already the smallest, in proportion to the population, of any OECD country, and it can ill afford to sustain dramatic reductions in its numbers. It is not at all clear that the private sector in Japan is necessarily more efficient than the public sector, or that it is able and willing to take over functions traditionally entrusted to the state. Nor is it clear exactly how the state's powers and responsibilities, including the crucial power to raise and allocate revenues, will eventually be redistributed among the slimmed-down national ministries and the (presumably) larger and more powerful prefectural bodies that are to take on some of the burdens of administration.

The "Japanese-Style Welfare Society" in Crisis

The very concept of "welfare" (*fukushi*) was not common in the Japanese language before the social reforms of the Occupation period (1945–52). Indeed, the government department known in English as the Ministry of Health and Welfare (MHW) retains in Japanese the name it was given when it was founded in 1938 – *Koseisho* – which has a strong connotation of "public welfare work" (*kosei jugyo*) rather than "welfare" in any broader, communal sense. Nevertheless, the terminology used in Japanese welfare discourses now has a firm foundation in the Occupation reforms, especially in the new Constitution of 1947, which emphasizes the principles of equality and welfare of all citizens (Takahashi pp. 33–75). These principles are widely accepted as core values of the postwar political system: the disagreements start when attempts are made to give these concepts more exact meanings.

Since Japan has been governed by the LDP for most of the postwar period, the "Japanese-style welfare society" (*Nihongata fukushi shakai*) created during these decades has taken on a fairly conservative significance. The opposition parties, especially the Japan Socialist Party (now the Social Democrats), used to be very vocal supporters of an extension of the "welfare society", and, notably during the 1970s, an influential body of LDP politicians also perceived a strong social and political need to improve welfare services. However, the power elite has tended to maintain an ideologically hostile attitude to over-intrusive state patronage in social policy, and most politicians in power since the end of the 1970s have been averse to strengthening the state's role in the process of building a welfare society. Thus, during the 1970s, when Japanese society reached new levels of affluence, existing systems of welfare were expanded and improved, and a comprehensive minimum welfare network was completed. However, in the late 1990s, with continuing economic stagnation, welfare services have come increasingly under threat from plans for extensive restructuring and reductions in costs and personnel. The political struggle now concerns, not the initiation of a Japanese-style welfare society, as in the 1950s, or its completion, as in the 1970s, but the course of its future development, and the costs of maintaining the current system or making modifications to it. Its current crisis is primarily the result of the national authorities trying to transfer responsibility to the local level, or to individuals and families.

Meanwhile, and perhaps more urgently from the point of view of the users of health and welfare institutions, there are specific problems in the ways in which they are run – whether they are formally public or private – especially in the attitudes of the people responsible for providing these services. For example, hospitals are notoriously places where doctors tend to exercise dictatorial authority over patients and nurses alike. Patients are expected to follow the rules and not to criticize anything, while the education of Japanese doctors is often rather superficial, most of them gaining their qualifications without doing any individual research work or dissertation. Instead of relying on scientific merit, loyalty to hospital hierarchies is the usual rule for determining careers. The result, all too often, is that doctors make no effort to explain illnesses or medication to patients; facilities are plagued by inefficiency and bureaucracy, being dirty and full of people; and the maximizing of profit can easily replace welfare as the main goal of the institutions. Hospitalization tends to last many times longer than in similar cases in the United States, while the monopoly that doctors and dentists enjoy in prescribing and selling most medication has guaranteed that the Japanese medical institutions have a stable source of income, without much need to make improvements in the efficiency or quality of care (see Ikegami and Campbell, Yamazaki, and Maruyama et al.).

The MHW itself has also come under increasing attack, partly because its spending has been the fastest-growing budget item for the past two decades. Its image has not exactly been enhanced by its association with a number of scandals, including the use of HIV-tainted blood products in Japanese hospitals for many years; the apparent failure to supervise greedy private institutions in both health and welfare services; and the disclosure of outright corruption in December 1996, when its Permanent Vice-Minister (its leading official) Okamitsu Nobuharu was arrested after receiving about ¥60 million in cash and other favors from the head of a private welfare foundation. In 1997 the MHW responded to its critics by publicizing plans to reform the nursing insurance system and the public pension system, thus drawing attention once again to one of the favorite themes of its officials, Japan's "aging society" (*koreika shakai*).

The Aging of the Population

In 1997, as in earlier years, the MHW cited the aging society as the main reason for radical reductions in benefits and increases in the contributions paid by working people. The change in average life expectancy in postwar Japan has certainly been significant: it has risen from 62.97 years for women and 59.57 years for men in 1950, to 83.59 years and 77.01 years, respectively, in 1996 – the highest average life expectancy of any country in the world. The number of people aged 65 years or more has risen over the same period of 46 years from 4,155,000 to 19,017,000, or 14.5% of the total population (*Asahi Shumbun Japan Almanac 1998* pp. 59 and 225). The aging of the population is expected to continue inexorably, since the birth rate and the death rate have both declined, with the result that, according to forecasts by the MHW, those over 65 will account for 27.4% of the population by 2025 and 32.3% by 2050 (Koseisho pp. 100–1). Of course, these figures are based on the assumption that there is not going to be any significant increase in immigration.

The "aging society" has become a familiar theme in the Japanese media and in everyday conversation. The emphasis is generally on the sacrifices that everyone will have to make in order to maintain welfare services, and on the threat that this major demographic shift appears to pose to Japan's economic competitiveness. One effect has been that reform of the pension system has tended to dominate political discussions about "welfare", leaving other issues to the MHW to take care of, more or less unobserved. In its turn, the MHW has devoted much of its policy output on the issue of aging to praising self-help and voluntarism, urging the elderly to participate in social activities in order to find meaning for their lives, and encouraging communication among the different generations (Koseisho pp. 102–5 and 124).

In these circumstances, the establishment of care facilities for the elderly has lagged behind their steadily rising needs. Much of the care of the elderly has been provided by their relatives, often daughters or daughters in law, in the past. However, since more women are going out to work, and most people now favor nuclear families over extended ones, it is becoming increasingly difficult to find people who are ready to dedicate their lives to providing such care. Because women live longer, and there are many women who are capable of taking care of their husbands at home, the proportion of elderly women who are in need of institutional care greatly exceeds that of elderly men, at 85.1% versus 14.9%. At the same time, most elderly people want to continue living at home as long as possible, which fits in well with the preferences of the MHW, and those of many local governments. The emphasis in the 1990s has therefore been on "normalization" (*nomaraizeshon*) and care provided at home (*zaitaku kea*), although practice falls far short of the rhetoric. About 60% of the elderly live with their relatives, 25% with their spouses, 10% alone, and 5% in institutions (Miura pp. 34–36).

Institutionalization usually takes place in the form of hospitalization under doctor's orders, which means in practice that many hospitals become places where patients spend extended periods waiting for death. In most localities there is a shortage of specialized institutions for the care of the elderly. Many local governments have begun to pay serious attention to the calls for more diversified services that would better meet the needs of the elderly, although the more cynical would observe that this may well be because hospitalization is relatively expensive. In addition to local initatives, which vary in quality and funding, as well as in motivation, the MHW's "Gold Plan" (*Gorudo Puran*) of 1989, and its modified version, the "New Gold Plan" (*Nyu Gorudo Puran*) of 1994, constitute a 10-year project to provide facilities and care for the aged. This is far from being ambitious in its aims. It is estimated that it will cover only 40% of those who will need such services; and it appears that the government is dragging its feet on implementing even the 1994 version (according to reports in the *Daily Yomiuri* newspaper for December 4 and 8, 1997; see also *Nihon no Ronten 1998* pp. 596–629). Both versions of the Plan called for a radical increase in the numbers of staff trained to care for the elderly, and for all kinds of institutional care, such as home help services, day care services, specialized nursing homes and health centers, and so on. In all these areas implementation is lagging vastly behind the stated goals of the MHW, and behind the needs identified in its own research (For more on the aging society, see Kaneko, Koseisho pp. 100–125, Miyajima Hiroshi, Shindo, and Takahashi pp. 180–202 and 230–1.)

Koreans Resident in Japan

Koreans were already regarded as a special case in 1876, when a friendship treaty signed by the governments of Japan and Korea first allowed them to live "freely" in Japan, outside the officially designated foreigners' settlements of the time. After Korea became a Japanese colony, in 1910, the number of Koreans resident in Japan started to rise: it has been estimated that there were more than 30,000 by 1920, and more than 800,000 by 1938. By then Koreans, as well as people from Japan's other main colony, Taiwan, were being brought into Japan as forced labor for the war effort. In addition, the numbers of Koreans and Chinese entering Japan illicitly in search of work rose steadily, especially after the economic boom at the end of 1910s. The wars of the 1930s and 1940s were particularly hard experiences for all Koreans, and the treatment of Koreans drafted into the Japanese Imperial Army, the laborers forced to work in Japan, and the Korean women who worked as "comfort women" (*jugun i'anfu*), remains a source of tension in relations between Japan and the two Koreas, even in the 1990s. The Korean minority in Japan itself inadvertently keeps these aspects of the two people's shared history in the minds of both the minority and the majority (see Kimu Chanjon, Nishinarita, Suzuki, and Weiner 1994).

After World War II a substantial portion of the Korean population residing in Japan at the defeat stayed on there, since the situation in Korea remained unsettled for a long time, and in any case many of these people had strong roots in Japan. In 1952 there were 535,065 Koreans registered as resident in Japan. Today there are about 1 million people from Korean backgrounds, including about 650,000 people who are citizens of one or other Korea, as well as naturalized Japanese citizens and their children. The community is growing increasingly heterogeneous, and it is difficult to make generalizations about shared identities. There are two main organizations that promote the interests of this minority in Japan, *Mindan*, which is closely linked to South Korea, and *Chongryun*, affiliated with North Korea. The vast majority of the first generation came from the southern parts of the peninsula, but identification with either Korea among Koreans resident in Japan is based more on social networks and political ideas than on ancestral origins. Some individuals have even tried to maintain some links with both camps, while others have found it difficult to relate to either, preferring to drop any interest in Korean politics once the elders who used to make them pay attention have gone. Koreans in Japan are constantly making choices, since there is no single model identity to adhere to; and among people born of mixed marriages the choice is usually even more complicated.

The rigid official treatment of residents of Japan according to their citizenship or lack of it has further complicated the choosing of identities among Korean residents. Japan recognized South Korea in 1965, since when those Koreans resident in Japan who have taken citizenship of South Korea have been treated as "permanent residents" of Japan, provided that they have lived there since 1945, or were born there. Since the Japanese government has not recognized North Korea, it has long regarded those who are loyal to that country, not as citizens of it, but as "resident aliens" of Japan. More recently, however, the government has liberalized the granting of permanent residence to those Koreans who are not citizens of the South, on an "exceptional basis", notably following Japan's ratification of the International Covenant on Human Rights in 1979. As a result, by 1990 there were more than 323,000 "permanent residents" who were citizens of South Korea, alongside another 268,000 who had permanent residence on an exceptional basis. Then, in 1991, the Foreign Ministers of Japan and South Korea signed a memorandum aimed at reforming Japan's Alien Registration Law and Immigration Control Law, and at changing practices which were seen as being discriminatory and had become a source of friction in bilateral relations. In 1992, all Korean permanent residents were re-designated as "special permanent residents", exempted from the fingerprinting which is still used to register other foreigners, and granted multiple re-entry permits to Japan for up to five years at a time.

This all too recent change in the legal status of permanent resident Koreans was not just a formal gesture towards marking the special circumstances of their presence in Japan. It was also needed to provide at least some guarantee against discrimination in their daily lives, for instance, in seeking and keeping employment, or when applying for pensions or welfare benefits. A fairly widespread lack of understanding of their status still causes problems. Some companies, for example, decline to rent electrical appliances or other goods to special permanent residents, claiming that, like other foreigners, they might take the goods with them "when they leave Japan". More pervasively, even though most of these people have lived all their lives in Japan, speak wholly or mainly in Japanese, and have only tenuous links with their forebears' home country, they can still be excluded from employment, education or business life for lack of formal Japanese nationality. It is not surprising, therefore, that there are now about 200,000 Koreans who have been naturalized as Japanese citizens. The fact that most Koreans resident in Japan now marry

Japanese citizens also contributes to an accelerating rate of naturalization. In 1991 82.5% of marriages involving Koreans in Japan were intermarriages with Japanese – 22.8% involving Korean bridegrooms, 59.7% involving Korean brides (Fukuoka pp. 64–75).

Yet another, much smaller group of Korean residents has been formed under special circumstances. During the late 1980s, the years of Japan's "bubble economy", a new wave of immigrant workers arrived from South Korea, notably after the Seoul government considerably liberalized travel restrictions in 1988, the year of the Seoul Olympics. As a result, citizens of South Korea have figured prominently on official lists of "illegal immigrants", numbering 5,534 in 1990 and 9,800 in 1993 (Tanaka pp. 182–4, and Yamanaka p. 83).

Whatever their official or marital status, the vast majority of Koreans resident in Japan have found more or less secure places in society, although lingering prejudices against the community as a whole can be revived whenever individuals from Korean backgrounds are associated with social problems in the media, or work (often alongside Japanese nationals) in employment which is not considered "respectable", such as owning and managing *pachinko* parlors. Discrimination against Koreans usually shows itself in Japanese individuals' avoidance of close relationships with members of this minority, but cases of discrimination in employment, housing, education and other social institutions are still widespread. Physical violence against Koreans and other minorities in Japan has been relatively rare, but it is not unheard of. Finally, the government has taken the uncompromising position that only Japanese citizens can exercise public authority, and, because of their numbers, and their generally greater commitment to living permanently in Japan, it is mostly the Korean minority, rather than any other group of non-Japanese, that suffers the effects of this policy. In the past few years, several municipal governments have started to test its limits by opening a wider selection of positions in the public sector to "foreigners", which in practice means to Korean residents. (On Koreans in Japan see Fukuoka, Iinuma, Kanai, Kimu Chanjon, Nakao, Ryang, and Sorano and Ko.)

Other Foreigners in Japan

Although a number of other aspects of social diversity in Japan could have been addressed here, for reasons of space we shall look at just one last example: the wide range of resident foreigners who are not of Korean descent. The Immigration Control Law, enacted in 1951, restricted the granting of work permits to those with specific skills needed or wanted in Japan. The number of registered aliens increased slowly over the ensuing decades, exceeding 1 million for the first time as recently as 1990, when the official figure was 1,025,911. By 1996 the figure was 1,415,136. The largest increases have been in the numbers of immigrants from China, Brazil and the Philippines, who now form the three largest communities of foreigners in Japan apart from the Korean "special permanent residents". In 1996 there were 234,264 Chinese, 201,795 Brazilians, and 84,509 Filipinos registered as alien residents (*Asahi Shimbun Japan Almanac 1998* p. 63, and Tanaka pp. 31–34).

Since the economic rewards of working in Japan have become known all over the world, the numbers entering the country have increased. In particular, it is fairly certain that the numbers of illegal immigrants have risen rapidly, although, given the penalties that can be imposed on these people, and the range of motives behind attempts to exaggerate or downplay the figures, all estimates of those numbers must be questionable at best. Thus, while it is on record that 2,339 such immigrants were arrested in 1983, 29,884 in 1990, and 67,824 in 1992, it is by no means clear that the numbers arrested correlate closely, if at all, with the numbers who enter illegally and evade arrest, in some cases for years on end. At the same time, it is also fairly clear that the number of people who have overstayed their visas has risen rapidly. There are reliable estimates that there are something like 450,000 foreigners who are either overstayers or working without authorization (Komai 1995 pp. 3–5).

Partly in response to the rising numbers of illegal immigrants, and partly also in reaction to scandalmongering by some sections of the mass media, the Immigration Control Law was subjected to a major revision in 1990. On the one hand, the entry formalities were simplified, and the number of categories of work permits was expanded from 18 to 28. On the other hand, specific sanctions were introduced against employers hiring illegal foreign workers, and against labor brokers and middlemen helping foreigners to work illegally, up to a maximum of three years in prison or a fine of ¥2 million (Yamanaka pp. 75–6).

The largest numbers of illegal immigrants, at least according to the official figures, are from South Korea, China, Bangladesh, Malaysia, the Philippines, Pakistan, Thailand and Iran (Tanaka pp. 182–4, and Yamanaka p. 83). Like illegal immigrants in western countries, they have often been at the center of attention from the mass media, not always with great regard for accuracy, and are always likely to break the law in many different ways, not always intentionally, precisely because they live on the margins of Japanese society. Some seek the protection of *yakuza* (gangsters), or work for them; others belong to criminal organizations based in their home countries, among which the Chinese gangs have acquired a notably poor reputation in Japan. Indeed, perceptions of criminality among immigrants, however distant they may be from the everyday reality of most foreign residents' lives, may have helped to harden attitudes toward foreigners in general during the 1990s (see Miyajima et al.).

It should also be noted, however, that some employers have now become dependent on foreign labor, and in some areas where the labor shortage is acute, including some small and medium-sized construction firms and factories, community control may actually work in favor of keeping illegal foreign workers in their jobs. In 1992 it was estimated that illegal foreign workers were especially likely to be drawn into construction, process work in manufacturing, or jobs as hostesses or hosts in restaurants, all fields in which it has become difficult to find willing Japanese labor (Tanaka and Ebashi p. 3).

The revised Immigration Control Law also in effect permitted foreign descendants of Japanese emigrants (*Nikkeijin*) to enter Japan and work there, by creating a new category of "long-term" residence available, usually for three years at a time, to members of *Nikkeijin* communities in such countries as Brazil and Peru (where they number about 1.28 million and about 80,000, respectively). A total of 27,500 *Nikkeijin* were given residence and work permits in 1991; an estimated 150,000 *Nikkeijin* were residing in Japan by the summer of 1992. They are employed primarily as unskilled laborers in crafts, manufacturing assembly and repair, and machine operation (Enari pp. 52–5, Kajita pp. 147–173, Sellek pp. 178–210, and Yamanaka pp. 73–9).

Another substantial group of foreigners comprises university and college students, who numbered about 53,000 in 1996, and pre-college students, at about the same number, most of whom are studying in small language academies. Of the university and college students, 23,300 (44%) were from China, 12,300 (23%) from South Korea, 4,700 (9%) from Taiwan, and 2,200 (4%) from Malaysia, while Indonesia, Thailand and the United States each sent about 1,000 (2%) (*Asahi Shimbun Japan Almanac 1998*, p. 251). In short, the great majority of such students come from neighboring countries in East Asia. However, in the past two years there has been a decrease in the numbers of foreign students, and the crisis sweeping East Asia in 1997–98, combined with the continuing recession in Japan, will probably lead to a further decrease, as many students from the region either discontinue their studies, or do not even take up places in Japanese institutions to begin with. In any case, it has become clear that it will not be possible for the Japanese government to meet its "internationalization" (*kokusaika*) target of having 100,000 foreign students at Japanese universities and colleges by the year 2000.

Still another major group of foreign residents to be considered comprises those academically educated foreigners who have relatively highly skilled positions in Japan, although even they are seldom employed on equal terms with Japanese citizens, and their numbers are relatively small. According to the official figures (see Somucho Tokei Kyoku), in 1994 they included 27,229 investors or business managers, 11,606 diplomats, and 14,066 other officials. In many cases, in Japan as elsewhere, such people live happily enough in their own ghettos and have limited contact with the society around them. Those who hold permits for "cultural activities" (7,189 people in 1994), religious activities (6,299), journalism (1,346), teaching (7,179), education (14,239), or the arts (291) are generally freer to seek the boundaries of their professions, but they too tend to be more or less dependent on the foreign organizations that brought them to Japan. This leaves such vaguely defined categories as "entertainers", "trainees", and "skilled laborers" – in all three of these cases, the overwhelming majority of permit-holders (and, probably, of illegal immigrants) are from countries in East Asia – as well as refugees, who have remained few in number.

In summary, foreigners have come to Japan from a variety of countries, most legally, some illegally; their total numbers have increased considerably in recent years, challenging notions of homogeneity and uniformity; and the presence of at least some among them, especially if it is sustained into another generation and beyond, could well have a significant impact on Japan in the 21st century. Almost all mainstream Japanese politicians and opinion-formers now pay at least lip-service to the concept of "internationalization" (*kokusaika*): the coming decades will show how much this really means in practice. They will also show what new forms of diversity are needed and can be generated, in a society that has never been as homogenous, or harmonious, as the myths have suggested.

Further Reading

Asahi Shimbun Japan Almanac 1998, Tokyo: Asahi Shimbunsha, 1997

A bilingual statistical yearbook.

Campbell, John C., *How Policies Change: The Japanese Government and the Aging Society.* Princeton, NJ: Princeton University Press, 1992

Campbell provides an analysis of Japanese welfare policies, using the policies and solutions of the United States as a contrast.

Denoon, Donald, Mark Hudson, Gavan McCormack and Tessa Morris-Suzuki, editors, *Multicultural Japan,* Cambridge and New York: Cambridge University Press, 1996

Enari Ko, "Raten amerika nikkeijin" [Latin Americans of Japanese Descent], in Komai Hiroshi, editor, *Shinrai-Teijugaikokujin ga Wakaru Jiten* [Dictionary for Understanding Foreign Newcomers and Residents], Tokyo: Akashi Shoten, 1997

Fukuoka Yasunori, *Zainichi Kankoku-Chosenjin: Wakai Sedai no Aidentiti* [Koreans Resident in Japan: The Identity of the Young Generation], Tokyo: Chuokoronsha, 1993

Iinuma Jiro, editor, *Ashimoto no Kokusaika: Zainichi Kankoku-Chosenjin no Rekishi to Genjo* [Grassroots Internationalization: The History and Present Condition of Koreans Resident in Japan], Osaka: Kaifusha, 1993

A book of 14 chapters by 12 researchers, analyzing Japan's problem of "grassroots internationalization" in the light of the lives and history of Koreans in Japan.

Ikegami Naoki and Campbell, John Creighton [Kyanberu J.C.], *Nihon no Iryo: Tosei to Baransu Kankaku* [Medical Treatment in Japan: A Sense of Control and Balance], Tokyo: Chuokoronsha, 1996

Ishida Takeshi, *Heiwa, Jinken, Fukushi no Seijigaku* [The Political Science of Peace, Human Rights and Welfare], Tokyo: Akashi Shoten, 1990

Ishida is among the most original and analytical of Japanese political scientists. This book provides a range of interpretations and analyses of Japanese political and social phenomena.

Kajita Takamichi, *Gaikokujin Rodosha to Nihon* [Foreign Workers and Japan], Tokyo: NHK, 1994

Kanai Yasuo, *Zainichi Korian Nisei, Sansei no Genzai* [The Present Condition of the Second and Third Generations of Koreans in Japan], Tokyo: Bakushusha, 1997

Kaneko Isamu, *Chi'iki Fukushi Shakaigaku: Atarashii Koreishakaizo* [The Sociology of Local Welfare: The Structure of the New Aging Society], Kyoto: Mineruba Shobo, 1997

Kaneko takes the analysis of the aging society to the local and regional levels, providing comparisons among different communities, primarily in Japan and Taiwan, and how the aging of the population affects them.

Kimu Chanjon (Kim Chan-jung), *Zainichi Korian Hyakunenshi* [History of 100 Years of Koreans in Japan], Tokyo: Sangokan, 1997

Kimu Kyondoku (Kim Kyeung-duk), *Zainichi Korian no Aidentiti to Hoteki Chi'i* [The Identity and Legal Status of Koreans Resident in Japan], Tokyo: Akashi Shoten, 1995

Komai Hiroshi, *Migrant Workers in Japan*, translated by Jens Wilkinson, London and New York: Routledge, 1995 [from *Gaikokujin Rodosha Teiju e no Michi*, Tokyo: Akashi Shoten, 1993]

Kosaka Kenji, editor, *Social Stratification in Contemporary Japan*, London and New York: Routledge, 1994

Koseisho [Ministry of Health and Welfare], editors, *Kosei Hakusho 1997: Kenko to Seikatsu no Shitsu no Kojo o Mezashite, Heisei Kyunenban* [Health and Welfare White Paper 1997: Aiming at Improvement in the Quality of Health and Life, Heisei Year Nine Edition], Tokyo: Koseisho, 1997

This is the 1997 edition of the MHW's major annual publication, an indispensable source for anyone studying health and welfare issues in Japan, which is unfortunately not published in any foreign language.

Kyogoku Jun'ichi, *The Political Dynamics of Japan*, Tokyo: University of Tokyo Press, 1987

This is an English-language version of Kyogoku's book, *Nihon no Seiji*, published by the same press in 1983, which has become one of the most widely used textbooks of Japanese politics at Japanese universities.

Macdonald and Maher, editors, *Diversity in Japanese Culture and Language*, London and New York: Kegan Paul International, 1995

This book demonstrates how much diversity there is in Japanese culture and shows how narrowly the Japanese mainstream is often defined.

Maruyama, Meredith Enman, Louise Picon Shimizu and Nancy Smith Tsurumaki, *Japan Health Handbook*, Tokyo, New York and London: Kodansha International, 1995

This book, written by three foreign nurses living and working in Japan, is a concise and practical introduction to the health and welfare system. It is very useful for anyone who needs to have practical information in English about Japanese health services.

Miura Fumio, editor, *Koreisha Hakusho 1996* [The Aging Society White Paper 1996], Tokyo: Zenkoku Shakai Fukushi Kyogikai [National Social Welfare Council], 1996

Miyajima Hiroshi, *Koreika Jidai no Shakai Keizaigaku* [The Sociology and Economics of the Era of Aging], Tokyo: Iwanami Shoten, 1992

Miyajima Ryu et al., *Nihon Kuroshakai* [Japan, Black Society], Tokyo: Takarajimasha, 1997

A collection of scandalous journalistic reports on foreign criminal activities in Japan.

Mouer, Ross, and Sugimoto Yoshio, *Images of Japanese Society: A Study in the Structure of Social Reality*, London, New York, Sydney and Henley: KPI, 1986

A highly analytical review of books and theories on Japanese society, challenging the *Nihonjinron* approach.

Nakao Hiroshi, *Zainichi Kankoku-Chosenjin Mondai no Kisochishiki* [Basic Knowledge about Koreans Resident in Japan], Tokyo: Akashi Shoten, 1997

Consisting of 55 explanatory answers to frequently asked questions related to Koreans in Japan.

Nihon no Ronten 1998 [Japan Talking-points], Tokyo: Bungei Shunju, 1997

This edition of a yearbook on contemporary social topics has 75 essays by eminent writers, each essay being supplemented with related data.

Nishinarita Yutaka, *Zainichi Chosenjin no Sekai to Teikoku Kokka* [The World of Koreans in Japan and the Imperial State], Tokyo: Tokyo Daigaku Shuppankai [University of Tokyo Press], 1997

This study of Koreans in imperial Japan has good sections on Japanese policies and on the economic role of Korean workers.

Oguma Eiji, *Tan'itsu Minzoku Shinwa no Kigen* [The Origins of the Myth of the Single Race], Tokyo: Shinyosha, 1995

A systematic historical analysis of the roots of the conception of the Japanese as a single "race".

Ryang, Sonia, *North Koreans in Japan: Language, Ideology, and Identity*, Boulder, CO: Westview Press, 1997

An exceptionally thoughtful and informative analysis of this section of the Korean community in Japan, written by a researcher who grew up among *Chongryun* Koreans.

Sellek, Yoko, "*Nikkeijin*: The Phenomenon of the Return Migration", in Michael Weiner, editor, *Japan's Minorities: The Illusion of Homogeneity*, London and New York: Routledge, 1997

Shimada Haruo, *Japan's "Guest Workers": Issues and Public Policies*, translated by Roger Northridge, Tokyo: University of Tokyo Press, 1994

Shindo Muneyuki, *Chiho Bunken o Kangaeru* [Considerations on Decentralization], Tokyo: NHK, 1996

Somucho Tokei Kyoku, editors, *Nihon Tokei Nenkan* [Japan Statistical Yearbook], Tokyo: Somucho Tokei Kyoku, 1996

Sorano Yoshihiro and Ko Chan-yu, *Zainichi Chosenjin no Seikatsu to Jinken* [The Lives and Rights of Koreans in Japan], Tokyo: Akashi Shoten, 1995

Sugimoto Yoshio, *An Introduction to Japanese Society*, Cambridge and New York: Cambridge University Press, 1997

A good general introduction to Japanese society, written within a multicultural framework.

Sugimoto Yoshio and Ross E. Mouer, editors, *Constructs for Understanding Japan*, London and New York: Kegan Paul International, 1989

A collection of chapters which share a critical attitude toward *Nihonjinron* literature.

Suzuki Yuko, *Chosenjin Jugun I'anfu* [Korean "Comfort Women"], Tokyo: Iwanami Shoten, 1991

Takahashi Mutsuko, *The Emergence of Welfare Society in Japan*, Aldershot: Avebury, 1997

This book places its focus on the postwar period, and especially on welfare policies and discourses from the early 1980s to the present.

Takita Sachiko, "Tan'itsu Minzoku Kokka Shinwa no Datsushinwaka: Nihon no Ba'ai" [Demythologizing the Myth of the Monoethnic State: The Case of Japan], in Kajita Takamichi, editor, *Kokusai Shakaigaku* [International Sociology], Nagoya: Nagoya Daigaku Shuppankai [Nagoya University Press], 1992

The book as a whole presents a comparative analysis of multiethnicity in different societies.

Tanaka Hiroshi, *Zainichi Gaikokujin: Ho no Kabe, Kokoro no Mizo* [Foreigners Resident in Japan: Legal Walls, Mental Trenches], Tokyo: Iwanami Shoten, 1991

Tanaka Hiroshi and Ebashi Takashi, editors, *Rainichi Gaikokujin Jinken Hakusho* [White Paper on the Rights of Foreigners Coming to Japan], Tokyo: Akashi Shoten, 1997

A review and handbook on the human rights and general situation of immigrants in Japan, including specific sections on work, nationality, residence, refugees, women, children, housing, law, health, crime, and discrimination).

Ventura, Rey, *Underground in Japan*, London: Jonathan Cape, 1992

Weiner, Michael, *Race and Migration in Imperial Japan*, London and New York: Routledge, 1994

Yamanaka Keiko, "New Immigration Policy and Unskilled Foreign Workers in Japan", in *Pacific Affairs*, Volume 66, number 1, Spring 1993

Yamazaki Fumio, *Dying in a Japanese Hospital*, translated by Yasuko Claremont, Tokyo: The Japan Times, 1996 [from *Byoin to Shinu Iu Koto*, Tokyo: Shufu no Tomo, 1990]

A book by a hospital doctor, describing and criticizing the inhuman conditions faced by terminally ill patients in Japanese hospitals.

Yano Tsuneta Kinenkai, editors, *Deta de Miru Kensei: Nihon Kokuseizukai Chi'iki Tokeihan*, dai shichi-han [Prefectures as Seen through Data: A Volume of Local Statistics on the State of the Japanese Nation, seventh edition], Tokyo: Kokuseisha, 1997

Dr Mika Merviö is Associate Professor of Political Science at Miyazaki International College, Miyazaki City.

Chapter Twelve

Urban and Regional Development

Peter J. Rimmer

Since 1950, Japan has had a Comprehensive National Land Development Law (*Kokudo Sogo Kaihatsu Ho*). The national land policy is designed to guide public investment in the construction of infrastructure, such as roads, railways, airports and seaports, to protect against natural disasters, and to conserve the natural environment. This public works policy has been variously applied to promote Japan's high economic growth, and to combat metropolitan problems, regional disparities, and deterioration of the national environment stemming from the concentration of industrial location and population in major metropolitan areas.

National Development Planning Since 1962

Regional spending by the central government on public works has been the cornerstone of Japan's postwar economic system, and has led to the country being described as a "construction state" (*doken kokka*) (see McCormack, and Sakakibara). Much of the early postwar public works planning in Japan was based on the creation of Tennessee Valley-style river basin projects emphasizing irrigation, transportation and the generation of electric power, but a distinct switch in emphasis occurred in 1962, with the introduction of the first of a succession of national development plans (see Table 12.1).

The Comprehensive National Development Plan (*Zenkoku Sogo Kaihatsu Keikaku* or *Zenso*) of 1962–68, also referred to as the First Plan, incorporated a "growth pole" strategy for the development of less developed areas. Fifteen new industrial areas and six provisional areas were designated for industrial reorganization. The New Comprehensive National Development Plan (*Shin Zenso*) of 1969–76, also known as the Second Plan, continued the policy of decentralizing industries from the major metropolitan areas by designating a fresh set of large industrial sites (see Figure 12.1). The idealistic Third Comprehensive National Development Plan (*San Zenso*) of 1977–86 sought to settle population in existing localities and further decentralize industries (see Figure 12.2). The Fourth Comprehensive National Development Plan (*Yon Zenso*) of 1987–97 aimed at developing Tokyo as a world city on a par with London and New York, while dissolving Tokyo's dominance by promoting decentralized development through a multipolar dispersed regional structure (*takkyoku busangata kokudo*), focused on four large cities (Sapporo, Sendai, Hiroshima and Fukuoka) (see NLA 1987).

Each plan has been subject to subsequent revision. Before the 1970s, the plans had to be modified because Japan's economic growth exceeded predictions. Subsequently, the plans were overhauled by additional legislation because they lagged behind or were ahead of forecasts. (It should be noted that only brief reference is made here to the relevant Japanese legislation, because it is written in general and ambiguous terms and phrases; and that, given the omnibus nature of the titles of individual laws, shortened versions are often preferred, as here.)

The Planning Structure: Regions and Cities

By the end of the Fourth Plan the key features of Japan's national land structure – the regions and its city system – had become entrenched (see Figure 12.3). For a variety of purposes, many Japanese institutions and organizations, including most government ministries, as well as the mass media, divide the country into nine regions: the three smaller "home" islands of Hokkaido, Kyushu, and Shikoku; the geographically distinct island group of Okinawa; and five regions of Honshu, the largest Japanese island. These five are:

- Tohoku (Northeastern Honshu);
- Kanto (the large plain which includes Tokyo);
- Chubu (Central Honshu, around the city of Nagoya), which is sometimes further divided into Tokai, on the Pacific coast, Tosan, the mountainous part of Central Honshu, and Hokuriku, on its Japan Sea coast;
- Kinki or, with different boundaries, Kansai (the region containing the cities of Kyoto, Nara, Osaka and Kobe); and
- Chugoku (Southwestern Honshu), which is sometimes further divided into San'yo, on its Inland Sea coast, and San'in, on its Japan Sea coast (see Association of Japanese Geographers).

The land policy has been focused on divisions which, in some areas, cut across these familiar regions, namely, the Pacific Belt; the three Metropolitan Areas of Tokyo, Nagoya and Osaka, linked together as the Tokaido Megalopolis; and the peripheries of Hokkaido, Tohoku, Hokuriku, Chugoku, Shikoku, Kyushu and Okinawa. The land policy is also based on a system of cities arranged in a five-level hierarchy: six metropolitan cities (Tokyo, Yokohama, Nagoya, Kyoto, Osaka and Kobe); four regional cities (Sapporo, Sendai, Hiroshima and Fukuoka); three other "designated" cities (Chiba, Kawasaki and Kitakyushu); seven sub-regional cities (typified by Kanazawa and Takamatsu); and six core cities.

Preparations for The Fifth Plan

In 1994 the National Land Agency (NLA) once again established the National Land Development Council to create another New National Land Policy. Undaunted by its past modest record in reshaping Japan's urban and regional development, the NLA sought to provide a new spatial structure which could respond to the anticipated era of high economic growth in East Asia. By the end of 1995, a special committee of 28 specialists had been formed by the Council's Planning Section to provide the first draft. The geographer Yada Toshifumi was vice-chairman of the technical committee handling the draft (see Yada 1996a). This draft plan, which became a blueprint for the 21st century, sought to balance the opportunities afforded by the information era and the constraints foreshadowed by critical environmental problems, international competition, Japan's declining population, and the imminence of the aging society.

In May 1996, the original committee of university professors and think tank specialists was enlarged to 50, in order to develop the interim report which was announced at the end of 1996. During the spring of 1997, the final report was discussed by the National Land Development Council. In mid-1997, the final report was formally accepted by the Cabinet. The report's broad aims are to provide good urban services and living environment, by constructing transportation and communications facilities, promoting medium-sized or small cities, and protecting the natural environment.

Before discussing the Fifth Plan (*Go Zenso*), which is being implemented from 1998 onwards, it is pertinent to address some of the issues that were considered by the Plan's five sub-committees (on environmental protection, the creation of "comfortable" cities and rural areas, the promotion of regional economies, the improvement of the quality of life, and the construction of transport and communication facilities). These issues included analyzing trends in the regional distribution of population (including the depopulation of the countryside and urban areas); developments in transportation and communications; proposals for new cities; and the impact of anti-pollution laws and policies on housing, construction, consumer protection and conservation.

Regional Distribution of Population

The most significant trend during the period of rapid economic growth, up to 1970, was the shift towards greater concentration of population and economic management functions in the large metropolitan cities – Tokyo, Osaka, and Nagoya – at the expense of nearly half of Japan's 47 prefectures. Industrialization since World War II has resulted in a decrease in the number of people engaged in primary industries and a rapid increase in the numbers of those in secondary and tertiary industries. While prefectural capitals grew rapidly, agricultural areas and their towns lost young people, even though the Japanese farmer's lot was improved by land reform, price support for rice, and mechanization. The government sought to stem migration. It increased investment in provincial areas by building industrial estates and improving the national highway system to attract firms from the metropolitan areas.

By the mid-1970s, the absolute population losses by prefectures had been stemmed, due largely to large corporations locating their factories in rural areas and branch offices in prefectural centers. This prompted some observers to identify a U-turn phenomenon: it was suggested that more people were migrating from large urban centers back to small urban centers and rural areas than in the reverse direction. However, this decline in economic disparity, evident from the first oil crisis of 1973 until the late 1980s, did not continue. The introduction, during a period of decreased government expenditure, of policies encouraging corporatization, deregulation and privatization favored Tokyo at the expense of other parts of the country. The Tokyo Metropolitan Area not only retained its small-scale machinery industries but developed as Japan's foremost center for research and development. A high-technology corridor across the Metropolitan Area linked Tsukuba and Yokohama via Kawasaki, home to Toshiba, NEC, Fujitsu and other leading high-technology producers.

Since the early 1980s, the regional distribution of population has reflected the transition from an economy dominated by manufacturing to an information and service economy. There was also a switch by Japanese corporations from viewing possible prospects within the Japanese archipelago to looking at opportunities overseas (see Chapters 9 and 18). Aided by a reversal in the pattern of public infrastructure investment, which has favored the capital since the mid-1980s, there has been a "monopolistic" or "unipolar" concentration in Tokyo (*Tokyo ikkyoku shuchu*) of organizations and industries (see Matsubara, and Yamamoto). In particular, there is a close association between the concentration of activities in the Tokyo Metropolitan Area and the internationalization of the Japanese economy. Befitting its status as a world city, Tokyo has attracted the headquarters of most of the important Japanese companies and the branch offices of many foreign companies, particularly those engaged in finance (see Abe).

Tokyo's new-found status has not only induced migration from provincial regions but also triggered an influx of highly skilled professionals engaged in international banking and finance (see Machimura 1992 and 1994). Most are legal foreign workers, but illegal foreign workers have also been attracted by the unskilled labor shortages (see Chapter 11). Although the majority of foreign workers are resident in the Tokyo Metropolitan Area, Koreans favor the Osaka area and the Brazilians are concentrated in major industrial areas throughout the Tokaido Megalopolis (as defined above), where most of them work in the automobile and electronics industries.

A more detailed examination of population changes between 1985 and 1995 needs to be interpreted in two stages: the economic boom (1985–90) and the period of stagnation (1990–95) that followed its collapse (see Figure 12.4). Population growth rates for the 47 prefectures during the first of these five-year periods varied between −2.7% and 9.2%, but the range narrowed in the second period to between −1.2% and 5.5%. When the prefectures are divided into quartiles, growth rates during the first period declined outwards from the core area (stretching from north of Tokyo to south of Nagoya) to peripheral areas. Elements of this pattern survived in the second period, but the core area was broken in two, and there was also a shift of prefectures in the lowest category towards western Japan. Other key features are: the "doughnut" phenomenon exhibited by Tokyo Prefecture, which is now experiencing declining population; the fall in the ranking of Osaka Prefecture; the precipitous decline of Hyogo Prefecture from the second to the fourth category, following the Hanshin Earthquake in January 1995; and the persistence of Okinawa and Miyagi Prefectures in the topmost rank of depopulated areas in both five-year periods.

These broad patterns, based on prefectural information, mask the imminence of Japan's aging society and the extent of depopulation. A survey by the NLA showed that "depopulated regions" (*kaso chi'iki*) accounted for 45% of the nation's land area in 1990 (see NLA 1991). Most of these areas are remote and difficult to reach: they include isolated mountainous regions, peninsulas, and former coalmining districts.

Depopulation of the Countryside

Japan's population is forecast to peak in 2011 and then decline. The number of municipalities in provincial areas which cannot sustain a natural population increase has risen from more than 20% in 1987 to almost 60% in the mid-1990s (MoC 1995). The contribution of outmigration due to "social factors" has declined, but depopulation still affects more than three fifths of municipalities in provincial areas. It is likely that by the early 21st century most of these municipalities will be undergoing decline. As it will not be possible for municipalities to increase their populations and amenities to the same extent, they will have to strengthen their outside links and cooperate more with each other to boost the number of visitors.

Signs of decay are already evident in provincial cities. These have stemmed from declining populations, changes in consumer behavior, and trends in the location of retail stores. More bulk shopping on weekends and holidays by working women is associated with the decline of neighborhood grocery stores and other outlets, and the increase in discount stores (MoC 1995). Shopping centers are also shifting their location from inner city areas, near railway stations, to roadside and other areas, where stores specializing in electrical appliances and men's clothing are popular.

These trends have been compounded by the "hollowing out" (*kudoka*) of the domestic production system which supported regional economies and employment. As companies have moved their production bases offshore, the processing and assembling of manufacturing in provincial regions has declined. This trend has not only weakened the economic base of provincial areas for renewed growth, but has also resulted in a decline in employment. The loss of employment at provincial plants has been much greater than in the main plants in the electrical machinery and precision machinery industries, although transportation machinery is a notable exception.

Urban Depopulation

Municipalities in the three metropolitan areas have also been affected by these trends. The proportion of municipalities in the Tokyo, Nagoya and Osaka areas that were experiencing natural increases in population had risen from below 10% in 1987 to 36% by 1994 (MoC 1995). Those experiencing a declining population due to outmigration declined from 43% to less than 40% in 1993. Most of these municipalities were within the inner cities. They were affected by the conversion of houses and apartments to offices; a fall in the numbers of families with children; and large areas of idle and underused land. Efforts were being made to identify new roles for the inner cities, and to restore their residential functions, through the supply of good-quality housing for middle-income people, together with the improvement of shopping and community facilities.

Much effort is being directed at creating housing supply within district planning projects, and directing the Housing and Urban Development Corporation and regional housing corporations to provide leasehold properties. Attention is also being given to the maintenance of collective housing and improving the urban environment. This is being done by updating and rebuilding existing "mansion" stock as part of redevelopment and infrastructure projects. The projects are associated with the creation of more convenient living space, involving the provision of barrier-free environments for the aged; various forms of child support; and facilities for health, social gatherings and waterfront redevelopment. These initiatives are based on experiments with carpooling and flexitime, and with advanced road transport systems that use optical fiber networks to improve communications between vehicles and road systems.

Transportation and Communications

The challenges posed by these population shifts and the advent of the aging society have led to major changes in transport and communications infrastructure, emphasizing high speed and comfort to meet the needs of consumers, such as more elevators at railway stations, and strategies to combat the disproportionate number of accidents involving senior citizens. These developments have been associated with the shift away from supporting industrial production and an export economy, to a greater emphasis by government ministries on the creation of an environment favorable to new Japanese and foreign business activities through privatization and deregulation. Japan National Railways (JNR) was divided in 1987 into a series of businesses being prepared for phased privatization, while much of its land was disposed of to retire long-term debts (see MoT). Six passenger companies were established – JR Hokkaido, JR East, JR Tokai, JR West, JR Shikoku and JR Kyushu – together with JR Freight. Government controls over trucking licenses, operation zones, freight charges and rates have also been deregulated. Supply and demand controls over new entrants into the taxi, bus and aviation businesses have been abolished. These initiatives, coupled with major technological developments, have significantly improved productivity in intra-city, interregional and international movements of people, goods and information.

Within Cities

Passenger transport in the three main cities has continued to increase within the "traffic ranges" of the three largest cities, although the rate of growth has slowed appreciably during the 1990s (see Table 12.2). Express railways have more or less held their own, but the patronage of buses, except in Tokyo metropolis, and that of taxis have both declined.

Serious traffic jams and accidents – those that cause more than 10,000 deaths a year – and congestion of commuter trains in large cities have led to a series of initiatives to increase traffic safety and transport capacity. As there are now more than 70 million motor vehicles in Japan, attention is being paid to the development of an advanced safety vehicle. New subway lines are also being constructed with tax exemptions to attract financial support. Existing tracks are being multiplied to reduce the congestion rate on commuter trains to 150% (with an interim target of 180% for Tokyo). Bus systems are also being revitalized, with newly designed vehicles, exclusive bus lanes, integrated control systems, and improved bus stop facilities. Where local bus routes are indispensable to the daily life of local areas they are being subsidized. Taxi pools are being encouraged for travel between housing developments in outlying suburbs and the nearest railway stations.

Recognition of the importance of logistics has led to increased attention to the movement of goods within cities. Various cooperative measures are favored by the government, including the establishment of joint collection and delivery points to smooth road traffic in urban areas; the transfer of privately owned trucks to commercial use; and, where possible, greater use of railways and river boats. By consolidating cargo transport in physical distribution bases it is possible to use more commercial trucks with higher load efficiencies.

Between Regions

There have been both an absolute increase and a marked modal shift in domestic passenger traffic between 1975 and 1995 (see Table 12.3). Automobiles have increased their share from 38% to more than 50%, largely at the expense of buses, the railways (both JR and non-JR), and coastal shipping. Even in passenger-kilometers, automobiles have increased their share from 35% to 47%, suggesting that the railways in particular have not retained their share of long-distance traffic. Other evidence suggests that the railways are still competitive at distances between 350 and 750 kilometers, with road transport dominating distances shorter than these, and air transport those that are longer.

As for interregional trunk passenger railways, the Ministry of Transport (MoT) is still promoting the construction of five sections on three lines of the Shinkansen ("bullet train") service. There are plans for extending the Tohoku Shinkansen from Morioka to Aomori, and adding sections of a Hokuriku Shinkansen and a Kyushu Shinkansen. Attention is also centered on speeding up connections between the Shinkansen and connecting lines. New rail lines will offer direct routes to ski fields, resorts and theme parks. Experiments with magnetically levitated ("maglev") railways are still continuing. Larger and faster ships are also planned, to provide intra-island ferry services capable of traveling more than 65 kilometers an hour, particularly between ports in Kyushu and Inland Sea ports in Shikoku and Honshu.

Interregional cargo distribution is becoming more diversified (see Table 12.4). With the rising standard of living and structural changes in industry, there are more requests by domestic shippers for services handling a greater range of items, smaller lots and more frequent services. Distribution costs in Japan are extremely high: the costs of shipping cargo from Kanagawa Prefecture near Tokyo to Hyogo Prefecture in western Honshu are almost the same as for shipping cargo from Japan to Rotterdam. Consequently, efforts are being made to improve the efficiency of long-distance trunk road transport by promoting "intermodal" transportation, involving better combinations of different modes of transport, and the greater use of palletized cargo in truck transport. Modal shifts are envisaged involving the use of container and "piggy back" rail services which can carry trucks; and more efficient vessels, such as the Techno Superliner, a new type of super-high-speed ship which runs at double the speed of conventional ships (50 knots) and has a capacity of 1,000 tonnes (see Ports and Harbors Bureau). These planned developments, giving rail and coastal shipping an edge at distances of more than 500 kilometers, are being dovetailed with the construction of strategic bases for physical distribution equipped with advanced information processors for labeling and other purposes.

Increasingly, information flows hold the key to urban and regional development within Japan as they involve both economic and social ties. Larger cities have a greater volume of telephone traffic than their population scale suggests, a disproportionate share of extra-regional traffic, and greater diversity in origins and destinations. A three-tiered structure is evident in the patterns of intraregional communications (see Murayama). First, prefectural core cities transmit more information to other prefectural cities than they receive. Second, six regional core cities – Tokyo, Osaka, Nagoya, Fukuoka, Sapporo and Sendai – transmit more information to the nearest prefectural core cities than they receive, with extraregional communications predominating in the Tokaido Megalopolis. Third, Tokyo, the national core city, transmits a greater volume of information to regional and prefectural core cities than it receives. The transmission of information from higher-order to lower-order centers is a distinctive feature of the Japanese national system of cities. It is this which underlies Tokyo's dominance over newspapers, television, publishing, and other mass media.

Between Japan and the Rest of the World

Ocean-going shipping accounts for 99.8% of total Japanese trade; Japanese imports and exports account for one fifth of the world's volume of freight. To stem the decline in Japanese-registered ships, the MoT has decided to treat them as international ships for taxation purposes (see MoT). Considerable efforts have been made to reduce the costs of container services, in order to prevent the shift of cargoes to ports in other parts of East Asia, by smoothing the linkages between the international and domestic distribution channels, and building deepwater international container terminals, with 15-meter draft, at gateway ports in Tokyo Bay, Ise Bay, Osaka Bay and northern Kyushu (see Ports and Harbors Bureau). The development of the international division of labor, and the opening of Japanese markets to imports of manufactured goods (particularly machinery and equipment) and of agricultural products, have also led to the development of local ports for international distribution. These include 22 "foreign access zones" (see Figure 12.5) and integrated air cargo facilities. International parcel and removal services are also growing quickly.

Cargo facilities have also been incorporated in the "Three Big Projects": the expansion of the New Tokyo International Airport at Narita Airport; the completion of the offshore expansion of the Tokyo International Airport at Haneda; and the Kansai International Airport, also offshore, opened in 1994. Studies are also being conducted for a Chubu New International Airport and a metropolitan airport as part of the program of internationalizing local airports for use on short-range and medium-range routes (for example, to and from Korea). Mega-Float, an ultra-large floating structure, has also been designed for the construction of earthquake-resistant airports offshore. These airports, however, are primarily designed to accommodate the increasing numbers of Japanese traveling overseas (now approaching 17 million a year) and of foreigners visiting Japan – a reflection of the fact that Japan accounts for around 10% of the world's "GNP" and is the world's largest creditor nation. The MoT is expanding airports, not only through bilateral agreements with other countries, but also by introducing competition between airlines, by permitting multiple entry onto international routes by Japan Air Lines and All-Nippon Airways.

Proposals for New Cities

The Japanese tradition of planning and building new towns was revived during the Meiji period (1868–1912), with the laying out of Sapporo and Ashikawa, both in Hokkaido, on the checkerboard pattern common in the United States. However, the establishment of full-scale new cities after World War II did not commence until after the introduction of the first Comprehensive National Development Plan in 1962. Since then, each new Plan has designated a series of new cities to correct regional imbalances throughout the country. At the same time, new cities have been developed to accommodate the spillover of population from the three metropolitan areas of Tokyo, Osaka and Nagoya.

Fifteen "new industrial cities" throughout Japan, and six special areas near the three major industrial centers, were designated under the first Plan (*Zenso*), but only three of the new cities – Mizushima (Okayama Prefecture), Oita (Oita Prefecture) and Kashima (Ibaraki Prefecture) – were successful in attracting investment. Simultaneously, new dormitory cities, in which public enterprises provided low-cost residences, were connected by commuter railways to the parent metropolitan cities. Tokyo was the mother city for Tama Denen (commenced in 1963), Kitachiba (1966), and Tama (1968). Kohoku (also 1968) was linked to both Tokyo and Yokohama. Osaka was the mother city for Senri (1961) and Semboku (1964), and Nagoya for Kozoji (1965). Specialist settlements were also established near Tokyo, including the research and university city of Tsukuba (1965), comprising laboratories and colleges (see Dearing), and Narita (1969), serving the New Tokyo International Airport.

In 1969, the Second Plan (*Shin Zenso*) designated six mega-industrial complexes (*kombinato*) on greenfield sites in remote areas, such as Mutsu-Ogawara in Aomori Prefecture. These schemes were reinforced by Prime Minister Tanaka Kakuei's unofficial "Plan for the Remodeling of the Japanese Archipelago" (*Nihon Retto Kaizo-Ron*), issued in 1973, which proposed taking industries from the major cities and relocating them in the regions (see Figure 12.6). Tanaka's Plan was undermined by the first oil "shock" in 1973, and the changeover from manufacturing to information and services. Reflecting the changed economy, plans were made in 1978 for Kansai Science City, a cultural, academic and research center started by academics and led by corporate investors.

In 1979, the Third Plan (*San Zenso*) established a set of "living spheres" in which local initiative and self-reliance were prominent.

As this concept was rejected by politicians, several high-technology cities were designated throughout Japan following the enactment of the Technopolis Law in 1983. Reflecting the Tsukuba prototype, 26 technopolises were established to provide strategic bases for regional economic development. They provided attractive urban settings for high-technology industries, such as semi-conductors, computers, and office and communications equipment, as well as universities, research institutes, and residential life. They were located near mother cities of 200,000 to 300,000 people, and were expected to attain populations of 40,000 to 50,000.

In 1987, the Fourth Plan (*Yon Zenso*) recognized the limitations of technopolises, and construction contractors sought a new component in addition to waterfront development, urban redevelopment, freestanding residential areas, and research parks. This led to the Resort Law, enacted in 1987, which permitted resorts within 24 prefectures; by June 1990 the government had approved 24 proposals for resort areas, out of 40 submitted (see Figure 12.7). Although many of the proposals for resorts or theme parks were speculative, a few had prospects of becoming resort cities, including the replica of a Dutch city at Huis Ten Bosch in Nagasaki Prefecture (see Rimmer 1992, and McCormack).

These resorts have been complemented by the implementation, by single ministries or groups of ministries, of policies seeking to develop future cities, in league with private developers. These schemes have included:

- "intelligent cities", proposed by the Ministry of Construction (MoC), which incorporate high-level information and communication networks – the Japan Regional Development Corporation is constructing Akita New City, Nagaoka New Town, Iwaki New Town, Tottori New City, Kibi Kogen City and Miyazaki Academic City (see MoC 1994);
- "research core", "high-tech" and "soft park" projects proposed by the Ministry of International Trade and Industry (MITI);
- "teletopias", "hi-vision cities" and "new media communities", proposed by the Ministry of Posts and Telecommunications (MPT), in which workers operate from satellite offices or their homes;
- local pilot schemes, proposed by the Ministry of Home Affairs (MHA), for diversification and core cities (*chukaku-toshi*) with populations greater than 300,000 and areas greater than 100 square kilometers;
- mega-commercial developments by MITI, the MoC and the MHA; and
- foreign access zones under the auspices of MITI, the MoT, the MoC, and the Ministry of Agriculture, Forestry and Fisheries.

Since 1992, following the collapse of the "bubble economy", there has been an emphasis by most ministries on integrated development (*chiho-kyoten-toshi-chi'iki no seibi-keikaku*). The aim is to create "regional node development for cities", based on converting existing regional industries into more advanced and valued enterprises, and attracting and fostering new high-technology industries. Collectively, MITI, the NLA, and the Science and Technology Agency have also placed emphasis on the role of research universities as urban bases for regional technological development. All of these efforts at developing new towns are intended to overcome the "Tokyo Problem".

Relocating the Capital

In November 1990, both Houses of the Japanese Parliament passed a resolution calling for the relocation of the capital, because without planning restraints Tokyo's exponential growth has produced a sprawling, frustrating and unsustainable city (see Cheung). As the linchpin of Japan's political, business and cultural infrastructure, Tokyo has arguably become too crowded to function efficiently. Increasingly, Tokyo has been geared to serving the world economy rather than Japan's regional areas. Instead of redeveloping Tokyo as a "super capital" (*kaito*) to accommodate the expansion of central government offices, the idea is to disperse the capital's functions.

Various suggestions have been put forward for this dispersal, and they may be classified into six categories (see Figure 12.8). First, the "capital development" theory involves leaving the main functions in Tokyo but relocating a set of connected functions within the National Capital Region (*tento*), as typified by the shift of research and technological activities to Tsukuba Science City. Second, advocates of the "dual capital" (*choto*) approach suggest that another city should be developed as a backup city for Tokyo in emergencies. Third, partial capital transfer (*sento*) would involve a piecemeal relocation of functions. Fourth, there is a proposal for the removal of all capital functions to a new capital city (*shinto*), more than 60 kilometers from Tokyo. Fifth, there is also a more radical proposal, known as *kakuto*, for the zoning of the capital to include Tsukuba, Tokyo, Kofu, Nagoya and Osaka, based on a super high-speed, central linear motor express offering a 60-minute trip between Tokyo and Osaka. Finally, others have suggested dividing up the capital city (*bunto*) by decentralizing political and administrative functions to regional core cities throughout Japan.

Among all these proposals, the relocation of the capital (*shinto*) is most widely favored. In 1992, the Investigative Committee for the Relocation of the Capital laid down a set of criteria for locating the new capital. They included:

- securing a location on the Japanese archipelago;
- choosing a site between 60 and 300 kilometers from Tokyo;
- providing a new international airport within 10 years; supplying 9,000 hectares of real estate;
- offering security from disasters, especially earthquakes, but also other disasters such as heavy snowfall;
- being at a low altitude above sea level; and
- being distant from other large cities.

Thus, a completely new capital is likely to be located in the northern part of the Kanto region or in the southern part of the Tohoku region, centered on the southern parts of Tochigi and Fukushima Prefectures.

This acknowledged solution will need to be combined with the reform of both central and local governments, which has also been the subject of various proposals. These include plans to reallocate certain powers of the central government, to existing prefectural and municipal authorities; to regional bureaus (*chiho-cho*) in the existing regions, to a reformed set of regions (*chiho-sei*); or to states in a federal system. The generally preferred option would be the devolution of central power to a reformed set of regions, as proposed by Ohmae Ken'ichi, as the basis for a "United States of Japan" by 2005 (see McCormack).

However, there is also much resistance to the plans for relocating Tokyo from local members of Parliament (especially in the governing Liberal Democratic Party) and from members of the Tokyo Metropolitan Assembly. Critics have argued that the location of the central government in Tokyo is not even a primary factor in Tokyo's growth. They argue that much of Tokyo's anticipated growth could be accommodated by the further reclamation of land from Tokyo Bay.

Measures Against Pollution

Progress has been made in combating the severe pollution generated during the period of high-speed growth in the late 1950s and 1960s, which created local nuisances, such as high noise levels, and severely affected people's health. This included pollution-linked illnesses such as mercury poisoning from the Chisso Corporation plant (Minamata disease), cadmium poisoning (*itai itai* disease) and pulmonary disorders (see Gresser et al.). As the extra-juridical settlement of disputes over pollution broke down, victims sought recompense under the Civil Code. Local citizens' activism on behalf of pollution victims, often led by women, resulted in the enactment of the Basic Law for Environmental Pollution Control in 1967, to establish environmental quality standards and draft pollution prevention measures for highly industrialized areas. The Ministry of Health and Welfare, however, continued to be responsible for industrial waste disposal.

The Environment Agency was formed in 1971, but its powers over environmental protection, air, water and noise pollution, and nature conservation are limited because it can only tender its legislation through ministries which are often growth-oriented. The Nature Conservation Law was enacted in 1972 to prevent the destruction of outstanding natural features and in 1973 the Law for the Compensation of Pollution-Related Health Injury was passed to ensure the polluter pays for any damage caused to the community.

This legislation had some success in mitigating the worst effects of pollution – for example, by decreasing sulfuric acid emissions – particularly when the measures were coupled with remarkable improvements in overall energy efficiency. These developments have included a range of technological innovations to control air pollution and permit continued economic growth. Energy consumption was drastically reduced by improved measurement and control devices, better waste heat collection systems and changed production systems, such as combined cycle power plants and increased automobile efficiency. During the 1970s, however, the energy crisis and Japan's efforts to maintain its prominence in international trade led to a retreat on the implementation of domestic environmental policies. This included a reduction of standards for nitrous oxides and the rapid development of nuclear energy.

Several intractable problems still remain, including air pollution by nitrous oxides, stemming from automobile exhaust emissions, in major urban areas; water pollution from household effluent; and waste disposal. Since the mid-1980s, there has been a slackening of interest in energy efficiency following the lowering of oil prices. There is continued reliance on oil, increased use of natural gas and coal, and, despite opposition, growth in use of nuclear power. The rash of resort developments during the late 1980s also created a new threat to the natural environment. These problems were encapsulated in the well-documented nature of golf course development during the "bubble" era (see Rimmer 1994, and McCormack). At best, these issues only generated weak national movements against the strong development lobby. Since the late 1980s, the construction of dams has generated more opposition from non-governmental organizations (NGOs), in particular for the Kamo River near Kyoto, the Shimanto River in Shikoku, and the Nagara River from the Japan Alps to Ise Bay.

There has also been a slowly growing recognition that Japan's environmental problems are not domestic but global in scope. They now encompass marine pollution, transboundary movement of acid rain and hazardous waste, global warming, and ozone layer depletion (Japan produces 10% of the world's volume of chlorofluorocarbons). These problems have led to cooperation with countries in the Asia-Pacific region which are in close proximity to Japan, and the Sixth Environment Congress for Asia and the Pacific (ECO ASIA 97) was held in Kobe. In December 1997, Japan hosted the third session of the Conference of the Parties (COP3) to the UN Framework Convention on Climate Change (UNFCC), popularly known as the Kyoto Conference, which was designed to stabilize greenhouse gases in the atmosphere at a level which would prevent anthropogenic interference with the climate system. In redressing this problem, the Japanese government has adopted a "green initiative" with developing countries. This program offers these countries both Japan's "green technology", which has had significant success in reducing greenhouse gas emissions (such as low emission vehicles), and "green aid" to integrate environmental considerations into economic plans.

Recognition of the transnational nature of Japan's environmental problems has also led to the revision of the Basic Law for Environmental Pollution Control and part of the Nature Conservation Law. In November 1993, a new Basic Environmental Law was enacted which embodied a new set of principles and directions for environmental policy. By December 1994, the Basic Environment Plan (*Kankyo Kihon Keikaku*) was introduced. This sought to address urban pollution, loss of the natural environment within urban areas, and degradation of forests and farmlands. The Plan's long-term objectives, however, refer to the importance of recycling, harmonious coexistence, participation, and international activities. In 1996, the Japan Council for Sustainable Development was established with representatives from government, industry and NGOs. In 1997, Japan enacted its first Law for Environmental Impact Assessment.

The Basic Plan embodies Japan's response to international criticism of the environmental impacts of its economic policies. Subsequently, Japanese overseas development aid has been increasingly devoted to construction projects related to the environment. Some of these projects have led to conflicts with local people. Consequently, NGOs in Japan have been lobbying for at least 20% of aid to be devoted to biodiversity, women's health care and empowerment, child health care, and basic human needs. Nevertheless, Japan is experiencing pressure from the international environmental community over driftnets, imports of ivory, whaling, and the exploitation of Southeast Asia's tropical rainforest (see Dauvergne).

Policies on Housing Construction

The immediate goal of housing policy after World War II was to complement investment in transportation and communications in order to help Japan recover from a state of devastation and to protect against natural disasters. Subsequently, the nature and direction of housing improvement linked to the four postwar National Comprehensive Development Plans was changed to mirror shifts in Japan's economic development and growing affluence.

The affluence is reflected in sewerage/utility diffusion ratios between 1963 and the mid-1990s. Ownership of electric washing machines increased from less than 70% to almost 100%; of private baths from 60% to 90%; of flush toilets from less than 10% to 70%; and of sewerage systems from less than 10% to 49% (see MoC). The average floor space per house increased from 74 square meters in 1968 to almost 92 square meters in the mid-1990s. Smaller cities lag behind larger ones in sewerage systems, sidewalk improvement and access to expressways.

The level of social capital improvement in Japan is still low compared with other industrialized nations. The state of housing and the living environment (such as parks), especially in the larger cities, has yet to be improved. Increasingly, investment in meeting this shortfall has shifted towards qualitative affluence, with more attention being paid to the environment, landscaping and social welfare.

Following the Hanshin Earthquake of January 1995, housing and social infrastructure policies were recast to accommodate heightened awareness of the vulnerability of large cities to disasters, particularly given their high concentration of wooden private rental houses. Much emphasis has been given to schemes which would ensure that cities have disaster-proofed urban structures.

One planned departure from existing policy is the abolition of the Housing and Urban Development Corporation. Originally established to provide low-income housing, the subsidized Corporation has been building houses which are more expensive and more inconveniently located than those provided by the private sector. The private sector, however, has been buffeted by the bad-debt problem of major banks left behind by the failure of seven housing loan companies (*jusen*). Public funds are being devoted over 15 years to writing off these debts. Ostensibly, the government's intention is not to safeguard the banks but to maintain Japan's financial settlement system, protect depositors, and avoid panic. Within this context, the government is planning a drastic reform of the banking system, modeled on Britain's liberalization of the securities markets in 1986.

Consumer Protection

In 2001, a "Big Bang" is scheduled for Japan (see Takenaka, and also Chapter 8). Phased in from April 1, 1998, the new policy will go beyond securities to liberalize all categories of financial services, so that Tokyo can transform itself into another New York and restore its status as one of three global financial centers. A policy to protect credit customers will be incorporated in the proposed Big Bang legislation. Hitherto, credit businesses have been regulated by a specific law because the authorities concerned are segmented. None of these laws is applied to banks. Consequently, consumers or customers are not legally protected but bound only by agreement through the law regulating banking businesses. Although the asset-inflated banks lent money during the "bubble economy" to those ineligible for loans, consumers were not protected against their losses. The new legislation proposed before the Big Bang will not be applied retrospectively to relieve past financial victims. Traditionally, the banks dominated customers, and clients had no real input into protecting themselves.

Consumer protection, and the delivery of products to consumers through wholesalers and retailers (including advertising), are coordinated by the Economic Planning Agency (see Rakugakisha). The Agency seeks to deal with growing complaints over the signing and canceling of contracts, sales methods (involving, for example, unjustifiable premiums and misleading representation), and product quality, especially goods with safety problems. However, it is very expensive for Japanese consumers to use the court system to vindicate their rights under the Product Liability Law 1994. There are a number of industry compensation schemes which provide an alternative to the judicial vindication of rights. Whenever the International Trade and Industry Bureau within MITI upholds complaints, MITI itself offers advice to the enterprise and industrial organization involved. This self-regulatory system may help to account for the low rate of litigation in Japan.

In 1995, a new Product Liability Law was enacted which lessened the burden on consumers seeking redress for injury or damage caused by defective products. Under the new law the plaintiff does not have to prove intent or negligence. The Law, however, does not apply to unprocessed foods. Generally, product liability laws in Japan are still less stringent than those in the United States.

Energy Conservation

Energy conservation has been a primary aim of Japanese governments (see also Chapter 4). Following the first oil crisis in 1973, both government and industry undertook conservation efforts which quickly reduced energy use ranging from better insulation to the development of alternative energy technologies. Since the late 1980s, however, there has been some relaxation of the laws on energy consumption and conservation laws, for example, through a reduction of tax on large automobiles. Efforts to increase consumption of imported goods, partly at the behest of the US government, has increased energy use, for example on electric bread-makers, full-sized refrigerators, and microwave ovens. There is also greater use of disposable goods, such as chopsticks and paper towels. Showers have been substituted for traditional baths, and larger houses are more costly to heat and cool. As the shift from labor-intensive industries to imported goods has been slower than anticipated, industrial growth and higher consumer spending have accelerated energy use.

The New National Land Policy

The Japanese government now faces a dilemma. Collecting funds centrally for expending on the regions has not changed. In the fiscal year 1996–97 expenditures by the prefectures exceeded central expenditure by a ratio of 2.3 to 1 (see Sakakibara). More than 36% of regional expenditure was earmarked for public works. Even efforts to protect the natural environment are still based on a catalogue of prospective construction projects, such as eco-roads, eco-ports, water-affinity green belts, marinas and esplanades. Having evolved as a "construction state" which makes extensive use of regional spending on public works, where should Japan go next? Given the devitalization of so many of the prefectures, the task of the land development planners was to find new ways of restoring them.

The Fifth Plan, "The Grand Design for National Land in the 21st Century", was introduced in 1998, and intended for implementation up to 2010, in the expectation of a period of high growth in East Asia. The underlying philosophy of the planners was that an improved quality of life, among all the regions of Japan, could only be achieved through equal opportunities in education, culture, employment, health and welfare. In adopting this philosophy the Fifth Plan seeks to counter: the agglomeration of 70–80% of the population, GNP, and cultural and educational facilities along the "First National Axis", the Pacific Belt stretching from the Tokyo Metropolitan Area to northern Kyushu, and encompassing both Osaka and Nagoya; and the disproportionate share of international exchange functions, notably international airports and seaports, and leading research and development functions, in the three metropolitan areas.

In meeting these twin goals the planners have devised a twofold strategy for creating a new regional structure (*chi'iki kozo*). The Plan's first strategy involves the construction of three new "National Axes" (*kokudo jiku*) in peripheral areas which will receive priority in public investment (see Figure 12.9). The three axes run from Hokkaido to the Tokyo metropolitan area; from Nagoya to Okinawa via the Ki'i Peninsula; and from Shikoku and the southern part of Kyushu to the Japan Sea. These axes will be complemented by "regional axes" (*chi'iki jiku*). Public investment in these corridors will be directed to improving city functions and intra-city highway networks, and to conserving the natural environment.

The second strategy is to form three "New Nodal Regions": Hokkaido-Tohoku, Chugoku-Shikoku, and Kyushu-Okinawa. The concept of the nodal region was fostered by Yada Toshifumi (see 1996b and 1996c). These regions will be the target of investments in new transportation and communications between local cities which are aimed at attracting international exchange functions, such as international airports, seaports, and research and development facilities. Once these New Nodal Regions have been established, people living in these peripheral regions will be able to engage more easily in interchanges with people in China, Korea and the Russian Far East.

Japan's land development planners have not forsaken their preoccupation with public works. Essentially, the Fifth Plan's aim is still to make the countryside an urban replica, by extending the Shinkansen networks and building more airports. In particular, the Linear Chuo Shinkansen is designed to strengthen the First National Axis, along the Pacific Industrial Belt. There is no hint that the "construction state" is about to be superseded by an "environmental state". The major conscious switch in policy is towards facilitating closer integration with East Asia. By exposing the Japanese to circumstances where they have to act internationally, the planners are presuming that they will rediscover their local identities.

Further Reading

Abe Kazutoshi, "The Status of Tokyo in Japan From the Standpoint of High-order Urban Function", in *Geographical Review of Japan*, Series B, Volume 63, number 1, 1990

Association of Japanese Geographers, editors, *Geography of Japan*, Special Publication No. 4 of the Association of Japanese Geographers, Tokyo: Teikoku-Shoin, 1980

A rather dated collection, but it still provides a series of useful articles on Japan's physical, historical and economic geography.

Cheung, C., "Decentralization of Power and Resources from Tokyo to the Regions: A Review of Issues and Proposals", in *The Review of Tokuyama University*, number 36, 1992

Chiba Tasuya, "Labour Migration to Japan on an International Scale from Asia and South America", in Flüchter, cited below, 1995

Dauvergne, P., *Shadows in the Forest: Japan and the Politics of Timber in Southeast Asia*, Cambridge, MA: MIT Press, 1997

A penetrating account of the effects of Japanese development on the Southeast Asian environment.

Dearing, J. W., *Growing a Japanese Science City: Communication in Scientific Research*, London and New York: Routledge, 1995

A solid account of the cloning of Tsukuba City as a planned scientific community.

East Asia Analytical Unit (EAAU), *A New Japan?: Change in Asia's Megamarket*, Canberra, ACT: Department of Foreign Affairs and Trade, 1997

An exhaustive compilation of material on Japan, originally aimed at providing assistance to Australian businesses interested in investment.

Flüchter, W., editor, *Japan and Central Europe Restructuring: Geographical Aspects of Socio-Economic, Urban and Regional Development*, Wiesbaden: Verlag Harrassowitz, 1995

A useful collection of papers by German and Japanese geographers. Most of the statistics refer to the early 1990s.

Gresser, J., Fukikura Ko'ichiro and Morishima Akio, *Environmental Law in Japan*, Cambridge, MA: MIT Press, 1981

A classic account of environmental pollution in postwar Japan.

Kokudocho [National Land Agency], *Kokudo Repoto 95* [National Land Report 95], Tokyo: Okurasho Insatsu-kyoku [Ministry of Finance Printing Office], 1995

McCormack, Gavan, *The Emptiness of Japanese Affluence*, Armonk, NY: M.E. Sharpe, 1996

A well-written critical account of contemporary issues in Japan.

Machimura Takashi, "The Urban Restructuring Process in Tokyo in the 1980s", in *The International Journal of Urban and Regional Research*, Volume 16, number 1, 1992

Machimura Takashi, *Sekai Toshi Tokyo no Kozotenkan: Toshi Risutorakucharingu no Shakaigaku* [World City – Restructuring of Tokyo: Sociology of Urban Restructuring], Tokyo: Tokyo Daigaku Shuppansha [University of Tokyo Press], 1994

A solid sociological account in Japanese of the nature of Tokyo's development.

Matsubara Hiroshi, "Internationalization of the Japanese Economy and its Impact on Japan", in Flüchter, cited above, 1995

Ministry of Construction (MoC), *White Paper on Construction in Japan 1994*, Tokyo: Ministry of Construction Research Institute of Construction and Economy, 1994; *Press Release: White Paper on Construction in Japan 1995: Highlights and Outlines*, Tokyo: Ministry of Construction, 1995

Ministry of Transport (MoT), *The Annual Report on Transport Economy*, translated by the Japan Transport Economics Research Center, Tokyo: Ministry of Transport, 1996

Mizuoka Fujio, "The Disciplinary Dialectics that Has Played Eternal Pendulum Swings: Spatial Theories and Discontructionism in the History of Alternative Economic and Social Geography in Japan", in *Geographical Review of Japan*, Series B, Volume 69, number 1, 1996

Murayama Yuji, "Regional Structure of Information Flow in Japan", in *Jimbun Chirigaku* [Human Geography], number 20, 1996

National Land Agency (NLA), *The Fourth Comprehensive National Development Plan*, Tokyo: National Land Agency, 1987; *The White Paper on the Depopulation of Japan*, Tokyo: National Land Agency, 1991

Ports and Harbours Bureau, *Long-term Port Policy of Japan: Ports in the Age of Global Exchange*, Tokyo: Ports and Harbors Bureau, 1996

Rakugakisha, *Yo no Naka Konatteru – Ryutsuhen* [What's What in Japan's Distribution System], Tokyo: PHP Institute, 1988; English-language edition, *Japan Times*, 1989

Rimmer, Peter J., "Urban Change and the International Economy: Japanese Construction Contractors and Developers under Three Administrations", in M. E. Smith, editor, *Pacific Rim Cities in the World Economy: Comparative Urban and Community Research*, Volume 2, London: Transaction Publishers, 1988

Rimmer, Peter J., "Japan's 'Resort Archipelago': Creating Regions of Fun, Pleasure, Relaxation and Recreation", in *Environment and Planning A*, number 24, 1992

Rimmer, Peter J., "Japanese Investment in Golf Course Development: Australia-Japan Links", in *International Journal of Urban and Regional Research*, Volume 18, number 2, 1994

Sakakibara Eisuke, "Moving Beyond the Public Works State", in *Japan Echo*, Volume 25, number 1, 1998

Statistics Bureau, *Japan Statistics Yearbook, 47th Edition (1998)*, Tokyo: Statistics Bureau, Management and Coordination Agency, 1997

This is the latest edition of the hardy annual which offers a useful starting point for studying Japan. It needs to be augmented by the annual reports of ministries and agencies, which can now be explored on the Internet (see below).

Statistics Bureau, *Wagakuni Jinko no Gaikan* [An Overview of the Population of Japan], 1995 Population of Japan Analytical Series No. 1, Tokyo: Statistics Bureau, Management and Coordination Agency, 1997

Takenaka Heizo, "Prospects for the Japanese 'Big Bang'", in *Japan Echo*, Volume 24, number 3, 1997

Tanaka Kakuei, *Building a New Japan: A Plan for the Remodeling of the Japanese Archipelago*, translated from *Nihon Retto Kaizo-Ron*, Tokyo: Simul Press, 1973

Yada Toshifumi (1996a), "Ajia no Jidai ni Okeru Nihon no Kokudo Seisaku" [Japan's National Land Policy in the Era of Asia], in *Keizai Chirigaku Nempo* [Annals of the Japanese Association of Economic Geographers], Volume 42, number 4, 1996

Yada Toshifumi (1996b), *Chiikijku no Riron to Seisaku* [Theory and Strategies of Regional Axes], Tokyo: Taimeido, 1996

Yada Toshifumi, editor (1996c), *Kokudo Seisaku to Chi'iki Seisaku* [National Land Policy and Regional Strategy], Tokyo: Taimeido, 1996

Yamaguchi Takeshi, "National Policies and the Urban System: The Japanese Experience", in *Proceedings of the Department of Humanities, College of Arts and Sciences, University of Tokyo*, number 84, Human Geography Series 9, 1986

Yamamoto Kenji, "Role of Government and Private Sectors in Regional Economic Development in Japan", in Flüchter, cited above

In addition, the annual reports of ministries and agencies, in Japanese and in English, can be explored on the Internet, through the site of the Official Residence of the Prime Minister of Japan (http://www.kantei.go.jp/index-e.html). Other useful sites are the Ministry Gateways (http://www.mha.go.jp/eng/indexb11.html), and the Japan Information Network (http://www.jinjapan.org/index.html).

Peter J. Rimmer is a Professor in the Department of Human Geography, the Research School of Pacific and Asian Studies, at the Australian National University, Canberra. The author is indebted to Professors Kumagai Keichi of Ochanamizu University, Matsubara Hiroshi of Tokyo University, Mizu'uchi Toshio of Osaka City University, and Okuno Shii of Tokuyama University, as well as to Dr Roger Farrell of the Japanese Embassy in Canberra and the Asian Studies section of the National Library in Canberra, for their assistance in locating relevant materials. Yeh Chien-wei has assisted with translation. Elanna Lowes has provided research assistance. The figures were drawn by Jenny Sheehan of the Cartographic Unit in the Research School of Pacific and Asian Studies at the Australian National University.

Table 12.1 National Land Planning Laws and Other Key Legislation, 1950–98

Year	Legislation
1950	Comprehensive National Land Development Law
1951	Special Region Development Planning Law
1962	Comprehensive National Development Plan (to 1968)
1962	New Industrial Cities Construction Promotion Law
1964	Law for Promotion of Special Areas for Industrial Facilities
1969	New Comprehensive National Development Plan (to 1976)
1972	Industrial Relocation Promotion Law
1974	National Land Use Planning Law
1977	Third Comprehensive National Development Plan (to 1986)
1983	Technopolis Law
1987	Fourth Comprehensive National Development Plan (to 1997)
1985	Capital Improvement Law
1987	Resort Law
1988	Regional Industrial Upgrade Law
1988	Multipolar Distribution Law
1992	Regional Base Cities Law
1998	Fifth Comprehensive National Development Plan (to 2010)

Source: Translated and adapted from Yada Toshifumi, "Ajia no Jidai ni Okeru Nihon no Kokudo Seisaku" [Japan's National Land Policy in the Era of Asia], in *Keizai Chirigaku Nempo* [Annals of the Japanese Association of Economic Geographers], Volume 42, number 4, 1996, p. 26

Table 12.2 Passengers Carried Within the Traffic Ranges of the Three Largest Cities, by Type of Transport, selected financial years 1975–94 (millions)

Tokyo Metropolis traffic range (within 50 kilometers of Tokyo JR Station)

	1975	1980	1985	1990	1994
JNR/JR[1]	4,066	3,938	4,283	5,109	5,373
Other railways	3,594	3,929	4,265	4,878	5,132
Subways	1,761	2,021	2,370	2,768	2,809
Buses	2,509	2,257	2,116	2,197	2,057
Hired cars and taxis	821	891	892	820	769
Private cars	3,074	3,965	4,731	6,936	7,664
Total	15,875	17,046	18,700	22,752	23,844

Chukyo traffic range (within 40 kilometers of Nagoya JR Station)

	1975	1980	1985	1990	1994
JNR/JR[1]	199	185	198	271	290
Other railways	493	485	478	496	500
Subways	260	304	319	352	377
Buses	579	491	421	388	346
Hired cars and taxis	142	151	144	146	116
Private cars	1,409	2,012	2,213	2,978	3,332
Total	3,085	3,639	3,781	4,637	4,966

Keihanshin traffic range (within 50 kilometers of Osaka JR Station)

	1975	1980	1985	1990	1994
JNR/JR[1]	1,148	1,086	1,074	1,229	1,308
Other railways	2,410	2,508	2,575	2,715	2,886
Subways	759	813	960	1,157	1,135
Buses	1,207	1,129	1,039	1,020	983
Hired cars and taxis	445	447	445	445	363
Private cars	1,858	2,457	2,745	3,677	3,710
Total	7,975	8,555	8,903	10,302	10,440

1 Japan National Railways up to 1987; regional Japan Railways companies thereafter

Source: Statistics Bureau, *Japan Statistics Yearbook, 47th Edition (1998)*, Tokyo: Statistics Bureau, Management and Coordination Agency, 1997, p. 382. No explanation is given in this source for the errors in the totals.

Table 12.3 Volume of Passenger Traffic, by Type of Transport, selected financial years 1975–95

	Buses	*Cars*	*JNR/JR*[1]	*Other railways*	*Ships*	*Air*	*Total*
Millions of passengers carried							
1975	10,731	17,681	7,048	10,540	170	25	46,195
1980	9,903	23,612	6,825	11,180	160	40	51,720
1985	8,780	25,899	6,941	12,048	154	44	53,866
1990	8,558	36,204	8,358	13,581	163	65	66,929
1995	7,619	37,777	8,982	13,648	149	78	68,253
Billions of passenger-kilometers							
1975	110	251	215	109	6.9	19	710.9
1980	110	321	193	121	6.1	30	781.1
1985	105	384	197	133	5.8	33	857.8
1990	110	576	238	150	6.3	52	1,132.3
1995	97	665	249	151	5.5	65	1,232.5

1 Japan National Railways up to 1987; regional Japan Railways companies thereafter

Source: Adapted from Statistics Bureau, *Japan Statistics Yearbook*, *47th Edition (1998)*, Tokyo: Statistics Bureau, Management and Coordination Agency, 1997, p. 374

Table 12.4 Volume of Freight Traffic, by Type of Transport, selected financial years 1975–95

	Motor vehicles	*Rail*	*Coastal shipping*	*Air*	*Total*
Millions of tons carried					
1975	4,393	181	452	0.2	5,026
1980	5,318	163	500	0.3	5,981
1985	5,048	96	452	0.5	5,597
1990	6,114	87	575	0.9	6,776
1995	6,014	77	549	1.0	6,643
Billions of ton-kilometers					
1975	130	47	184	0.2	360
1980	179	37	222	0.3	439
1985	206	22	206	0.5	434
1990	274	27	245	0.8	547
1995	295	25	238	0.9	559

Source: Statistics Bureau, *Japan Statistics Yearbook*, *47th Edition (1998)*, Tokyo: Statistics Bureau, Management and Coordination Agency, 1997, p. 374

Figure 12.1 Japan as the Spatially Integrated National Economy Envisaged Under the New Comprehensive Development Plan (*Shin Zenso*), 1969

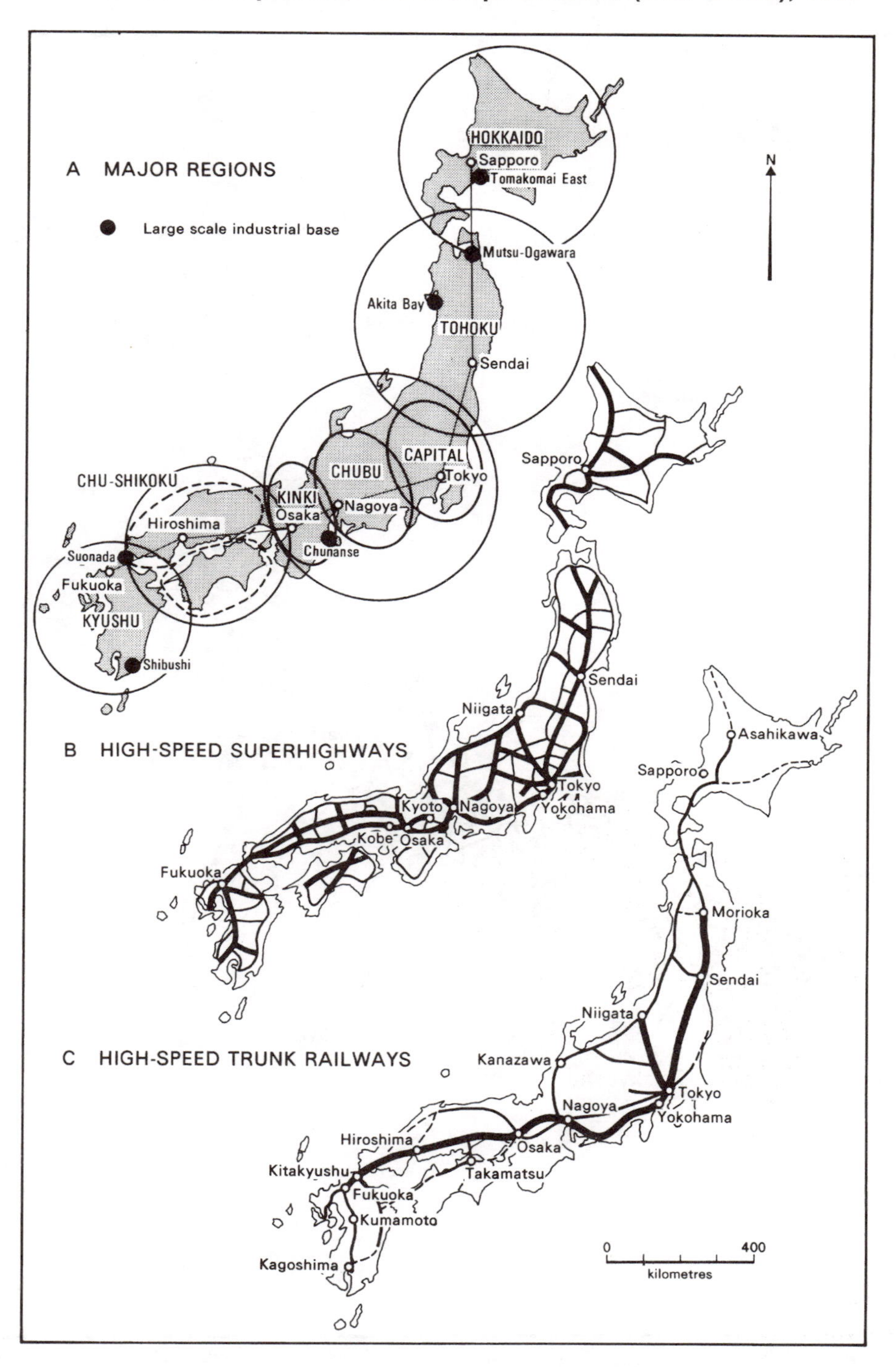

Source: Peter J. Rimmer, "Urban Change and the International Economy: Japanese Construction Contractors and Developers under Three Administrations", in M. E. Smith, editor, *Pacific Rim Cities in the World Economy: Comparative Urban and Community Research*, Volume 2, London: Transaction Publishers, 1988

Figure 12.2 Self-governing Units and Major Core Regions Established Under the Third Comprehensive National Development Plan (*San Zenso*), 1977

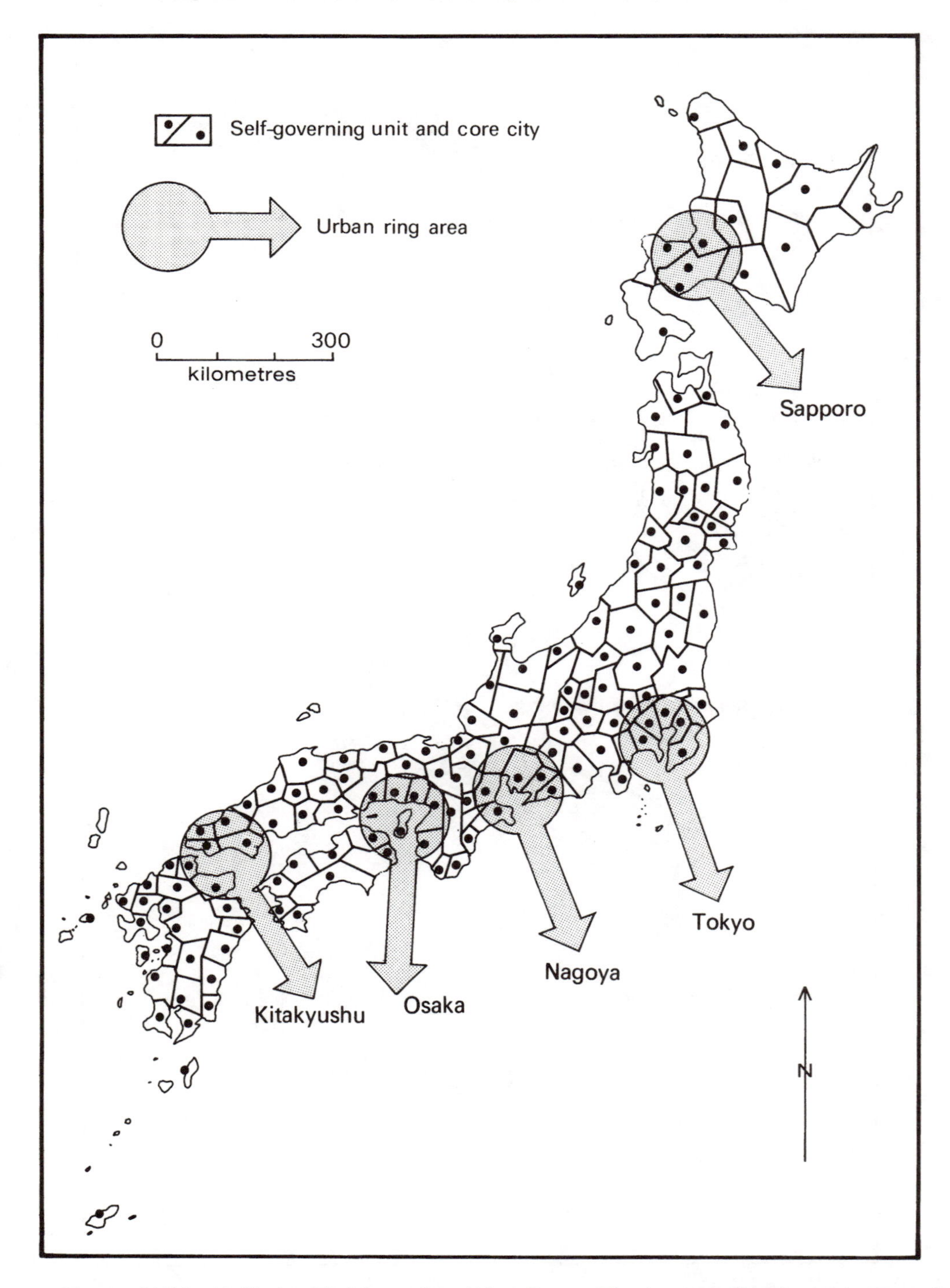

Source: Yamaguchi Takeshi, "National Policies and the Urban System: The Japanese Experience", in *Proceedings of the Department of Humanities, College of Arts and Sciences, University of Tokyo*, number 84, Human Geography Series 9, 1986

Figure 12.3 Model of Japan's National Land Planning Structure

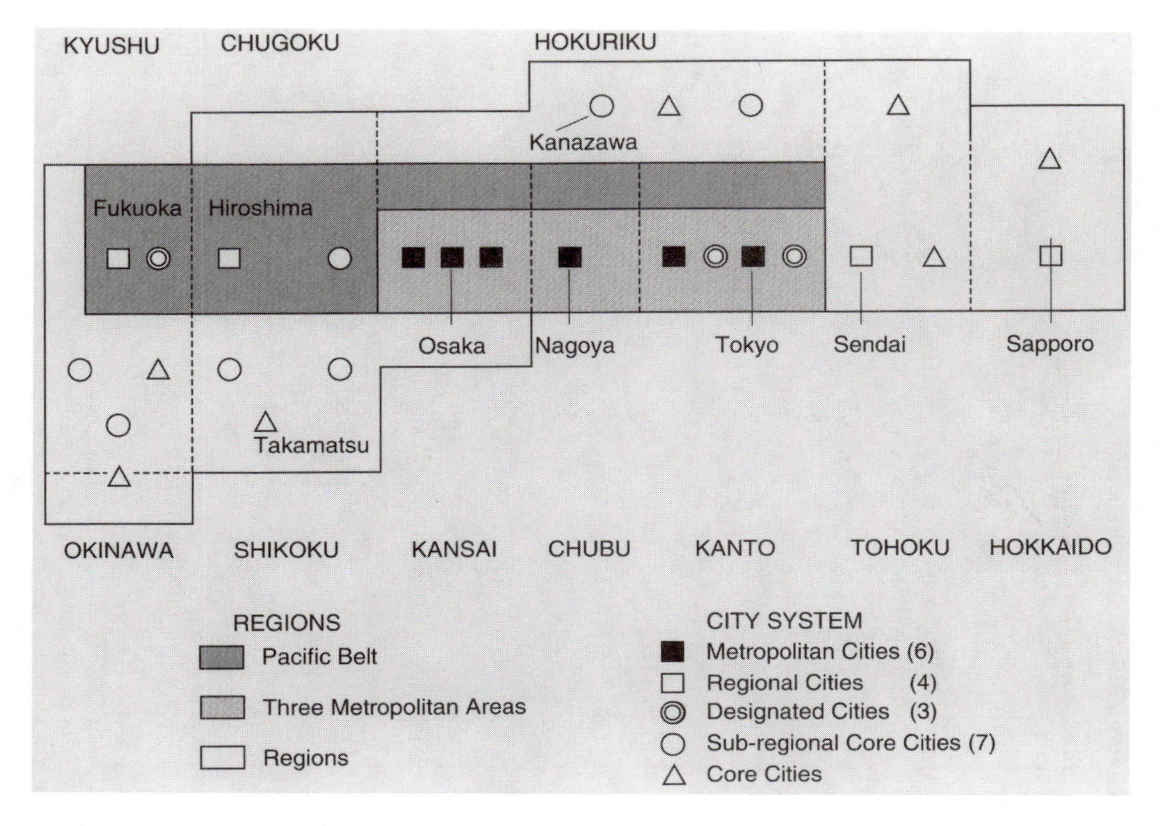

Source: Yada Toshifumi, editor, *Kokudo Seisaku to Chi'iki Seisaku* [National Land Policy and Regional Strategy], Tokyo: Taimeido, 1996

Figure 12.4 Changes in Prefectural Populations Between 1985–90 and 1990–95

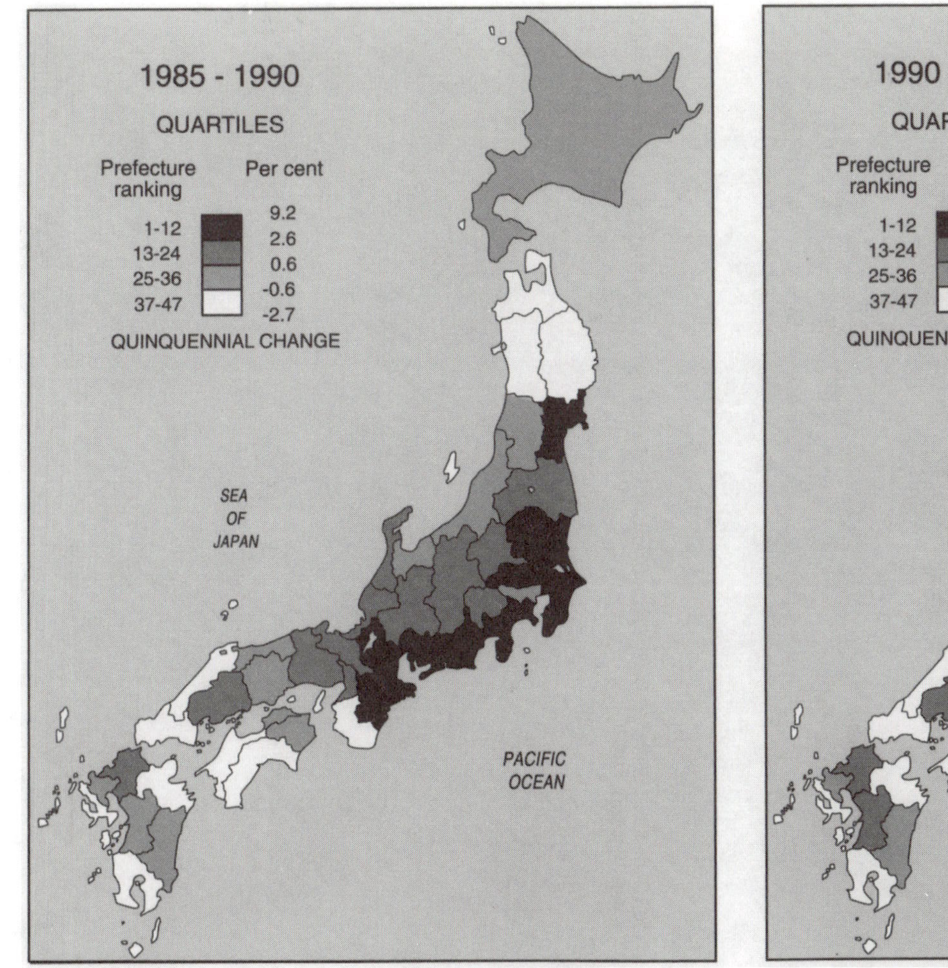

Source: Based on data from Statistics Bureau, *Wagakuni Jinko no Gaikan* [An Overview of the Population of Japan], 1995 Population of Japan Analytical Series No. 1, Tokyo: Statistics Bureau, Management and Coordination Agency, 1997

Figure 12.5 Designated Foreign Access Zones, as of 1996

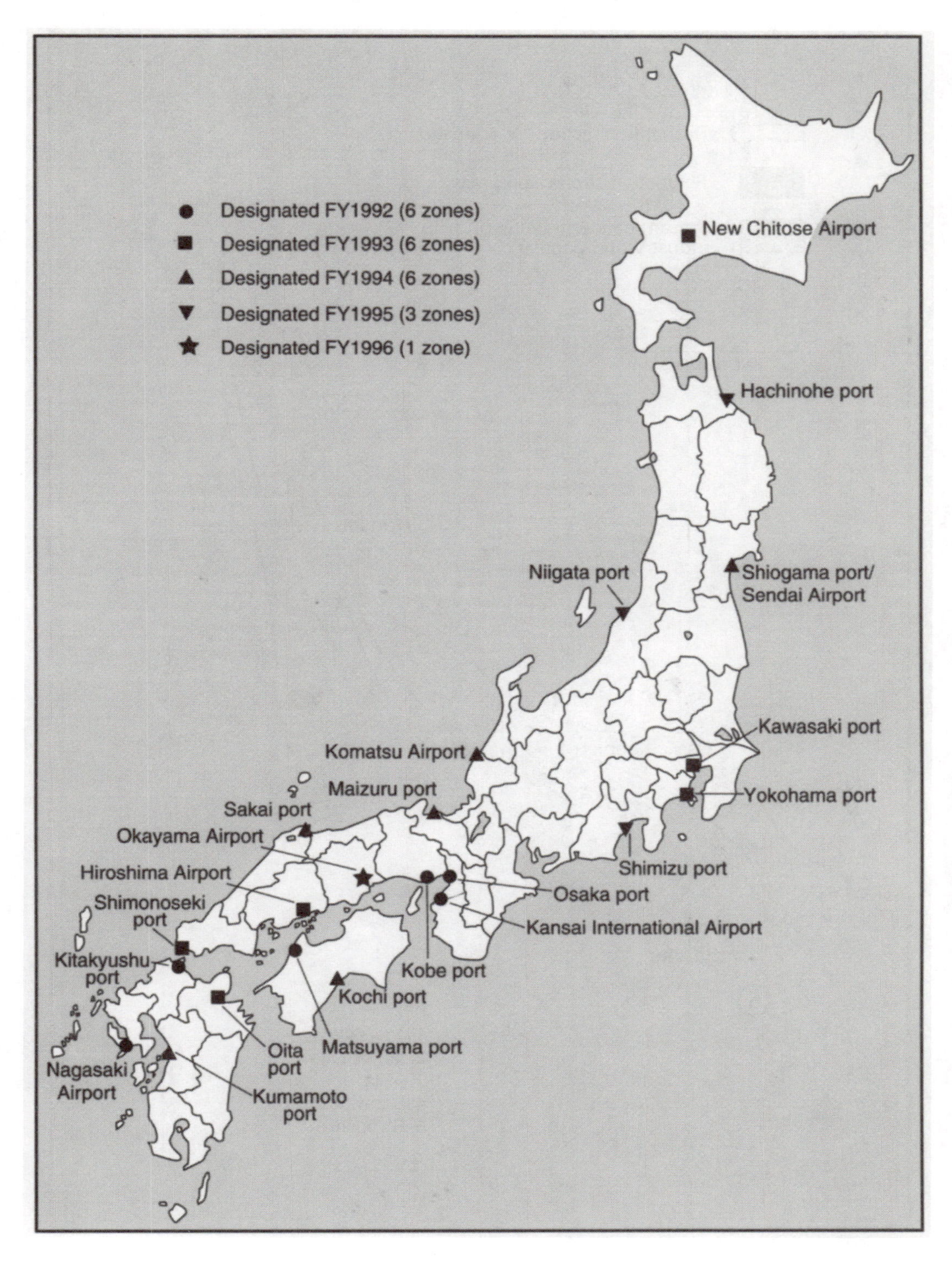

Source: Ministry of Transport, *The Annual Report on Transport Economy*, translated by the Japan Transport Economics Research Center, Tokyo: Ministry of Transport, 1996

Figure 12.6 Tanaka Kakuei's Proposed New Industrial Pattern, 1973

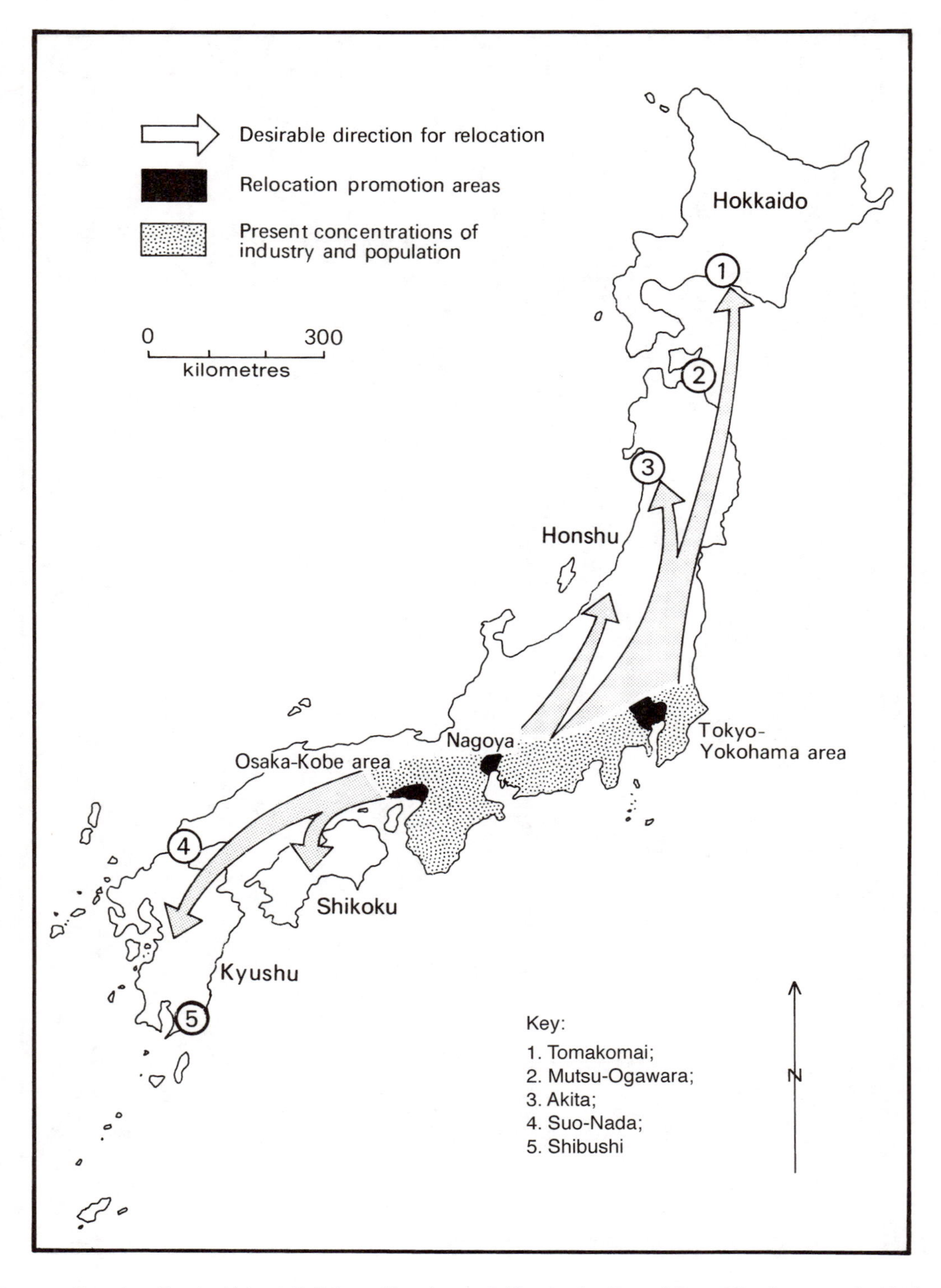

Source: Based on Tanaka Kakuei, *Building a New Japan: A Plan for the Remodeling of the Japanese Archipelago*, translated from *Nihon Retto Kaizo-Ron*, Tokyo: Simul Press, 1973

Figure 12.7 Distribution of the 24 Resort Areas Designated up to June 1990

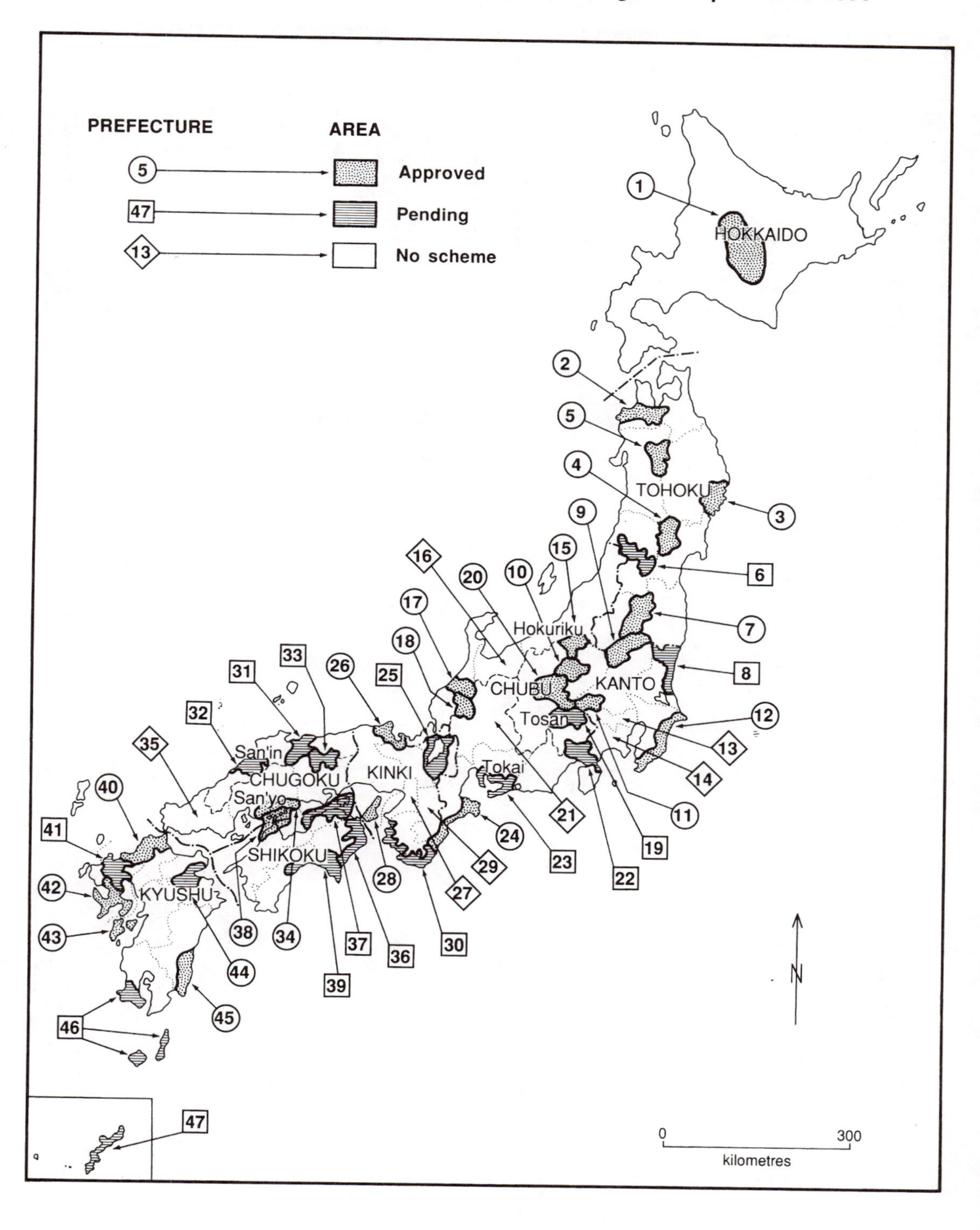

Source: Rimmer, Peter J., "Japan's 'Resort Archipelago': Creating Regions of Fun, Pleasure, Relaxation and Recreation", in *Environment and Planning A*, number 24, 1992

Figure 12.8 Options for Relocating the Capital

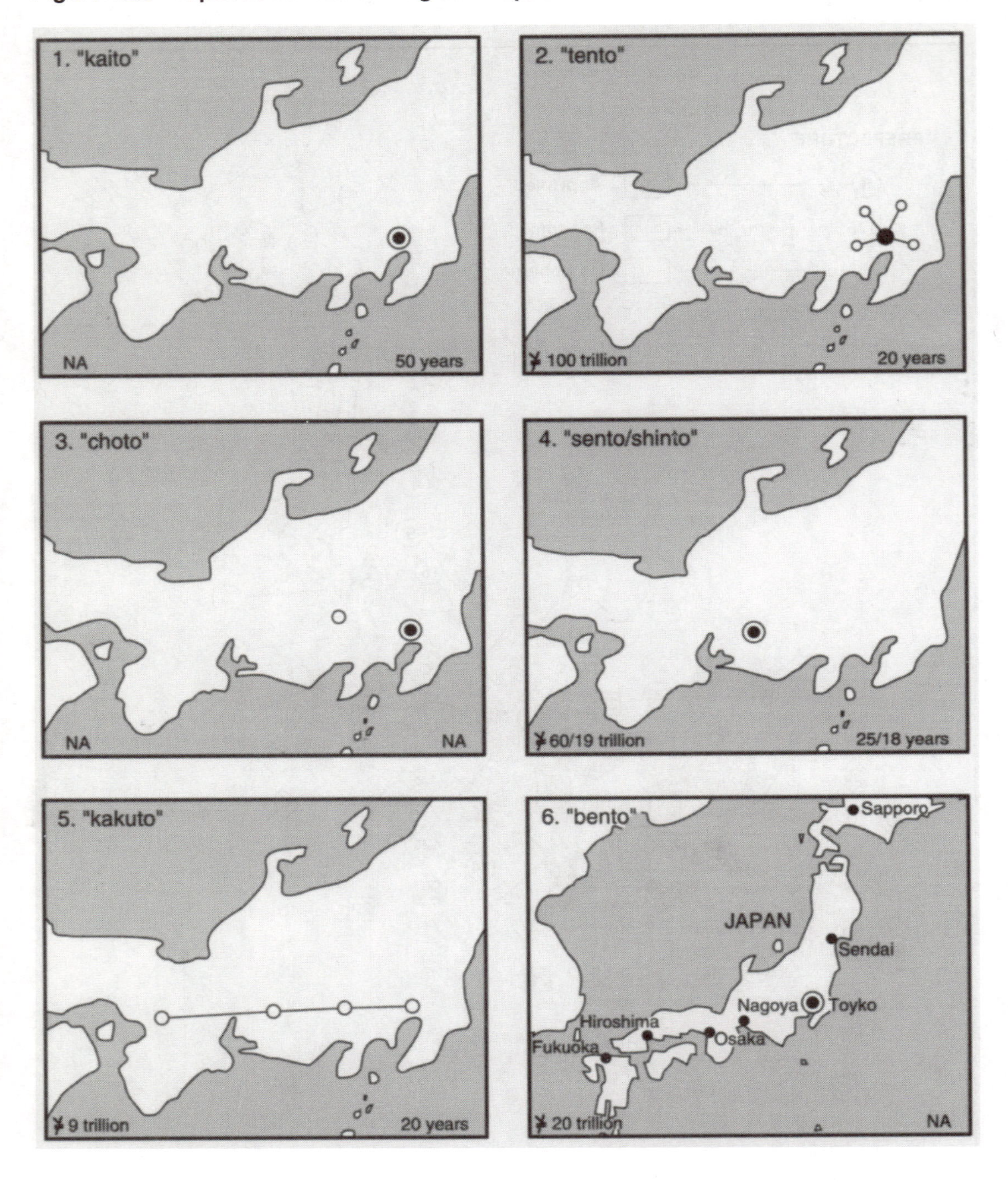

Key: 1. Complete redevelopment of Tokyo to contain new capital; 2. Spread of government offices to fringe of Tokyo; 3. Establishment of a second or emergency capital; 4. Move capital to a new location at least 60 km from Tokyo; 5. Spread capital along the one hour Tokyo-Osaka route linked by the planned Linear Express; 6. Move political and administrative functions to regional core cities throughout Japan

Source: Cheung, C., "Decentralization of Power and Resources from Tokyo to the Regions: A Review of Issues and Proposals", in *The Review of Tokuyama University*, number 36, 1992, p. 100

Figure 12.9 Schematic Diagram of Multiple Axes and Nodal Regions Under the Fifth Plan (*Go Zenso*)

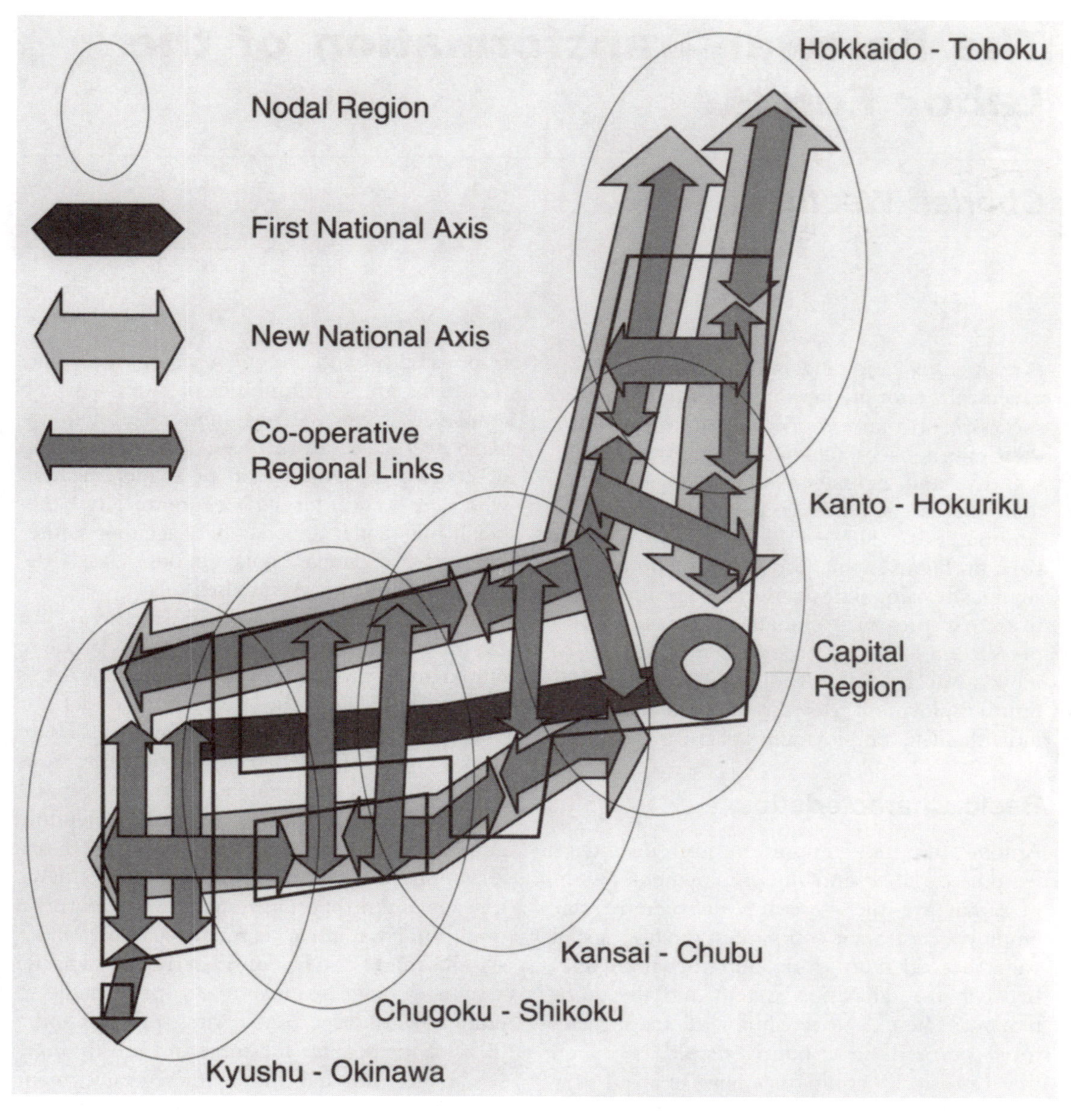

Source: Kokudocho [National Land Agency], *Kokudo Repoto 95* [National Land Report 95], Tokyo: Okurasho Insatsu-kyoku [Ministry of Finance Printing Office], 1995, p. 171

Chapter Thirteen

The Postwar Transformation of the Labor Force

Charles Weathers

A cooperative and diligent labor force has been a central factor in Japan's postwar economic success. For many workers, the rewards for their efforts have included strong employment security and management's recognition of their status as full members of "enterprise communities". However, critics have emphasized problems such as long working hours, and significant inequalities between large and small firms. At present, "globalization" and other pressures are leading businesses and workers to move away from the traditional ideal of "lifetime employment" toward more flexible and individualistic employment practices.

Basic Characteristics

Among the most important and distinctive features of labor and the employment system in Japan are the close ties between regular employees and their companies; the high social value accorded to work; and the close links between the education system and the labor market. These factors cannot all be satisfactorily covered here, but it should be noted that historically companies have enjoyed very high prestige, as agents of modernization and as sources of social mobility. To a greater degree than in almost any other country, an individual's status has been closely linked to that of his or her company – in general, the larger the firm, the higher the status. As a result, class consciousness has been relatively weak and status distinctions have been pronounced. Managers have had considerable success in equating loyalty and work with social responsibility, and in encouraging the belief that firms are "communities of fate" in which employees have a strong material and moral stake (see Clark, and Rohlen). Many Japanese, of course, recognize that close identification with corporate interests is economically beneficial, but that it also has had negative consequences for social policy-making and the quality of life (see Watanabe).

The close ties of workers to firms are expressed in the concept of "the Japanese employment system", which, under various names, has been the dominant model for understanding employment practices. There has been heated debate about whether it is beneficial to, or exploitative of, employees. The three main features of the system are lifetime employment, more accurately termed long-term employment; strong seniority components in wage and promotion systems; and enterprise unionism. Extensive company-based welfare systems might also be considered integral to the system. These features apply principally to employees of larger enterprises, perhaps some 30% of the regular full-time and mostly male workforce. Although the Japanese employment system has important roots in prewar and wartime practices, it basically took its contemporary form in the early 1950s (see Gordon 1985), and became firmly established under the favorable conditions of high growth, from 1955 through the early 1970s.

As Nakamura Keisuke and Nitta Michio have suggested, the lifetime employment and seniority systems "are normative assumptions mutually shared by management and workers rather than a set of objective rules and practices"; and these practices have endured largely because they are "flexible or vague enough to be adjusted to [a] changing environment". In fact, only about 20% of employees spend their entire careers with one employer. However, strong social norms and legal precedents bring pressure on firms to protect livelihoods, and, when "employment adjustment" becomes necessary, to use "soft" measures, such as hiring freezes or overtime cutbacks, over the "hard" measures of (so-called) voluntary early retirement or outright dismissal.

In principle, the seniority-based wage and promotion systems function as follows, especially in production industries. The promise of steadily rising compensation and status motivates employees to remain with their firms, despite the low pay received by younger workers. Since firms feel confident that most workers will remain over a long term, they can bear the costs of continuous training and skill development, which in turn raises the productivity and job satisfaction of the workforce. (Voluntary departures are concentrated among workers with perhaps three years of experience or less; their wages are lowest and accumulated benefits fewest.) Since most job training is conducted by firms, skills are largely company-specific, making it difficult for employees to increase their wages by changing employers. It is often overlooked that employers have steadily increased the importance of merit pay and evaluations since the 1950s, so that seniority and merit pay systems essentially coexist.

Most unions in Japan are enterprise unions. An enterprise union includes only employees of a single firm, but all regular, non-managerial employees, including both blue-collar and white-collar workers, may join. Part-time and temporary employees are excluded. Membership is quasi-mandatory for regular employees in many enterprises, a situation which effectively locks out competing unions. Almost all large firms are unionized, but the organization rate is very low and falling in small and medium-sized enterprises (SMEs), defined as companies which have fewer than 300 employees, or which are capitalized at less than ¥100 million (see Table 13.1). Since unions are company-based, collective bargaining or consultation procedures usually take place at the level of the firm.

Enterprise unionism has been at the center of the debate on whether Japanese industrial relations practices are primarily beneficial, or exploitative of, workers (see Gordon 1993, Shirai 1983, and Shimizu). A major advantage of enterprise unions is that they are well-positioned to understand both the needs of employees and the financial position of the company when conducting collective bargaining. They are also effective at promoting communication between labor and management. On the other hand, enterprise unions are inherently vulnerable to management pressure, since they are inclined to make the well-being of the company their main priority. They rarely enjoy significant external support since industrial unions in Japan mainly provide coordinating functions for their enterprise union affiliates. It should be noted that the debate on industrial relations has been more or less superseded by a debate on Japanese production practices (see, for example, Elger and Smith), in which the context is global, and emphasis has shifted from unions to management.

There are distinct differentials in income and working conditions between large firms and SMEs (as defined above). Compensation in SMEs is generally around 70% of that in large firms, the hours are longer, and working conditions are generally inferior. Many SMEs are subcontractors, their employees being often subject to the strains of working odd hours to produce to the demanding specifications of larger firms. Generally, improvements in working conditions in larger firms gradually influence those in SMEs, but the timelag can be considerable. The 40-hour working week, for example, will be implemented in all SMEs in April 1999 at the earliest.

Despite these differentials by size of firm, Japan's pay structures are relatively egalitarian, largely because differentials between employees and managers are relatively small. The major determinants of an individual's income are the size of the firm, that individual's level of education, and his/her gender. Although wage-setting takes place yearly within enterprises, many firms coordinate their settlements within the *shunto* ("spring offensive") framework. In *shunto*, the settlements of large firms serve as unofficial guidelines for the rest of the economy. The *shunto* wage determination system has provided the advantages of formally coordinated systems (as in Germany) but is more flexible. Japan's nominal wages are among the world's highest, but the relatively high cost of living means that purchasing power is not satisfactory.

Japan has probably the tightest connection between education system and job market in the world (see Dore and Sako). The ideal course for a firm has been to select incoming employees in their last year of school and then to nurture the development of their skills over the long term. The seniority wage system has made it practical for employers to hire relatively cheap school-leavers in order to provide long-term training. Competition to hire good school-leavers is often fierce. Until recently, schools and firms often forged close ties which provided the firms with favored access to the school's best students, and helped the schools build reputations for effective job placement. At the university level, major firms have tended to judge applicants in large part by the prestige of their universities, although hiring practices are becoming more flexible.

1945–55

Encouraging the development of a labor union movement was one of the pillars of the strategy of the Occupation authorities for democratizing Japan. The prewar labor movement had been weak, but the new ideal of democracy, combined with the loss of prestige by traditional elites, including managers, meant that the social and political environment of the early postwar years favored the rapid growth of unionism. The number of union members grew from 5,000 in October 1945 to 5 million in December 1946. The enterprise proved to be the natural base for union organization, partly because difficult conditions made it imperative for workers to support their companies to protect their jobs. At first, unions won numerous concessions, including strong job guarantees and wage systems determined largely by the workers themselves. However, the employers began to take the initiative in 1947, as economic conditions worsened and the Occupation authorities wearied of aggressive, sometimes ideological, unionism. In 1948, the government revised labor legislation to ban strikes in the public sector and generally weaken protections for unions.

In the difficult economic and political environment of the late 1940s and early 1950s, most unions in the private sector either moderated their policies, or were replaced by more cooperative unions. Often, militant "first unions" were replaced in the course of industrial disputes. One consequence of aggressive management tactics was that the organization rate, which reached a peak of 56% in 1949, declined to 46% in 1950, before stabilizing in the 35% range in the mid-1950s. It was after these years of disruption that the contemporary elements of the Japanese employment system, including new pledges of job security, began to emerge, especially in large firms. Improved job security was partly a price paid by employers to secure workers' cooperation, but it also reflected strong social norms. Relations between labor and management calmed steadily from the early 1950s, as indicated by the falling number of strikes (see Table 13.2 – and note that many strikes in the 1960s and 1970s were rather ritualistic). Much of the public sector, however, remained contentious for years.

In 1950, most major unions participated in the founding of a new labor federation, *Sohyo*. It soon became caught up in Cold War politics and turned to the left, prompting conservative (or right-wing) unions to leave and form competing federations. Although Sohyo remained Japan's largest labor federation until its dissolution in 1989, it suffered continuous decline throughout its existence. Perhaps the most decisive blow to Sohyo came with the Mi'ike mine strike in 1960. The original union confronted managers for several months, in a dispute so heated that large-scale violence was barely averted. The strike ended in total defeat for the union, but some observers believe that the near-tragedy prompted policy-makers to take more effective measures to assist the re-employment of redundant workers in declining industries.

1955–73

During Japan's era of high growth there was a dramatic rise in the numbers, educational attainments and skills of the workforce. The blue-collar workforce was reshaped by the intensive efforts of managers to bolster efficiency and competitiveness in manufacturing. Consequently, productivity in production industries increased rapidly, but surprisingly little effort was made to increase productivity elsewhere, including white-collar and service industries. As a result, there have been marked differences between the intense, competitive working conditions in manufacturing workplaces and those of offices.

The era of high growth also saw the consolidation of the Japanese employment system and of cooperative unionism. The Marxist-oriented Sohyo steadily lost private sector affiliates, becoming by the late 1960s a federation of primarily public sector unions. The more conservative federations merged to form *Domei* in 1964. Committed to cooperation with managers, Domei became a bitter ideological rival to Sohyo. The two federations were also divided in the political realm. The Japan Socialist Party (now known in English as the Social Democratic Party of Japan) was closely aligned with Sohyo, and public sector unions have contributed many of its politicians. However, the right wing of the Socialist Party seceded in 1960 to form the Democratic Socialist Party (DSP), which came to be aligned with Domei. Both parties stagnated electorally, however, so that labor has never enjoyed adequate political representation.

While Sohyo and Domei have garnered the most attention, probably more important has been the International Metalworkers Federation-Japan Council (IMF-JC). Founded in 1964, the IMF-JC soon included all the major unions from the metalworking industries – steel, shipbuilding, vehicles, and electronic and electrical products – which drove high growth in Japan and many western countries after World War II. It also included affiliates from Sohyo, Domei, and two smaller labor federations, and thus served as a symbol of unity among unions based on "pragmatic" rather than "ideological" grounds. The IMF-JC has no formal authority, but it has served as an important platform for the coordination of union policies, notably in wage-setting. It supports the principle of cooperative, enterprise-centered relations between labor and management (see Miyata).

The *shunto* system (described above) was launched in 1955 by Sohyo as a means of mobilizing workers, and overcoming the political and economic weaknesses of enterprise unionism. The workers were less militant than hoped, but both companies and unions found that *shunto* was a useful means of coordinating wage determination. Although millions of employees participated in yearly *shunto* strikes and demonstrations, it was primarily the combination of high growth, high inflation, and severe shortages of young workers which ensured high wage rises, as well as improved working conditions and job security. Following the contentious 1950s, fast-rising incomes and improving conditions helped to legitimize the new industrial relations practices, and to undermine the appeal of left-wing unionism.

Workplaces and productivity

The keys to expanding production during the era of high growth were heavy investment and massive imports of new technology; the development of new production and workplace practices; and a rapidly growing and increasingly better-educated young workforce. A large domestic market and global postwar prosperity made it possible for manufacturing firms to conduct continuous rationalization, and yet increase the overall number of employees, although it should be borne in mind that "employment adjustment" was a common occurrence even in prospering industries.

Manufacturing industries borrowed heavily from western, especially US, managerial and production techniques, often refining those techniques extensively. One of the most important examples was the quality control movement, guided in Japan by the Japanese Union of Scientists and Engineers (JUSE). JUSE's activities helped lead to the development of quality control circles (QCCs) and other forms of small group activities by the early 1960s. In addition to improving productivity, small group activities are noted for helping to promote job satisfaction among many employees (see Cole 1989). The Japan Productivity Center, established in 1955 with considerable assistance from the United States, has played a prominent role in encouraging cooperation between labor and management. Around 1960, managers also began to place increasing emphasis on merit in deciding wages and promotions in production workplaces.

Foremen have combined the roles of supervisors and personnel managers on the shopfloor. In principle, they are eligible for promotion to managerial positions, but in practice most have remained in unions as regular members and officials. Foremen were important in managing rapidly changing workplaces from the late 1950s and 1960s, as new technology transformed production and new labor techniques, such as QCCs and the "zero defects" (ZD) movement, were introduced.

The rising levels of education in Japan also contributed to the transformation of the labor force (see Table 13.3). From the late 1950s, manufacturing firms for the first time began to hire large numbers of high school graduates, rather than primarily junior high school graduates, for blue-collar jobs in manufacturing. The better educations of the workers made it possible for firms to make full use of the new technologies being rapidly introduced into factories, as well as to train foremen in the demanding new methods of labor supervision and management.

Rising education also meant higher aspirations, and high turnover among young workers became a major problem in production industries, such as vehicle manufacturing, where working conditions were difficult. Shortages of young workers became a perennial economic problem during the 1960s. Firms responded in part by steadily improving benefits and working conditions. In addition, the ability of agriculture to supply inexpensive temporary workers contributed to the very high flexibility of the labor market during the era of high growth. With increased urbanization, this flexible labor source no longer exists.

1973–91

The "oil shock" of 1973 brought the era of high growth to an abrupt end, and the subsequent deep recession created the first major challenge to the lifetime employment system. Firms managed to minimize dismissals largely through massive transfers, often to affiliated enterprises. In addition, the close cooperation of unions made it possible for managers in production industries to step up the introduction of microelectronic technology into workplaces. Consequently, firms were able to maintain productivity without hiring large numbers of expensive new workers. However, the gains in productivity generally helped Japan, unlike Europe, to maintain the level of employment in manufacturing industries. Overall unemployment rose significantly after the oil shock, to about 2.0%, but remained well below European levels (see Table 13.4).

The recession also brought a major transformation of wage-setting practices. Double-digit wage increases had become routine during the high-growth, high-inflation 1960s, but in the mid-1970s, the IMF-JC and large manufacturing firms began to emphasize wage restraint (see Shinkawa; also Table 13.5). As metalworking industries are highly sensitive to international economic conditions, the unions sought to protect jobs by restraining wage demands and strengthening competitiveness. The major metalworking firms and unions have continued to act as the main pattern-setters in *shunto* since 1975. Their preference for rather tight wage restraint has helped Japan to maintain its strong economic competitiveness, but the relatively low wage rises have, arguably, depressed internal demand and made the economy overly reliant on exports for growth. Unions in service industries have criticized IMF-JC wage policies, and long-term wage restraint has also reinforced the widespread impression that unions have little effect on wages.

Unable to secure high wage demands, unions have sought new means of protecting workers' incomes and interests since 1975. At the level of the firm, they have gained a greater voice in managerial decision-making, although the outcome is usually closer to consultation rather than actual influence. At the national level, unions began to participate actively in economic and social policy-making following the 1975 *shunto*.

The unions' participation in policy-making encouraged progress toward unifying the union movement. The major steps were taken in the private sector: the establishment by 16 major industrial unions in October 1976 of the Trade Union Council for Policy Promotion (*Seishin Kaigi*), a forum for coordination and discussion, representatives of which met regularly with ruling Liberal Democratic Party (LDP) and government officials; the establishment in December 1982 of the Japanese Private Sector Union Council (*Zenmin Rokyo*), representing around 4.2 million workers; and the transformation of Zenmin Rokyo, in 1987, into *Rengo*, initially known in English as the Japanese Private Sector Trade Union Council. Finally, in 1989 the left-wing federation Sohyo disbanded and its unions entered Rengo, the English name of which was revised to the Japanese Trade Union Confederation, completing the unification of most of the union movement. The absorption of public sector unions brought Rengo's roster to 78 industrial unions. In 1997, according to the Ministry of Labor, they had a total membership of around 7.58 million workers, who account for 62% of the nation's organized employees, but only 14% of total employees. Rengo is essentially recognized by the government as labor's chief representative.

Two smaller federations, the National Confederation of Trade Unions (*Zenroren*) and the National Trade Union Council (*Zenrokyo*), were also formed in 1989, in large part by leftists from Sohyo who were unhappy with the creation of the enlarged Rengo from a merger between the left and right wings of the union movement. In 1997, these two federations had 840,000 members and 270,000 members, respectively. Both are virtually ignored by the government and receive little attention from the media. Also in 1997, unions representing around 3.59 million workers remained unaffiliated with Rengo, Zenroren or Zenrokyo, while 41.97 million workers, or 77.4% of the workforce, were unorganized.

Japan's massive trade surpluses in the early 1980s led to the Plaza Accord of 1985, in which the Japanese government acceded to demands by irritated western trade partners to let the yen appreciate. The result was the *endaka* ("yen-high") recession of 1985–87, and another round of "employment adjustment", especially in industries with high fixed costs, such as steel. The vehicle and electronics industries began to shift sizable amounts of production offshore for the first time, although the effect of overseas investment on manufacturing employment was mitigated for several years by the heavy demand of the transplanted factories for capital equipment from Japan.

The employment structure changed substantially between 1970 and 1985, reflecting rationalization in manufacturing, the shift in economic structure toward service industries, and demographic changes. The share of employment in manufacturing remained fairly constant, at around 35%, while the share of the services sector rose from 47% to 57%. Change was most pronounced in 1980–85, with the beginning of the so-called distribution revolution and increasing "informationalization" (*johoka*) of the economy.

The two largest state-owned corporations, Nippon Telegraph and Telephone (NTT), and Japan National Railways (JNR), were nominally privatized in 1985 and 1987, respectively. The restructuring of JNR led to the near-destruction of a major Sohyo union, and signaled the virtual end of significant labor militancy in Japan. As a result of restructuring, the public sector accounted for only 6% of total employment in 1994, compared to 14% for the United States, 14.4% for Britain, and 24.8% for France (according to OECD statistics).

The 1990s

During the late 1980s, Japan enjoyed an era of tremendous prosperity. Steady growth and labor shortages pushed up real incomes and helped break down barriers to women in a number of business and professional fields. It became increasingly difficult to hire well-qualified people for "3K" work (*kiken*, *kitanai*, *kitsui*; or in English, the 3 Ds, "dangerous, dirty, difficult"), forcing many firms to reduce working hours and otherwise improve working conditions and benefits, or, in some cases, to pass on

"3K" work to immigrant laborers (see Chapter 11). There was a great deal of discussion about placing more emphasis on improving the quality of life, and reducing long working hours became a major policy issue, although one that produced little change in the long run. Working hours have indeed fallen, but largely because of labor market factors (see Foote 1997). However, the boom was followed by the deep Heisei recession, lasting from 1990 until late 1995; further, the value of the yen surged in 1995. These economic problems raised fears that Japan's high wages and its high cost structure threatened economic welfare. As a result, the national mood has shifted towards concern for protecting competitiveness and jobs.

Surprisingly for an economy deeply concerned by looming shortages of labor (primarily young workers), unemployment has hovered around 3.4% in the mid-1990s, record highs for Japan since the statistics started being kept in 1953. The ratio of openings to applicants, which is closely watched in Japan, peaked at around 1.4 in 1991 just before the boom ended, but has since been in a worryingly low range around 0.7. This is partly because companies overhired during the late 1980s and have since struggled to adjust, but it also partly reflects "structural mismatch", meaning in part the mismatch between young people's expectations or qualifications and employers' needs. Companies in white-collar industries used to spend years nurturing university-educated but non-specialized young workers. Now more specialized skills are in demand, and firms want to see results faster.

The workforce was very young in the 1950s and 1960s, which made the seniority system attractive to managers, but now that it is much older, the high costs of older workers, who are usually less productive despite continuous training, make them a liability for many firms. Not surprisingly, workers over the age of 45 bore the brunt of "downsizing" in the three major recessions, in the mid-1970s, in 1985–87, and in the early 1990s.

The Labor Market

Compared to other countries, the distinctive feature of job security in Japan is not comprehensiveness but flexibility of application. Companies are under strong pressure from social norms and legal strictures to retain personnel, but they enjoy great freedom in how they allocate those personnel (see Foote 1996). This means that transfers, for example, are regarded as standard practice, even if spouses have to live apart for extended periods (*tanshin funin*), as occurs frequently since families are reluctant for their children to switch schools. In general, EU countries provide stronger guarantees than Japan, including higher unemployment insurance and stricter rules for justifying layoffs. Most EU countries also strengthen protections for workers in line with seniority, whereas older workers are the most vulnerable in Japan. There are similarities, however, in methods used for cutting workforces in the EU, the United States, and Japan. These include cutbacks on new hires and overtime work, and, when companies are in deep distress, induced early retirements, euphemistically termed "voluntary" early retirements in Japan. In general, compared to firms in other industrialized democracies, Japanese firms reduce their workforces quite slowly in response to changes in economic conditions, and they are about average in the speed of adjustment of total working hours, but the extensive use of temporary and part-time workers enables them to adjust total real wages very flexibly (see Chuma).

In the mid-1970s, some 20% of firms, including many large firms, had to resort to the "hard" measures of voluntary early retirement and dismissal. During the *endaka* recession of 1985–87 and the Heisei recession of 1990–95, there was decreasing resort to hard measures. This probably indicates, not the strengthening of job security as such, but the refinement of personnel policies which allow flexible adjustment, including greater use of transfers, which sometimes become permanent, and increased use of irregular workers, whose employment can be easily terminated.

In recent years, the ideology of lifetime employment has lost favor with managers and many unions, who now prefer to emphasize that Japan needs greater flexibility and mobility in the labor market. This attitude reflects in part the desire of many managers to escape the burdens of providing long-term job security, but also the growing belief that practices such as seniority wages, elaborate corporate welfare systems, and retirement pay simply make the labor market rigid. Substantial retirement allowances, for example, are an incentive for employees to stay with a firm throughout their careers, and constitute a key component of the lifetime employment system. However, such changes as the increasing importance of knowledge-intensive skills are expected to make regular job changes both desirable and inevitable for many employees. Companies and unions in the vital electronics and electrical industries are moving toward major modifications of pay and work structures to accommodate the new conditions. Matsushita Electric, traditionally a pace-setter in personnel practices, is experimentally introducing a new pay system which substitutes a higher monthly wage for retirement pay. Such innovations should make it easier for employees to change jobs, and are being closely observed.

Such terms as "globalization" or "the age of super-competition" are often heard in Japan these days, as businesses strive to reduce personnel costs and improve productivity. At least two major trends are currently reshaping labor and personnel practices in this regard. First, employers are attempting to boost productivity in white-collar industries, where productivity has been notoriously low in contrast to manufacturing workplaces. Thus a shift toward more flexible pay and hiring practices is occurring. The rising importance of specialized skills means somewhat less emphasis on recruiting school-leavers, and more opportunities for changing jobs. Firms are steadily changing compensation systems, notably by making greater use of various forms of merit pay. This should make white-collar workplaces more competitive, and raise the compensation of key specialists, such as software designers, who will increasingly be "rewarded for performance" rather than tied to relatively inflexible seniority-plus-merit systems. "Yearly salary" (*nemposei*) is one widely heralded pay system, although far fewer than 10% of firms have actually introduced it so far. It basically means an emphasis on evaluation-based pay over the traditional regular monthly wage. Changes in pay systems are so far primarily incremental rather than revolutionary, but compensation is certainly becoming more individualistic and "market-oriented." This is one reason why the *shunto* wage-setting structure is quickly losing importance; this is reflected in stagnant wage settlements, despite the somewhat improved economic conditions of the mid-1990s (see Table 13.6). It should be noted that about 2% of each wage settlement is tied to seniority and is more or less automatic, so recent wage raises have approached zero, but they do not include raises tied to promotions or skill acquisition, which are considerable for many workers.

Second, deregulation of the labor market and rationalization of working conditions are proceeding rapidly, as businesses seek to reduce personnel costs and enhance flexibility. The use of part-time employees, who are mostly women, is growing. Regulations on temporary workers, contract workers, and discretionary labor (*sairyo rodo*) are being liberalized. Discretionary labor means that working hours and other conditions will no longer be regulated for an expanding range of white-collar occupations. As with changes in pay systems, this will increase corporate flexibility, and will mean heavier demands on at least some employees. Finally, corporations are being permitted to create holding companies (see Chapter 10); how employees will be protected under the new holding company structures has yet to be determined.

Women

Women have traditionally been relegated to relatively menial jobs with low pay and benefits. It was long expected of many female employees, especially in office jobs, that they would quit upon getting married, although the practice has been recognized as illegal since the 1960s. It was also long the custom for women to serve as "flowers of the workplace", making work environments more pleasant, though this practice is fading as competitive pressures force companies to cut costs. The classic model of female labor force participation is the M-shaped curve. Rates of participation are high for young women and older women, but sharply lower for women aged between 25 and 35, the main years for child-bearing and child-raising. The curve began to flatten out in 1987 because women started to marry later and the number of working mothers rose.

Career opportunities have improved, and it appears that discriminatory practices are being eroded as social values and economic conditions change. This is partly because labor shortages in the late 1980s created new professional opportunities for women. On the other hand, female university students seem to have been the most adversely affected by changes in work rules. The Equal Employment Opportunity Law of 1985 was a largely symbolic measure with little real impact, while the less noticed Worker Dispatching Law, enacted one month later, legalized private temporary employment agencies for the first time, and made it easier for companies to hire women as temporary instead of permanent employees. The number of female part-timers has also grown steadily: the ratio of part-timers among female employees rose from 31.6% in 1995 to 34.0% in 1996.

In 1992, the average female employee's wages were only 50.7% of those of male colleagues, compared to differentials of around 70–80% in other industrialized nations. There is little difference in starting wages for new young male and female employees, so the differential is due largely to relatively short terms of employment for women and the importance of seniority pay. Many women clearly prefer the freedom that comes even with lower-paying jobs, but others are still simply subject to various forms of gender discrimination.

The Aging Population

The rapid aging of the population (also discussed in Chapter 11) is one of Japan's most pressing economic concerns. The low birth rate, which in 1996 dropped to a historical low of 1.42 for the average woman over the course of her lifetime, coupled with the world's longest life expectancy, means that Japan faces long-term labor shortages, as well as high and rising welfare costs. The situation is a reversal from the era of high growth, when the population was young, and large numbers of increasingly well-educated school leavers provided high-quality, low-cost labor.

People aged 65 and over are expected to account for about 27.4% of the population by 2025. The speed of increase in the rate of aging has been far faster than in other OECD countries. The populations of Germany and Sweden are still older than that of Japan, for example, but by around 2000 the aggregate age of the Japanese population will surpass them; and it already surpasses those of the United States, Britain, and France. Policymakers are searching for means to employ more of the elderly in order to resolve the labor shortage and to ease the strain on the pension system. However, the labor force participation rate of the elderly is already the highest among industrialized nations, even after three decades of decline. Increasing the participation of the elderly will also require adjustment to the employment system, which has generally protected the jobs of younger workers at the expense of older workers. The pension system is to undergo incremental adjustment to raise the age of eligibility to begin drawing pensions from the present 60 to 64, between 2001 and 2013 (it is already about 65 in most countries). In addition, more than 93% of firms have now shifted their mandatory retirement age from the previous standard of 55 to 60, and the government hopes to extend it to 65.

Labor Representation

The establishment of the labor federation Rengo was intended to help overcome the limitations of enterprise unions in areas such as new organizing and collective bargaining, but its resources are few, despite its high profile, and most unions are not exerting themselves in these areas. The federation's main function is to guide policy-making efforts, and here it has enjoyed modest success. Rengo and its major union affiliates maintain extensive contacts in the political parties, employers' associations, and the national ministries, and they send hundreds of representatives to the deliberative councils (*shingikai*) attached to ministries, which often play important roles in formulating new policies. Rengo is credited with helping to institute legislation in areas such as child care, and assistance for the elderly and disabled, but its impact is probably limited to areas with minimal budget implications. Nonetheless, since unions do not mobilize workers or enjoy much political support, they rely heavily on policy-making participation to promote labor interests.

The major unions, although nominally united in Rengo, remain divided more or less along the old Sohyo/Domei (left/right) lines. At present, the IMF-JC and many former Domei unions generally support business on economic deregulation, which tends to boost national economic competitiveness and lower the cost of living. However, such deregulation often threatens the jobs and working conditions of employees in the public and services sectors, and sometimes in SMEs.

Almost all unions traditionally oppose the LDP, and Rengo has openly proclaimed its desire to support a progressive party able to challenge it. In 1994, Rengo played an important role behind the scenes in establishing the coalition government which briefly replaced the LDP in power, and which included both the Social Democrats and the DSP. However, the coalition soon collapsed. Presently, the former Sohyo unions maintain uneasy ties with both the Social Democrats, in support of the LDP government, and the new Democratic Party, in opposition to it. The DSP, comprising politicians who are mostly former Domei union officials, also remains in opposition. In short, the unions' program of electoral politics has been a stark failure, their participation in policy-making has not been much more successful, and neither actvity elicits much interest from workers.

Unions have suffered steady decline for two decades. The organization rate held fairly steady, at around 35%, from 1955 until 1975, but has since fallen continuously (see Table 13.7). The organization rate was 22.6% in December 1997, down from 23.2% from a year earlier. The absolute number of union members has started to decline as well, by 162,000 to 12,451,000 between 1995 and 1996. One problem is that the manufacturing workforce peaked at 15.7 million in 1992, but decline has also occurred in industries which are not being restructured, indicating that the problem is not simply one of economic transformation.

Few unions actively recruit new members, since most members of enterprise unions enter quasi-automatically, and few employees are motivated to join unions independently because they do not believe that unions have an effect on wages or working conditions (see Tsuru and Rebitzer). They are probably correct, as researchers have consistently failed to find any such union effect. Some scholars believe that this is because unions generate a strong "spill-over" effect, while others think that working conditions and unions themselves are primarily characteristics of firms. As bad as the decline in the organization rate is, more worrying to union officials is distancing from unions (*kumiai-banare*), meaning members' losing interest in union activities. One of the ways unions have tried to maintain influence and employee support has been by seeking a stronger voice in managerial decision-making, although, as mentioned above, actual influence on core policies remains limited.

The steady decline of unions in Japan has prompted a debate on the merits of mandating some form of non-union employee representation, along the lines of Germany's works councils. Two forms of non-union employee representation, joint consultation and employee associations, are already common. Joint consultation involves more or less formal committee meetings of labor and management representatives, to discuss business conditions, working conditions, and fringe benefits. Unions often gain access to confidential information during the course of consultation, but the process often leads employees to suspect collusion between unions and managers. Joint consultation is most common in unionized firms. Employee associations are widespread in both unionized and non-union firms, and, unlike unions, they may include managers as well as regular employees. About two thirds of employee associations are oriented toward recreational activities, and the remaining one third are "voice-oriented" organizations which discuss matters such as working conditions. The widespread existence of employee associations seems to indicate the importance placed by managers and many employees on maintaining a voice for workers, and a sense of participation in decision-making, but they do not appear to exert much influence over final decisions.

Conclusion

The labor market is undergoing rapid changes, as the population ages, and technology- and information-intensive industries steadily replace mass-production manufacturing. "Lifetime employment" will not exactly disappear: job security remains legally and socially entrenched, and many workers and companies will continue to value the stability it confers. However, the concept of lifetime employment will continue to lose importance as ideology or symbol, to be replaced by some emerging vision of a more flexible employment structure adapted to the increasingly knowledge-intensive and "individualistic" social and economic environment of the 21st century. There will be increasing differentials in compensation and diversity in working situations, and probably continued expansion of the non-regular workforce. The challenge for Japan, especially in the absence of assertive unionism, will be to manage these changes without allowing excessive marginalization of less favored workers, or abandoning efforts to improve working conditions.

Further reading

Aoki Masahiko, *Information, Incentives, and Bargaining in the Japanese Economy*, Cambridge and New York: Cambridge University Press, 1988

A well-known thesis on the employment system from the conservative viewpoint, using some formal modeling and emphasizing the important benefits of communications between labor and management in Japanese firms.

Brown, Clair, Yoshifumi Nakata, Michael Reich, and Lloyd Ulman, *Work and Pay in the United States and Japan*, New York and Oxford: Oxford University Press, 1997

A new comparative analysis of US and Japanese industrial relations and human resource management.

Chuma Hiroyuki, *Nihon-gata "Koyo Chosei"* [Japan-style "Employment Adjustment"], Tokyo: Shueisha, 1995

Japanese employment adjustment examined through 40 case studies, plus comparative and historical analysis.

Clark, Rodney, *The Japanese Company*, New Haven, CT: Yale University Press, and Rutland, VT, and Tokyo: Charles E. Tuttle, 1979

A close analysis of employment relations from a social perspective; excellent on the importance of the company in Japan.

Cole, Robert E., *Japanese Blue Collar: The Changing Tradition*, Berkeley, Los Angeles: University of California Press, 1971

An excellent participant-observer study of conditions in an auto parts subcontractor.

Cole, Robert E., *Strategies for Learning: Small-group Activities in American, Japanese, and Swedish Industry*, Berkeley, Los Angeles: University of California Press, 1989

A comparative analysis of small-group activities and why they are hard to emulate.

Dore, Ronald, *British Factory – Japanese Factory: The Origins of National Diversity in Industrial Relations*, London: Allen and Unwin, and Berkeley, Los Angeles: University of California Press, 1973

Dore's influential conception of convergence between Japanese and western practices; also important for its insights into factory-level industrial relations and enterprise consciousness.

Dore, Ronald, and Mari Sako, *How the Japanese Learn to Work*, London and New York: Routledge, 1989

The links between the education and employment system analyzed.

Elger, Tony, and Chris Smith, editors, *Global Japanization? The Transnational Transformation of the Labour Process*, London and New York: Routledge, 1994

A recent collection of work criticizing Japanese production practices (by both Japanese and non-Japanese firms).

Foote, Daniel H., "Judicial Creation of Norms in Japanese Labor Law: Activism in the Service of – Stability?", in *UCLA Law Review* Volume 43, number 3, February 1996

An investigation of the nature of job security.

Foote, Daniel H., "Law an Agent of Change? Governmental Efforts to Reduce Working Hours in Japan", in Harald Baum, editor, *Japan: Economic Success and Legal System*, Berlin and New York: Walter de Gruyter, 1997

An excellent case study on an important labor issue, with insights into the policy-making roles of unions and government.

Gordon, Andrew, *The Evolution of Labor Relations in Japan: Heavy Industry, 1853–1955*, Cambridge, MA: Harvard University Press, 1985

Perhaps the most widely cited source on Japanese industrial relations, both before and since 1945, although it deals primarily with the earlier period. Gordon's forthcoming book on the postwar steel industry will revise and deepen the analysis of the postwar era.

Gordon, Andrew, "Contests for the Workplace", in Andrew Gordon, editor, *Postwar Japan as History*, Berkeley, Los Angeles: University of California Press, 1993

A study of management's efforts to control postwar workplaces and shape workplace values, by a scholar sympathetic to the left; a recommended starting point.

Inoki Takenori, *Gakko to Kojo: Nihon no Jin-teki Shigen* [School and Factory: Japan's Human Resources], 1996

An historical overview of the important school-to-workplace connection.

Koike Kazuo, *Understanding Industrial Relations in Japan*, translated by Mary Saso, London: Macmillan, and New York: St Martin's Press, 1988

One of Japan's most prominent labor economists lays out his influential theses on skill formation and other topics.

Koshiro Kazutoshi and Rengo Sogo Seikatsu Kaihatsu Kenkyusho, editors, *Sengo 50-nen Sangyo-Koyo-Rodo Shi* [50-year Postwar History of Industry, Employment, and Labor], Tokyo: Japan Institute of Labor, 1995

A comprehensive source on all aspects of postwar industrial relations, and an excellent starting point for basic information and further reading in Japanese.

Kumazawa Makoto, *Portraits of the Japanese Workplace: Labor Movements, and Managers*, edited by Andrew Gordon; translated by Andrew Gordon and Mikiso Hane, Boulder, CO: Westview Press, 1996

An insightful criticism of employment and industrial relations practices by a prominent leftist scholar.

Miyata Yoshiji, *Kumiai Zakkubaran* [Union Frankness], Tokyo: Toyo Keizai Shimposha, 1982

Miyata, arguably the most influential labor leader in postwar Japan, on "cooperative unionism".

Nakamura, Keisuke, and Michio Nitta, "Developments in Industrial Relations and Human Resource Practices in Japan", in Richard Locke, Thomas Kochan, and Michael Piore, editors, *Employment Relations in a Changing World Economy*, Cambridge, MA: MIT Press, 1995

A very useful overview of the evolution of postwar industrial relations and human resources management.

Orii Hyuga, *Romu Kanri Nijunen* [Twenty Years of Personnel Management], Tokyo: Toyo Keizai Shimbunsha, 1973

One of the essential works on the formation of postwar industrial relations.

Otake Hideo, "The *Zaikai* under the Occupation: The Formation and Transformation of Managerial Councils", in Robert E. Ward and Sakamoto Yoshikazu, editors, *Democratizing Japan: The Allied Occupation*, Honolulu: University of Hawaii Press, 1987

Perhaps the best secondary source on the important, but little studied, employers' association Nikkeiren.

Rodo Sogi-shi Kenkyu-kai, editors, *Nihon no Rodo Sogi (1945–80)* [Japanese Labor Disputes (1945–80)], Tokyo: Tokyo Daigaku Shuppansha [University of Tokyo Press], 1991

A collection of articles analyzing the processes and significance of Japan's major industrial disputes.

Rohlen, Thomas P., *For Harmony and Strength: Japanese White-collar Organization in Anthropological Perspective*, Berkeley, Los Angeles: University of California Press, 1974

An excellent participant-observer analysis of work and values in a Japanese company.

Sako Mari and Sato Hiroshi, editors, *Japanese Labour and Management in Transition: Diversity, Flexibility and Participation*, London and New York: Routledge, 1997

A new collection of articles (of varying quality) on industrial relations and human resource management, for specialists.

Shimada Haruo, "Japan's Industrial Culture and Labor/Management Relations", in Kumon Shumpei and Henry Rosovsky, editors, *The Political Economy of Japan*, Volume 3, *Cultural and Social Dynamics*, Stanford, CA: Stanford University Press, 1992

A good overview of postwar industrial relations by a prominent conservative scholar; notable for its mild criticism of unions' excessive commitment to their firms' economic goals.

Shimizu Shinzo, editor, *Sengo Rodo Kumiai Undo Shi Ron* [On the Postwar History of the Labor Union Movement], Tokyo: Nihon Hyoronsha, 1982

An important collection of articles on industrial relations by leading leftist scholars.

Shinkawa Toshimitsu, "1975-nen Shunto to Keizai Kiki Kanri" [The 1975 *Shunto* and the Management of Economic Crisis], in Otake Hideo, editor, *Nihon Seiji no Soten: Jirei Kenkyu ni yoru Seiji Taisei no Bunseki* [Issues in Japanese Politics: Analyses of the Political System through Case Studies], Tokyo: Sanichi Shobo, 1984

An excellent case study of the 1975 *shunto*, a critical juncture for the Japanese industrial relations system.

Shirai Taishiro, editor, *Contemporary Industrial Relations in Japan*, Madison: University of Wisconsin Press, 1983

A widely cited collection, generally expounding the conservative view.

Shirai Taishiro, *Gendai Nihon no Romu Kanri* [Labor Management in Contemporary Japan], second edition, Tokyo: Toyo Keizai Shimposha, 1992

A useful account of the development of industrial relations and human resource management.

Tsuru Tsuyoshi and James B. Rebitzer, "The Limits of Enterprise Unionism: Prospects for Continuing Union Decline in Japan", in *British Journal of Industrial Relations*, Volume 33, number 3, September 1995

A data-based analysis of the causes of union decline.

Watanabe Osamu, "*Yutaka na Shakai*" *Nihon no Kozo* [The Structure of Japan, an "Affluent Society"], Tokyo: Rodo Jumposha, 1990

A best-selling semi-academic work studying and decrying the development of Japan's company-centered society.

Periodicals

Useful English-language periodicals include the *Japan Labor Bulletin*, a monthly publication from the semi-governmental Japan Institute of Labor, and the *Nikkei Weekly*. Among the most useful Japanese-language periodicals are *Shukan Rodo Nyusu* [Weekly Labor News], the *Rodo Hakusho*, and its abridged English-language version, *Labor White Paper*, annual publications from the Ministry of Labor. Some useful information can be found in the *Keizai Hakusho* [Economic White Paper], published by the Economic Planning Agency. The major newspapers, including *Asahi Shimbun* and *Nihon Keizai Shimbun*, are indispensable for tracking labor issues; the English-language newspapers published in Japan cover these issues in less detail. The best source for further general information is the Japan Institute of Labor. Its various publications in both English and Japanese can be investigated on line.

Dr Charles Weathers is a Postdoctoral Fellow in the Institute of Economic Research at Hitotsubashi University, Kunitachi City, Tokyo.

Table 13.1 Union Density, in Total and by Size of Firm, 1985–94 (%)

		Size of firm (by number of employees)		
	Total	*1–99*	*100–999*	*More than 1,000*
1985	24.4	2.5	28.3	64.5
1986	23.9	2.4	27.3	65.6
1987	24.0	2.3	27.4	68.0
1988	23.5	2.2	26.8	66.1
1989	22.5	2.1	25.7	62.0
1990	21.9	2.0	24.0	61.0
1991	21.4	1.8	23.3	58.7
1992	21.8	1.8	22.5	57.2
1993	21.3	1.8	22.0	58.2
1994	21.2	1.7	21.6	59.8

Source: Sako Mari and Sato Hiroshi, editors, *Japanese Labour and Management in Transition: Diversity, Flexibility and Participation*, London and New York: Routledge, 1997, p. 303

Table 13.2 Strikes Lasting Half a Day or Longer, selected years 1946–96

	Number of strikes	*Thousands of workers involved*	*Thousands of working days lost*[1]
1946	702	517	6,266
1948	744	2,304	6,995
1950	584	763	5,486
1955	659	1,033	3,467
1960	1,063	918	4,912
1965	1,542	1,682	5,669
1970	2,262	1,720	3,915
1975	3,391	2,732	8,016
1980	1,133	563	1,001
1985	627	123	264
1990	283	84	145
1991	308	53	96
1992	261	109	231
1993	251	64	116
1994	229	49	85
1995	208	38	77
1996	189	23	43

1 These figures include days lost through lock-outs by employers.

Source: Rodosho [Ministry of Labor], *Rodo Hakusho* [Labor White Paper], various years

Table 13.3 Numbers Leaving Education and Taking Jobs (and Their Relative Proportions, by Type of Institution), selected years 1955–90 (thousands and %)

	All	*Junior high schools*	*High schools*	*Universities*	*Two-year colleges*
1955	1,123	698 (62.2)	340 (30.3)	70 (6.2)	15 (1.3)
1960	1,373	683 (49.7)	572 (41.7)	100 (7.3)	18 (1.3)
1965	1,499	624 (41.6)	700 (46.7)	135 (9.0)	36 (2.4)
1970	1,370	271 (19.8)	817 (59.6)	188 (13.7)	87 (6.4)
1975	1,038	93 (9.0)	591 (57.0)	233 (22.4)	111 (10.7)
1980	1,100	67 (6.1)	600 (54.4)	285 (26.0)	136 (12.4)
1985	1,086	70 (6.4)	564 (52.0)	288 (26.5)	148 (13.6)
1990	1,212	54 (4.5)	622 (51.3)	324 (26.7)	189 (15.6)

Source: Inoki Takenori, *Gakko to Kojo: Nihon no Jin-teki Shigen* [School and Factory: Japan's Human Resources], 1996, p. 122

Table 13.4 Unemployment Rates in the Group of Seven Economies, selected years 1975–96 (% of workforce, as measured by national standards)

	Japan	*United States*	*United Kingdom*	*Germany*[1]	*France*	*Italy*	*Canada*
1975	1.9	8.5	4.3	3.6	4.0	5.8	6.9
1980	2.0	7.2	6.4	2.9	6.2	7.5	7.5
1985	2.6	7.2	11.5	7.1	10.1	8.4	10.5
1990	2.1	5.6	7.1	4.8	9.0	9.1	8.1
1991	2.1	6.8	8.8	4.2	9.5	8.8	10.4
1992	2.2	7.5	10.1	4.6	10.4	9.0	11.3
1993	2.5	6.9	10.5	7.9	11.7	10.3	11.2
1994	2.9	6.1	9.6	8.4	12.3	11.4	10.4
1995	3.1	5.6	8.8	8.2	11.7	11.9	9.5
1996	3.4	5.4	8.2	8.9	12.4	12.0	9.7

1 West Germany up to 1990, unified Germany thereafter

Source: OECD, *Quarterly Labor Force Statistics*, various issues

Table 13.5 *Shunto* Wage Rises, 1965–88

	Rise in monthly wage (yen)	*Nominal rise (%)*
1965	3,150	10.60
1966	3,403	10.60
1967	4,371	12.50
1968	5,296	13.60
1969	6,865	15.80
1970	9,166	18.50
1971	9,727	16.90
1972	10,138	15.30
1973	15,159	20.10
1974	28,981	32.90
1975	15,279	13.10
1976	11,596	8.80
1977	12,536	8.80
1978	9,218	5.90
1979	9,959	6.00
1980	11,679	6.74
1981	14,037	7.68
1982	13,613	7.01
1983	8,964	4.40
1984	9,354	4.46
1985	10,871	5.03
1986	10,146	4.55
1987	8,275	3.56
1988	10,573	4.43

Source: Rodosho [Ministry of Labor], *Rodo Hakusho* [Labor White Paper], various years

Table 13.6 Average Monthly Wages and *Shunto* Wage Rises, for Major Firms and for Samples of Small and Medium-sized Enterprises, 1989–97

	Major firms				*Small and medium-sized enterprises*[1]		
	Average wage	*Rise (yen)*	*Nominal rise (%)*	*Real rise (%)*[2]	*Average wage*	*Rise (yen)*	*Nominal rise (%)*
1989	246,549	12,747	5.17	2.2	192,378	9,061	4.71
1990	252,752	15,026	5.94	2.6	199,668	11,050	5.53
1991	264,082	14,911	5.65	2.8	207,406	11,447	5.52
1992	276,275	13,662	4.95	3.3	214,885	10,707	4.98
1993	284,444	11,077	3.89[3]	2.7	222,699	8,699	3.91
1994	291,694	9,118	3.13[3]	2.7	227,280	6,902	3.04
1995	296,006	8,376	2.83[3]	3.0	229,919	6,184	2.69
1996	305,066	8,712	2.86	2.3	233,178	6,148	2.64
1997	308,106	8,927	2.90	1.3[4]	n. a.	n. a.	n. a.

1 On the basis of annual surveys of around 8,000 firms
2 Nominal rises deflated by annual rises in the consumer price index
3 The lowest *shunto* wage rise recorded by that date
4 According to estimates by the Ministry of Labor

Source: Official figures published by the Ministry of Labor, July 8, 1997

Table 13.7 Union Membership and Organization Rate, 1945–97

	Thousands of members	Organization rate (% of workforce)		Thousands of members	Organization rate (% of workforce)
1945	381	3.2	1972	11,889	34.3
1946	4,926	41.5	1973	12,098	33.1
1947	5,692	45.3	1974	12,462	33.9
1948	6,677	53.0	1975	12,590	34.4
1949	6,655	55.8	1976	12,509	33.7
1950	5,774	46.2	1977	12,437	33.2
1951	5,687	42.6	1978	12,383	32.6
1952	5,720	40.3	1979	12,309	31.6
1953	5,927	36.3	1980	12,369	30.8
1954	6,076	35.5	1981	12,471	30.8
1955	6,286	35.6	1982	12,526	30.5
1956	6,463	33.5	1983	12,520	29.7
1957	6,763	33.6	1984	12,464	29.1
1958	6,984	32.7	1985	12,418	28.9
1959	7,211	32.1	1986	12,343	28.2
1960	7,662	32.2	1987	12,272	27.6
1961	8,360	34.5	1988	12,227	26.8
1962	8,971	34.7	1989	12,227	25.9
1963	9,357	34.7	1990	12,265	25.2
1964	9,800	35.0	1991	12,397	24.5
1965	10,147	34.8	1992	12,541	24.4
1966	10,404	34.2	1993	12,663	24.2
1967	10,566	34.1	1994	12,699	24.1
1968	10,863	34.4	1995	12,614	23.8
1969	11,249	35.2	1996	12,451	23.2
1970	11,605	35.4	1997	n. a.	22.6
1971	11,798	34.8			

Source: Rodosho [Ministry of Labor], *Rodo Hakusho* [Labor White Paper], various years

Table 13.8 Significant Labor Events in Japan, 1945–97

Year	Month	Event
1945	October	Widespread "production control" disputes begin, irritating the Occupation authorities, which ban them in June 1946.
	December	The Labor Union Law, the first of the "three fundamental labor laws", is passed by Parliament.
1946	August	Two labor federations are founded: the moderate Sodomei and the leftist Sanbetsu Kaigi.
	September	Parliament passes the Labor Relations Adjustment Law.
	October	Unions win a labor dispute in the electric power industry and establish a livelihood wage system.
1947	January 31	A general strike led by left-wing unions, planned for February 1, is banned by the Occupation authorities.
	April	Parliament passes the third "fundamental labor law", the Labor Standards Law.
	September	The Ministry of Labor (*Rodosho*) is established.
1948	July 31	A revision of the Labor Union Law and other new legislation prohibit public sector strikes and weaken unions.
	August	The Federation of Employers' Associations (Nikkeiren) is founded.
1949		Japan's unionization rate peaks at 55.6% of the labor force. Starting with Toshiba, firms begin large-scale lay-offs.
	March	The "Dodge Line" austerity program begins, enabling employers to weaken the union movement.
1950	July 11	The labor federation Sohyo is founded.
	Summer	The "Red Purge", aimed at union activists, spreads from the public to the private sector .
1951	March	Sodomei unions secede from Sohyo.
1952	July	The Labor Standards Law is revised to provide better protection for women workers.
1953	August	Nissan strike
1954	February	Nikkeiren issues its "Three Wage Principles" to promote competitiveness.
	April	The Japan Federation of Labor Unions (*Zenro Kaigi*) is founded.
	May	Omi Kinshi strike by mainly women workers
1955	February	The Japan Productivity Center (JPC) begins operations.
	Spring	Sohyo launches the first *shunto*.
1956	Spring	Public sector unions participate (illegally) in *shunto* for the first time.
1957	Fall	A major Steelworkers Federation strike fails, shifting power from left to right in the steelworkers' unions.
1958	September	Yawata Steel Corporation introduces a new foreman (*sagyocho*) system at its state-of-the-art Tohata plant.
1960	January	A major split in the Japan Socialist Party, formalized by the founding conference of the Democratic Socialist Party, reinforces the division of the labor movement into left and right wings.
		The Mi'ike coalminers' strike begins, becoming the longest strike in Japanese history before ending later in the year in total defeat for the union.
1963	May	The Japanese Union of Scientists and Engineers (JUSE) convenes the first conference on quality control circles.
1964	April	Ota Kaoru of Sohyo and Prime Minister Ikeda Hayato meet to compromise on public sector labor issues.
	May	The International Metalworkers Federation-Japan Council (IMF-JC) is founded.
	November	The Japanese Confederation of Labor (Domei) is founded.

Table 13.8 Significant Labor Events in Japan, 1945–97 (Continued)

Year	Month	Event
1965		The International Labor Organization publishes its report on Japan's contentious public sector labor relations, though it has little effect.
1967	Spring	The first *shunto* led by the IMF-JC
1969	March	Nikkeiren announces its "Three Merit Principles" for wage-setting.
1971		The failure of a productivity campaign (*marusei undo*) in Japan National Railways is widely taken as a humiliation for public sector management. *Sanrokon*, a tripartite forum, improves contacts among moderate unions, business leaders, and the government.
1973	April	A massive demonstration over the pension system marks the peak of labor activism in Japan.
1974	Spring	Unions win a 32.9% wage increase in *shunto*; Nikkeiren proclaims a 15% wage guideline for the 1975 *shunto*; and manufacturing union leaders call for an economically rational *shunto*.
1975	Spring	The *shunto* rate falls to less than 14%.
	November	The failure of a "strike for the right to strike", launched by Sohyo public sector unions, gravely weakens the left in the labor movement.
1976	October	The founding of the Trade Union Council for Policy Promotion (Seishin Kaigi) marks the start of active union participation in the policy-making process.
1981	March	The establishment of the Administrative Reform Commission paves the way to the restructuring of JNR, NTT and other state-owned corporations.
1982	December	The Japan Private Sector Trade Union Council (Zenmin Rokyo) is established.
1985	April	The first phase of the privatization of NTT creates further impetus to unification of the organized labor movement.
	May	Parliament enacts the Equal Employment Opportunity Law.
	June	The Dispatch Law is enacted to promote the use of temporary employees.
1987	April	The division of JNR and the beginning of privatization of the resulting JR companies virtually eliminates significant militant left-wing unionism.
	September	Parliament passes the first set of labor law revisions intended to reduce working hours.
	November	The labor federation Rengo is founded; Domei is dissolved.
1988		*Karoshi* (death from overwork) becomes a prominent issue.
1989	November 21	Rengo (now known in English as the Japanese Trade Union Council) expands to nearly 8 million members with the addition of public sector unions; Sohyo is dissolved.
1990	June	Revision of laws on foreign labor
1991	Spring	The average *shunto* settlement falls for the first of five consecutive years.
	May	The Child Care Leave Law is passed.
1993	June	The Part-time Labor Law is passed.
	August	Rengo begins to play an important role behind the scenes in the Hosokawa coalition government (continuing up to its fall in April 1994).
1995	Spring	The average *shunto* wage settlement drops for a fifth straight year.
1997		The 40-hour working week officially takes effect in small and medium-sized enterprises. Deliberations continue on major revisions to the Labor Standards Law.

Chapter Fourteen

The Education System

Robert Aspinall

Japan's rapid economic growth in the latter part of the 20th century owed much to the existence of an education system that produced a highly skilled and hard-working labor force. It has now become clear, however, that continued economic development into the 21st century will not be sustainable without the modernization and continuous improvement of this system. With the changing economic environment, schools that were once required to turn out armies of loyal factory workers and highly disciplined white-collar workers will soon be required to produce a more flexible, creative and varied workforce. A debate is now under way in Japan about how exactly to reform the country's educational institutions in order to meet this challenge.

The Historical Background, 1868–1945

Once Japan's national leaders, in the second half of the 19th century, had decided to embark on the road of competition with the western powers, it became clear that in order to be successful Japan, a land of few natural resources, would have to make maximum use of the one resource it had in reasonable abundance, its people. Thus Japan, ever since the Meiji Restoration (1868), has always taken the creation and continuous improvement of a universal and modern national education system very seriously. The first modernizers were helped by the existence of a widespread network of schools and teachers, and a comparatively high level of literacy throughout the population. Further, members of the ruling class in 19th-century Japan were less concerned about the possible dangers of educating the masses than were many of their counterparts in Europe in the same period, because of their Confucian belief that one of the most important roles of education was to teach the populace proper respect for authority. Obedience to the father as head of the family and to the Emperor as head of the nation were cornerstones of this philosophy, and the latter was used by Japan's modernizers to instill a new nationalism into the population.

Japan's modernizers therefore had two parallel goals in mind when they drew up the blueprint for their new education system. First, they wanted to "catch up" with western technology, industry and machinery of modern government, by teaching the necessary practical skills in new schools and universities. Second, they wanted to create a new nationalism by instilling reverence for the Emperor into the entire population, through a program of compulsory moral education. Japan's leaders believed that both goals had to be achieved if Japan was to avoid the fate of other Asian nations which had been colonized by western powers. However, the nationalism and Emperor worship they nurtured later came to be exploited by the military as they mobilized the population for total war from the early 1930s onwards. A generation of boys was indoctrinated with the belief that their greatest honor would be to give their life for the Emperor on the battlefield. Girls, meanwhile, were taught to be "good wives and wise mothers". The prewar education system consisted of near-universal provision of elementary education; a smaller, mainly male secondary sector; and a very exclusive, elitist university system, consisting of a handful of government-controlled "imperial" universities and some prestigious private institutions. The system was very hierarchical and centralized, and lent itself to abuse by those with their hands on the levers of power. The tragedy of this system, therefore, was that it helped Japan's militarists lead their country into the disaster of World War II.

The Postwar Education System and its Contribution to Japan's Economic Success

When the Allied forces arrived in a defeated Japan in 1945 with a mission to democratize the Japanese people, they regarded Emperor worship and blind obedience to authority as evils that had to be rooted out and destroyed. From the occupiers' point of view, the education system had been mainly responsible for instilling these traits into people, and therefore the education system would have to be reformed in order to remove these beliefs and replace them with democratic ones. With hindsight, the optimism with which the reformers set about their democratizing mission seems a little naive. However, the trauma of defeat led many ordinary Japanese people to turn against the militarism that had brought their nation along such a disastrous road. They turned against their wartime leaders and welcomed General Douglas MacArthur, the head of the Occupation forces, as a new Shogun who would help to rebuild the shattered country. Thus the occupiers found that they had a lot of domestic support for their reform program.

Education reforms during the Occupation era included the establishment of a "6–3–3" system of schools (six years for elementary school, three years for junior high school, and three years for senior high school); the introduction of coeducation; the legalization of teachers' unions; and the removal of military instructors and portraits of the Emperor from schools. In spite of the initial popularity of these reforms, there were some Japanese conservatives who were uneasy about what they regarded as foreign intrusions into their education system. After the occupiers left in 1952, conservatives who were in positions of power – some of whom had been regarded as war criminals by the Occupation authorities, but had been able to return to their old jobs when the Occupation ended – began a reverse course to remove some of the "excesses" of the Occupation years. Thus the years following the Occupation bore witness to prolonged and bitter struggles between the left and the right, over such issues as government control over textbooks, the right of teachers to strike, and the election of local boards of education. On the left, the strongest and most militant organization fighting against the reverse course in education was *Nikkyoso*, the Japan Teachers' Union (JTU), which in the 1950s represented about 90% of the nation's schoolteachers. The leadership of the union was dominated by Socialists and Communists, and for many years it was one of the strongest and most important elements making up the "progressive camp" (*kakushin jin'ei*) in Japanese politics. This loose alliance of unions, political parties and their allies acted as a counterbalance to conservative dominance of the national government. In this way, the battle over education came to be subsumed by the broader ideological conflict between left and right that characterized Japanese politics until the 1990s.

Although the postwar education system was overshadowed by perpetual conflict between left and right, it was still able to deliver, very successfully, an education for the entire population that played a major role in making the Japanese "economic miracle" possible. Thomas Rohlen and other scholars have shown that this paradox came about because the bitter ideological conflict between the two sides was largely confined to the higher echelons of the education system, while at the level of the individual school – the place where education is actually delivered – teachers and school managements would usually unite around a shared loyalty to the school and the children in its care (see Rohlen 1984). Further, although the leadership of the JTU was very politically motivated, this ideological enthusiasm was shared by only a minority of the ordinary membership. Most teachers continued to support the union, although national membership figures showed a slow but persistent decline, and they looked to it to defend their economic interests. They were also glad that the JTU was acting as a defense against those who would turn the clock back to prewar totalitarian days. However, this did not mean that they shared the Marxist or socialist beliefs of their leaders. Ordinary teachers, therefore, tended to be unenthusiastic about responding to national calls for action made for political purposes. Thus, within individual schools conflict was the exception rather than the rule and, while national leaders carried on a more or less permanent war of words with one another, ordinary teachers and school principals got on with the job of teaching. It was this teaching, carried out in thousands of schools the length and breadth of Japan, that is widely seen by commentators at home and abroad as having made Japan's economic miracle possible.

The postwar Japanese education system served the nation's economy well by providing it with a hard-working, loyal workforce, well-versed in basic skills and equipped with the habits of learning that would facilitate continued training once formal education was over. Kindergartens and elementary schools instilled basic numeracy and literacy, as well as the importance of working as part of a group. Junior high schools (for students from 12 to 15) continued with the work of teaching basic skills, as well as stressing the importance of hierarchy and discipline. Senior high schools (for students from 15 to 18) began the job of segregating children according to their future careers. For example, those destined to go to elite universities would go to elite academic high schools, while those who were destined to become secretaries or factory workers would go to commercial high schools or industrial high schools, respectively.

The role of the private schools is interesting, as it differs from level to level. Private schools as a group make up only a tiny part of the compulsory sector of Japan's education system. In 1992, only 0.7% of elementary and 5.5% of junior high schools were private. However, in the same year 24% of senior high schools were private. Parents gain little by sending their children to private elementary or junior high schools, since the curriculum followed is the same as in the public schools, but there is a greater demand for private senior high schools, partly because of the inability of all parents to get their children into "academic" (non-vocational) public senior high schools. There is also a small number of very elite private senior high schools that draw students from all over the country.

Throughout Japan the population tends to emphasize the importance of education, and to regard school and university entrance examinations as the key determinant of a child's future career prospects. Employment structures and company cultures that would not allow women the opportunity to advance very far in most middle-class careers encouraged the creation of the phenomenon known as *kyoiku mama* ("education mother"), referring to the tendency for competitive, middle-class women to look to the educational achievements of their sons as their main source of social status. A *kyoiku mama* would dedicate her time to helping with her son's homework, organizing his private lessons (either at home or at evening and weekend cram schools known as *juku*), taking part in school or PTA events, and so on. Mothers with less lofty ambitions for their children were also greatly influenced by a culture which put such a great emphasis on the importance of education. It is still a loss of face for a family to have one of their children fail to graduate from senior high school, even though compulsory education in Japan only goes to the end of junior high school. This helps to explain why Japan has one of the highest staying-on rates in the world. In 1994, a typical year, 95.4% of all children in the relevant age group completed senior high school.

The success of the Japanese education system in producing a hard-working, loyal and highly skilled workforce drew attention from abroad. The governments of the United States and the United Kingdom, where the education systems came under attack from the late 1970s onwards for their perceived failure to teach basic skills and instill proper discipline, were especially interested. The argument was made that poor educational performance was linked with poor economic performance, and that lessons could be learned from the Japanese model. Thus, during the 1980s a steady stream of British and US educators, academics and government officials visited Japan in search of the secret of its success (see Goodman). It is quite ironic, therefore, that the 1980s was also a decade when a "great education debate" was started within Japan, based on the widespread belief that its education system was in need of full-scale reform. Why was a system that had been so successful in helping the Japanese economic miracle now the subject of widespread criticism?

Criticisms of the System and the Debate on Reform

The postwar education system had been very successful in providing the economy with the workforce that it needed, but this had not been achieved without a price. Many fundamental features of the system contained their own intrinsic problems and drawbacks. Take, for example, the system's stress on conformity and group harmony. This was very successful in instilling the virtues of teamwork and group loyalty into future workers. The downside, however, was that it also stifled individual creativity and spontaneity, and could often make school bullying problems worse by encouraging children to look harshly on those who were different from the group norm. In fact, by the mid-1980s school bullying (*ijime*) had become such a problem that it was rarely out of the headlines of national newspapers. Attention was especially focused on the tragic increase in suicides by apparent victims of bullying, although by international standards the overall rate of teenage suicides never became especially high. Bullying, in turn, was linked with the issue of violence in schools. The beating of children by teachers in some junior and senior high schools was almost routine, and when it resulted in serious injury or death it would usually hit the national headlines. This teacher-centered violence could be regarded as the downside of the system's stress on obedience and discipline.

Another issue that became a focus of public debate starting in the 1980s was "examination hell" (*juken jigoku*). Entrance examinations for high schools, universities and government departments had been introduced in the Meiji period (1868–1912), in a deliberate attempt to overthrow the class-based favoritism of the old regime. They had been very successful in doing that, and the massive expansion of the secondary and higher sectors of education that followed World War II opened opportunities for advancement to ever wider sections of the population. Thus, by the 1970s the competition for entrance to elite senior high schools and universities had resulted in the necessity of making entrance examinations more and more demanding for those who took them. This in turn spawned the creation of a private industry of *juku* (as mentioned above), to help children prepare for examinations. By the 1980s an ambitious child, egged on by an equally ambitious *kyoiku mama*, would have to sacrifice most of his/her adolescence to non-stop preparation, in the hope that this would yield a place at an elite university, and the lifetime elite career and elite status that this would inevitably bring with it. Stress related to this system resulted in misery, breakdowns and even suicides for those who tried but failed, while even those who were successful had to sacrifice the broader experiences and personal development enjoyed by adolescents in the less demanding education systems of other countries.

Alongside widespread public concern about school violence, bullying and stress, a more specialized debate has been going on among academics and concerned parties about what they see as the excessive uniformity and rigidity of the system. It has been argued that a more flexible and diverse education system would not only ease the pressures on children and families, but would also be more appropriate for Japan's economy as it moved from the "catch up" phase of development to take on the more prominent and innovative role that would properly suit its recently acquired status as an economic superpower. From about 1980 onwards, business and industry started putting pressure on the governing Liberal Democratic Party (LDP) and the Ministry of Education to overhaul the education system, in order to adapt it to changing times. In response, Prime

Minister Nakasone Yasuhiro in particular made education reform one of the main priorities of his government. In 1984 he established an advisory body directly under his office, the Ad Hoc Council on Education (*Rinkyoshin*). Nakasone believed that it was unlikely that anything substantial would be achieved if reform were left up to the Ministry, because of its bureaucratic inertia and its defense of its discretionary powers.

Rinkyoshin was made up largely of university professors, government advisors, and representatives of the business world. It was hoped that it would bring a fresh approach to education reform unhindered by bureaucratic interests. True to its remit, between 1984 and 1987 Rinkyoshin scrutinized the education system from top to bottom. By the time that it was disbanded, it had issued about 500 separate recommendations for reform, which, for the purposes of clarity, can be divided into those related to "old issues", which had been the subject of debate and controversy since the Occupation, and those concerning "new issues" related to the changing needs of the economy at the end of the 20th century.

In the "old issue" category, recommendations from Rinkyoshin included proposals to reform the rigid 6-3-3 system, by introducing a greater variety of types of school, including six-year secondary schools, and proposals to bring back traditional moral education. The right in Japan have been trying to bring about a return to moral education that emphasizes "Japanese values" ever since the Occupation authorities removed all the trappings of Emperor worship from the schools in the 1940s. One of the main obstacles to the restoration of this kind of teaching was the fact that it would have to be taught by ordinary classroom teachers, most of whom, until the 1980s, were members of the JTU. A split in the union's ranks in 1989 seriously weakened it further, which in theory should have made it easier for the Ministry of Education to implement its desired policy. One sign that this has happened is its recent victory in the struggle to force all schools to fly the national flag and play the song *Kimigayo* during school ceremonies. The JTU was opposed to this, on the grounds that the flag and the song are symbols of militarism and imperialism. After a bitter struggle in the 1980s that saw many teachers and school principals disciplined by the authorities, the JTU has now dropped its opposition to the flag and the song. The smaller breakaway union, the All-Japan Teachers and Staff Union, maintains its absolute opposition to these symbols, but it has now become established practice in the vast majority of schools to comply with the Ministry's wishes. This retreat by the left came shortly after the split between the unions. Both events are indicative of the decline of the progressive camp's influence in education.

The decline of the left inevitably brought about a realignment of Japanese educational politics. The old issue of moral education had always divided left from right. Such ideological clarity, however, was (and remains) largely absent from the debate over most of the "new educational issues". Rinkyoshin's recommendations in this category included references to three of the main education buzzwords of the 1980s: liberalization, "flexibilization", and internationalization (*jiyuka, junanka,* and *kokusaika*). Every group that was a party to the education debate claimed to be in favor of these three trends, although there continues to be considerable confusion about what they actually mean in practice. Liberalization mainly involves increasing competition and choice in the compulsory education sector (elementary and junior high schools). Under existing arrangements, parents have no choice about where to send their children up to the age of 15, and the curriculum of all schools is uniform and strictly controlled by the Ministry of Education. The related concept of flexibilization involves increasing the variety of schools within the education system, and allowing more student choice within schools. By comparison to school systems in other developed countries, the Japanese school curriculum is very inflexible, children having no choice whatever over around 90% of the schedule. Everyone agrees that this is too rigid, but there is disagreement about how far to go in increasing choice for parents and students.

The Ministry of Education jealously guards its powers to control the curriculum, the textbooks and the minutiae of school life, including even the size of classrooms and the height of desks. Commentators from both left and right are highly critical of this centralization of power, which they regard as a throwback to the 19th century's drive to create a strong state. The problem remains, however – as the Occupation authorities also found – that in order to initiate real change in the education system it is very difficult to sidestep the Ministry. In other words, if a government wants to initiate change it has no choice but to make use of, and thereby legitimize, the very centralized ministerial power that it seeks to undermine. Of course, Nakasone's main reason for forming Rinkyoshin in the first place was precisely to try to avoid this Catch-22 situation. The Ministry, for its part, woke up to the fact that there was a tide of opinion against it, not only from its traditional enemies on the left but also from business people and free-market ideologues on the right. It has therefore joined in the call for liberalization and flexibilization, but it is trying to keep the scope and pace of change as limited as possible, in order to minimize any threat that there may be to its own bureaucratic interests.

Confusion also surrounds the concept of internationalization. No actors in the education policy-making process say that they are opposed to it. More than any other concept, it seems to represent the arrival of the education system in a world in which Japan must play its part as a global economic power. On one level, internationalization represents little more than efforts to increase exchanges between Japanese students, and teachers, and their counterparts in other countries. It has also been used to help promote better foreign language teaching (meaning, overwhelmingly, the teaching of English) in Japanese schools. In 1988, the Council of Local Authorities for International Education (CLAIR) was established jointly by the Ministries of Education, Home Affairs and Foreign Affairs, to promote internationalization at the local level. One of CLAIR's main achievements has been the expansion of the Japan Exchange and Teaching (JET) Program, which sends foreign nationals into Japanese schools to assist with language teaching and various international exchange projects organized at the local level. There is a consensus of support for this and other programs, although there is still dissatisfaction with the generally poor level of spoken English achieved by students in the formal school and university systems, in spite of the large sums of money spent on efforts to improve language teaching there. However, at the ideological level internationalization can also be a cause of conflict. For example, the right has used it as justification for an increase in "patriotic" education, the argument being that children cannot be taught to respect the cultures of other countries before they respect their own. In this way the new issue of internationalization can be brought in to add weight to arguments in favor of a return to "old issue" policies.

Nakasone himself was disappointed with the lack of major concrete changes resulting from Rinkyoshin's deliberations: it seemed that bureaucrats and the opponents of change had won. Although it is true that no major education reforms were implemented in the immediate aftermath of Rinkyoshin, it is now being argued by Japanese education experts that this does not mean that Nakasone's efforts were in vain. From the perspective of the 1990s, it can be seen that the great debate started by Rinkyoshin has been carried forward by politicians, teachers, academics and the general public. Further, the debate on education has now managed to shed much of the ideological point-scoring that characterized it in the past. The consensus is now that change must take place, and that it must be somehow centered around liberalization, flexibilization and internationalization. Political ideologues of both left and right still fight their own corners, but they are fast being drowned out by calls for pragmatic change to a school system that has served its purpose of helping Japan catch up with the West, but which is now in need of a serious overhaul. Such an overhaul must inevitably also encompass higher education, which shares many of the general problems outlined above, but also has some particular problems of its own.

Problems with Higher Education

Although western commentators have often been full of praise for Japan's school system, especially its compulsory sector, similar praise has not been forthcoming for Japan's universities. The three main functions of a modern nation's university system – the teaching of undergraduates, the supervision of graduate study, and the carrying out of fundamental and applied research – have all been found wanting in the case of Japan. In order to understand the reasons for these deficiencies, it is necessary to review the recent history of higher education.

As with the rest of the education system, the university sector was transformed by the Occupation authorities, and, in a similar fashion to the way in which postwar schools developed, postwar universities were influenced by a combination of foreign (mainly US) and domestic pressures and influences. Following the Occupation, Japan's economic expansion was accompanied by a parallel expansion in the number of (four-year) universities and (two-year) junior colleges, so that by the 1970s going to an institution of higher education was no longer for the privileged few, but had become a normal progression for anyone considering a white-collar or skilled blue-collar career. In fact, the "qualification inflation" that accompanied this huge expansion meant that young people had to get some kind of post-secondary qualification if they were to stand any chance of avoiding a lifetime of insecurity and low pay (see Table 13.3). This harsh fact helps explain the rise of the *kyoiku mama* and her obsession with the academic success of her children. Further, the high financial cost involved in getting a modern child into and through university has been cited as one of the main reasons for the decline in Japan's birthrate. Ordinary families simply cannot afford to get more than one or two children through the whole process, and failure to get them through would condemn them to a life as second-rate citizens.

The reforms of the Occupation era were aimed at creating a "democratic" university sector that would be the opposite of the prewar elitist system, centered as it had been around a handful of imperial universities, the most important of which was Tokyo. Although they were successful in opening up higher education to a far greater proportion of the population, the reformers failed to touch the elitist credentials of the leading institutions. After the war the prefix "imperial" was removed from Tokyo and Kyoto Universities but they still continued to dominate the university sector. In particular, the University of Tokyo (*Tokyo Daigaku*, or *Todai*) is so venerated an institution that, in the words of one western journalist: "a diploma from this school is practically a ticket into the ruling class" (van Wolferen p. 111). This arrangement is damaging for the education system as a whole, because it is the name of the university that is all important, not the content of the work that is carried out there. Since graduation is almost guaranteed to all students who have been in reasonably regular attendance, the key feature of Japan's elite universities is not how difficult they are to graduate from, but how difficult they are to get into. Thus, once students have overcome the considerable hurdle of the entrance examinations, they are able to "rest" for the following four years and await the next stage of the process, which is usually entry into a leading company or a government ministry, where they are guaranteed secure employment and access to the elite career track. It should be noted that, while the time spent at institutions of higher education by those in training for a specific profession such as medicine or engineering is by no means restful, even for these people the name of the institution that they attended – or, in some notorious cases, falsely claim to have attended – is very important for their future careers.

Companies and ministries value the graduates of elite universities not only for their ability to work hard – albeit not actually while they are at university – but also for the connections they will have built up while at university with other future members of the elite. Thus, social life for undergraduates is almost more important for their future career than any studying that they may do. Students in non-elite universities, meanwhile, have little to gain from exerting themselves, since all the openings to elite career paths are reserved for those above them in the university hierarchy. It is not surprising, therefore, that foreign lecturers at Japanese universities complain that undergraduates there are poorly motivated, and lack the intellectual curiosity expected of their counterparts in other countries. This has been put down to the content of the entrance examinations, almost all of which are based upon multiple-choice factual questions. These are good tests of candidates' mastery of huge amounts of detailed information, but are not designed to elicit, among other things, original answers or interpretations of conflicting arguments. In other words, the university system does not reward original or creative thinking, nor does it reward novel approaches to analysis, problem-solving or self-expression.

Just as company recruitment practices have contributed to the deficiencies of undergraduate study in Japan, so the employee-training policies of the same companies have stifled the growth of graduate education. The "lifetime employment" system run by large companies has come under pressure in recent years, but still hires recruits upon graduation from university or college, on the understanding that they will serve with the company until retirement. In order to maintain this system, companies do not hire people in mid-career except in exceptional circumstances. This system has enabled companies to invest significant resources in training their employees, in the secure knowledge that the employees will not then be poached by another company. The ideal of company loyalty reinforces the companies' preference for taking care of as much of their own training as possible, since if employees train with their colleagues it helps maintain group cohesion and loyalty. Thus, Japanese companies have almost always preferred to provide the equivalent of graduate training themselves. An undergraduate who wants to work for a large company therefore has a disincentive to embark on a graduate study program at university. He or she will be much better off joining a company straight away and getting the training later, at company expense.

Since the training of graduates is mainly carried out by companies, it is no great surprise to learn that important new research is also more likely to be carried out at a company research facility than in a university. In 1994, a team from the British government's Office of Science and Technology found that in Japan company laboratories and company-staffed "National Project" laboratories were much more important than university laboratories in basic, innovative research related to new technology (see Chisolm). As well as a shortage of graduate students, Japanese universities are also hindered by bureaucratic restrictions and a lack of public money earmarked for research. Thus, not only applied but also basic research is more likely to be carried out by private industry than by the universities.

From the above discussion, it can be clearly seen that postwar Japan has failed to create a university system that can rival those of other advanced nations, and that the reasons for this failure can be traced to the function performed by Japanese universities within the wider economy. Rather than developing as centers for teaching and research, universities have taken on the function of acting as clearing houses that allocate young people to their future career tracks.

The criticisms of Japanese higher education are not new. Government officials and university professors have been painfully aware of adverse international comparisons for quite some time, and have been trying to address the various problems. Rinkyoshin's recommendations included many far-reaching proposals that were aimed at higher education reform, and its buzzwords of liberalization, flexibilization and internationalization were just as applicable to higher education as they were to secondary education, if not more so. Indeed, the last-named concept has a more direct relevance to universities than it has to other areas of education, since the increasing trend for Japanese students to go abroad for their higher education has threatened universities at home with a decline in "customers". If this trend is taken in tandem with the inevitable downturn in the size of the teenage population after the year 2000, then the threat to the existence of some universities becomes very real. Private universities in particular are in danger, because they depend almost entirely on student tuition fees for their funding. Thus, demographic factors have added another impetus to the drive for reform.

Following Rinkyoshin's recommendations, the Ministry of Education has implemented some specific reforms in higher education.

However, in doing so it is handicapped by the same problem that reduced the effectiveness of its secondary education reforms, its unwillingness to shed any of its discretionary powers. Thus, although it favors increasing the diversity of educational institutions, and increasing the choice offered to their students, it is still determined to maintain an iron grip on the regulations that determine the exact parameters within which these changes can be made. It is also determined to maintain its total centralized control of the public purse strings. Within these constraints, actual, concrete reforms have been minor. The main reform came in 1991, when the Ministry removed the distinction between "liberal arts" subjects – those compulsory, general subjects, such as foreign languages and physical education, studied by undergraduates in their first two years – and the specialized subjects that are taken by third-year and fourth-year students. This reform allows universities more freedom in drawing up their curriculums, and more choice for students in what subjects they study. In a limited number of cases, students can even transfer their credits from one university to another. As a response to the problems of graduate study, the Ministry has started to encourage the diversification of graduate schools. It is also promoting existing financial support programs for distinguished students, as well as a research assistant program, under which graduate students are financially rewarded for their contributions to research. However, the number of bursaries and scholarships available to students at Japanese universities remains very small.

As with reforms in the junior and senior high schools, the failure of the Ministry of Education to respond adequately to the problems of higher education is drawing criticism from business and industry. The Japan Federation of Employers' Associations (*Nikkeiren*) now believes that the only way that progress can be achieved is to grant universities more freedom from the Ministry's control. The established certainties of the university system are also under threat from Japanese industry's transformation of its lifetime employment system. If Japanese companies respond to commercial pressures by ending the practice of recruiting all their employees fresh out of university, and reduce the amount of training they carry out themselves – in other words, if they become more like western companies – then the burden will be on universities to offer their students more relevant and demanding curriculums.

Conclusion

Some Japanese commentators have recently started to talk about three educational "revolutions". The first was the creation of a modern, centralized education system in the Meiji period. The second was the era of democratization and expansion that accompanied the Occupation. The third revolution, it is argued, is the one that is currently under way, the one that is the child of Nakasone's Rinkyoshin and its massive review of the entire system. Some argue that this third period will result in little more than a lot of talk, accompanied by a few minor changes. The strength of vested interests and the paralysis of the policy-making process will ensure, so this argument goes, that major reform is postponed indefinitely. The decade of the 1980s provided abundant evidence in favor of this negative verdict (see Schoppa). However, it can now be argued that some important events, right at the end of the 1980s and in the first years of the 1990s, have increased the chances of Japan's education reformers achieving significant results in the near future. The ending of the Cold War, the collapse of Japan's "bubble economy", and the LDP's loss of its monopolistic grip on political power (albeit temporary) have all contributed to a profound alteration of the postwar certainties that influenced most people's thinking on education, among many other topics. For example, the world of business and industry, which used to regard the Ministry of Education as a vital ally against the perceived threat of the left to the education of Japan's youth, now regards it as an obstacle on the road to extensive education reform. With the near-disappearance of the former Japan Socialist Party, which for most of the postwar period was the main opposition party, and the splitting into two of its ally, the JTU, business leaders are now less worried about left-wing influence on education than they are about the failure of schools to produce workers who will be more suited to an advanced, globally competitive economy. For its part, the Ministry of Education has had serious problems coming to terms with an age that calls for more flexibility and diversity. In its official reports and pronouncements it recognizes the need for this kind of change, but in its unwillingness to relinquish the power that it has over the whole education system it acts as a barrier to real reform. Whether the education system can be adapted to face the needs of Japan in the 21st century, or remains stuck in the past, will be decided by the outcome of the struggle between reformers, of all political shades, and the established education bureaucrats.

Further Reading

Chisolm, John, *Turning Research into Wealth: The Japanese Way* (Report of the Technology Foresight Team, Office of Science and Technology), London: Her Majesty's Stationery Office, 1994

This report is the result of a British government team's visit to Japan, carried out with a view to learning from Japan's "virtuous circle" of investment in research and development producing returns that lead to more investment, despite comparatively low public funding of research. The report identifies a "national consensual process", which includes the willingness of Japanese companies, unlike British companies, to work together on pre-competitive research.

Cutts, Robert L., *An Empire of Schools: Japan's Universities and the Molding of a National Power Elite*, Armonk, NY: M. E. Sharpe, 1997

This is a critical account of how the Japanese university system is dominated by a few elite institutions, especially the University of Tokyo. Because of entrenched interests, Cutts is pessimistic about the chances of the current wave of reform achieving much real change.

Dore, Ronald, and Mari Sako, *How the Japanese Learn to Work*, London and New York: Routledge, 1989

This is a very thorough and comprehensive account of vocational education and training in Japan, examining the roles of both public and private sectors.

Goodman, Roger, "Japan – Pupil Turned Teacher?", in David Phillips, editor, *Lessons of Cross-National Comparison in Education*, Oxford Studies in Comparative Education, Volume 1, Wallingford, Oxon: Triangle Books, 1992

This is an excellent analysis of how, in the 1980s, Japan's role apparently changed, from being a student and copier of western knowledge and institutions to being a teacher from which the West could learn. Goodman shows how scholars and politicians in the United States and the United Kingdom set up the Japanese educational system either as a "straw man" or as an "idealized model" off which to bounce their own ideas and beliefs.

Marshall, Byron K., *Learning to be Modern: Japanese Political Discourse on Education*, Boulder, CO: Westview Press, 1994

This is an historical analysis of the evolution of the Japanese education system between 1800 and 1989, concentrating on political, social and ideological developments. Marshall covers a lot of ground in a very informative and accessible way.

Ministry of Education, Science and Culture, *Education in Japan: A Graphic Presentation*, Tokyo, 1994; and *Japanese Government Policies in Education, Science and Culture 1994*, Tokyo, 1995

These are annual publications in English, providing statistics and descriptions of how the education system is administered, and a detailed overview of government policies in the field.

Okano Kaori, *School to Work Transition in Japan*, Clevedon, Avon, and Philadelphia, PA: Multilingual Matters, 1993

Here Dr Okano, an educational sociologist, closely examines the process by which Japanese vocational high school students decide on their future careers. The study describes in detail the process by which students, teachers and employers interact during the transition from school to employment.

Rohlen, Thomas P., *Japan's High Schools*, Berkeley, Los Angeles: University of California Press, 1983

Rohlen describes and analyzes all aspects of Japan's senior high school system. Based, in part, on a thorough anthropological study of five of these schools, it outlines clearly all the most important strengths and weaknesses of the system.

Rohlen, Thomas P., "Conflict in Institutional Environments: Politics in Education", in Krauss, Ellis S., Thomas P. Rohlen and Patricia G. Steinhoff, editors, *Conflict in Japan*, Honolulu: University of Hawaii Press, 1984

The most important contribution of Rohlen's article to our understanding of Japan's postwar education system is its account of how persistent conflict at the national level of the system did not prevent teachers and administrators lower down the educational hierarchy from getting on with the job of providing very effective education to the majority of Japan's children.

Schoppa, Leonard, *Education Reform in Japan: A Case of Immobilist Politics*, London and New York: Routledge, 1991

This is an excellent analysis of the policy-making process involved in education reform in Japan in the 1970s and 1980s. Schoppa argues that the failure of Nakasone and others to bring about any major reform was caused by splits and disagreements within the "conservative camp", some of which could be exploited by the government's opponents in the progressive camp. The ability of the Ministry of Education to block change at the implementation stage is also described.

Shields, James J., editor, *Japanese Schooling: Patterns of Socialization, Equality and Political Control*, University Park: Pennsylvania State University Press, 1989

This is a very useful collection of articles by distinguished American and Japanese writers about contemporary issues in Japanese education. Contributors include Catherine C. Lewis, Ikuo Amano, Hidenori Fujita, Benjamin Duke, and Richard Rubinger.

Tsuchimochi, Gary H., *Education Reform in Postwar Japan: The 1946 US Education Mission*, Tokyo: University of Tokyo Press, 1993

This is a detailed account of a fascinating chapter in the history of the Japanese education system. It charts the interactions between the occupiers and Japanese reformers that took place during the Occupation. The actions of these people have had a lasting impact on the structure of education in Japan.

Van Wolferen, Karel, *The Enigma of Japanese Power: People and Politics in a Stateless Nation*, London: Macmillan, 1988, and New York: Knopf, 1989

This controversial book provides a very critical analysis of Japan's major political, social, cultural, economic and educational institutions, and how they are held together by a web of informal or semi-formal contacts, connections and "understandings", most of which are not codified in law. The book includes a critical study of schools and universities, and the attacks made by politicians and bureaucrats against the Japan Teachers' Union, which was perceived by them as a threat to what van Wolferen calls the "System". Many scholars disagree with van Wolferen's conclusions, but most agree that the book makes fascinating reading.

Dr Robert Aspinall is a Lecturer at Nagoya University, Japan.

International Relations

Chapter Fifteen

Japan's Relationship with the United States

Chitta R. Unni

The key events in the shaping of the postwar relationship between Japan and the United States took place on September 8, 1951. On that day Japan signed a Peace Treaty with 48 other nations at a conference organized by the United States in San Francisco, preparatory to the ending of the Allied Occupation. In a private ceremony on the same day, Japan also signed a Treaty on Mutual Cooperation and Security with the United States. As a result of these two treaties, Japan regained her sovereignty, but, as John W. Dower has pointed out, it was "sovereignty without autonomy", and was not what Japan's leaders had wanted for their country. The combined effects of the two treaties obligated Japan to become a junior partner of the United States. Much against the wishes of many of its policy-makers and opinion-formers, Japan had to consent to become the main bulwark against the expansion of Communism in East Asia for almost all of the postwar period. It can be argued that the consequent subordination to the United States continues even now to have a negative impact on the policy-making of Japanese governments and interest groups.

The Occupation Years, 1945–52

The signing and implementation of the two treaties represented the completion of the policies pursued under the Allied Occupation (1945–52). Extensive political and economic reforms had been carried out in the immediate aftermath of Japan's defeat, but as the Occupation went on its main purpose shifted, from securing internal reconstruction to building a close relationship between Japan and the United States.

From Reform to the "Reverse Course"

The policies of the Occupation authorities, known as SCAP, originally had three objectives: to liberalize, demilitarize and democratize Japan (see Halliday). Accordingly, the reforms included the dismantling of the *zaibatsu* (industrial conglomerates), the massive reallocation of landholdings, and the encouragement of labor unions. Among their immediate consequences, however, were severe inflation and a troubling downward movement of other macroeconomic indicators. These trends, combined with international events such as the ascendancy of the Communist forces in China in 1949, the war on the Korean peninsula from 1950, and the perceived threat of a worldwide Communist movement spearheaded by the Soviet Union, caused a reversal of Occupation policies in Japan.

The crucial change, which has affected relations between the United States and Japan ever since, was that Japan came to be viewed as a key player in the overall strategy pursued by the United States and its allies, rather than as a separate and special case. The US foreign policy expert George Kennan envisaged a coordinated use of military, economic and political forces to stem Communism, and Japan was now expected to play its part in this strategy (see Borthwick). Reformist zeal abruptly gave way to a "reverse course" in

1949–50, as US efforts were redirected to strengthening Japanese industry, and rebuilding Japan as the forward defense against international Communism. In particular, in order to rebuild its economy, Japan needed to secure supplies of inexpensive labor and to accelerate capital accumulation. Thus, for example, SCAP abandoned plans to remove factories and equipment in payment of reparations claims, and mandated revisions in the Labor Union Law of 1945, restricting the bargaining rights of public sector workers. Further, in 1953, soon after the Occupation ended, the Anti-Monopoly Law of 1947 was relaxed to permit the formation of cartels, enabling Japan to move into a phase of rapid economic growth and technological upgrading.

The Formulation of a Security Policy for the New Japan

Japanese policy-makers continued to favor a strategy of economic nationalism, which, among other things, implied that they would seek to avoid any entanglement in collective security arrangements with the United States beyond the immediate areas surrounding the four "home islands". Ever since 1950, Japanese governments have resisted US efforts to build a regional defense organization for Asia and the Pacific, along the lines of the North Atlantic Treaty Organization (see Yoshitsu), and from the very beginning of the Occupation the objective of Yoshida Shigeru, Minister of Foreign Affairs and later Prime Minister, was to concentrate on rebuilding the economy, while leaving security problems to the United States. In exchange for bases on Japanese territory, Yoshida expected the United States to guarantee the security of Japan in such a way that Japan would not have to take any direct part in US-led military activities.

Yoshida and other Japanese policy-makers used many arguments to prevent the United States from drawing Japan into any collective, multilateral security system during the Cold War. Thus, for example, they would habitually cite Article 9 of the 1947 Constitution, which declared that Japan had renounced war. From the US point of view, the standard Japanese interpretation of Article 9 was political rather than juridical, since there was nothing in the article to prevent Japan from participating in collective defense activities. Indeed, according to Charles Kades, one of the US officials who took part in the drafting of an early version of the Constitution, Article 9 had been written in order not to contradict Article 43 of the UN Charter, which requires all member countries to make armed forces available to the UN Security Council for the purpose of maintaining international peace and security (see Kades). In other words, the United States already envisaged, as early as 1946, that Japan would soon become a member of the UN (see Chapter 19).

It was in 1947, after SCAP had successfully worked with Japanese leaders to implement the new Constitution, that General Douglas MacArthur, the head of the Occupation forces, decided that the time had come to end the Occupation. The United States had already started to change its attitude towards Japan, and MacArthur's announcement that the objectives of the Occupation had been fulfilled is indicative of this change. The US State Department was beginning to look to Japan as a potentially powerful ally of the United States in its emerging Cold War strategy of containing Communism, just at a time when the idea of signing both a multilateral Peace Treaty and a bilateral Security Treaty was being broached, and relations with the Soviet Union were progressively deteriorating. The United States was emboldened to make arrangements to end the Occupation on terms that would favor its own global political strategy; and, despite lingering memories of wartime alliances, neither the Soviet Union, nor the People's Republic of China, nor even the Chinese Nationalist regime in Taiwan, took part in the San Francisco Peace Conference.

However, the alternative of making Japan neutral as between the two Cold War camps was also being floated in the late 1940s and early 1950s, notably by the Japan Socialist Party (JSP), which included neutrality among its four "political principles". Its other principles at that time were demands for a peace settlement with all of Japan's former enemies, including the Soviet Union and China; for a ban on the maintenance of any foreign troops on the home territory of Japan; and for a pledge never to rearm the Japanese state. A fifth principle, reflecting anti-nuclear attitudes, was added later. The United States, and the increasingly dominant conservative politicians in Japan, obviously disagreed with all five principles. In particular, following the outbreak of the Korean War in 1950, Prime Minister Yoshida and US President Truman's representative, John Foster Dulles, agreed on the necessity of keeping US forces in Japan for the foreseeable future. They therefore finalized the plan to sign a bilateral Security Treaty at the same time as the overall Peace Treaty.

The Security Relationship Since 1952

The San Francisco Peace Treaty formally ended the Allied Occupation and provided for the withdrawal of all foreign military forces from Japan. However, it also left open the possibility of maintaining such forces if their presence was required under a bilateral treaty – such as the Security Treaty Between Japan and the United States (*Nichibei Anzen Hosho Joyaku*, or, in abbreviated form, *Ampo*), which went into effect at the same time as the Peace Treaty, on April 28, 1952. A revised version, the Treaty on Mutual Cooperation and Security Between Japan and the United States (*Nichbei Sogo Kyoryoku Oyobi Anzen Hosho Joyaku*, also known, again, as *Ampo*), was signed in Washington, DC, on January 19, 1960, and went into effect from June 23, 1960.

The 1952 Security Treaty

The original Security Treaty had five articles, which may be briefly summarized as follows. First, the United States was to keep its forces in Japan to secure it against attacks from abroad, as well against insurrection by domestic groups. Second, Japan agreed not to grant the right to maintain military forces on its territory to any other country apart from the United States. Third, the details of the joint security policy to be implemented under the Treaty were to be laid out in a separate administrative agreement between the United States and Japan. Fourth, the Treaty would expire only if both parties agreed, and only after alternative security arrangements, possibly involving the UN, had come into effect. Finally, the Treaty would come into force after both parties had ratified it and the instruments of ratification had been exchanged in Washington. In general, US policy-makers viewed the Treaty as a great foreign policy success, while their Japanese counterparts tended to view it more pragmatically and soberly, as part of the inevitable political and diplomatic price that had to be paid, not only to regain independence and sovereignty, but also to secure the basis for achieving economic self-sufficiency. It is noteworthy, for example, that when the Japanese Self-Defense Forces (SDF) were organized in 1954, largely in response to US pressure stemming from the provisions of the Security Treaty, the House of Councillors, the upper house of Parliament, passed a resolution prohibiting their deployment anywhere outside Japan. For the Japanese, the Security Treaty was a strictly bilateral arrangement, and was not to be used to draw their country any deeper into the Cold War.

The 1960 Security Treaty

The 1960 version of the Security Treaty originated from proposals put forward by the Japanese Prime Minister, Kishi Nobusuke, for certain changes in the original text, which he hoped would make the relationship between the two countries somewhat more equal. By the end of the 1950s, Japan was no longer a war-shattered country. It had made striking progress in economic reconstruction and development; it had been admitted to the UN and several other international organizations; and it had strengthened its own military capabilities, in the form of the SDF.

Accordingly, the officials who drafted the new Security Treaty explicitly placed economic cooperation among the objectives of the bilateral arrangements; incorporated a greater emphasis on the necessity of following the principles of peaceful international relations set out in UN Charter; and removed some of the elements that Kishi and other Japanese leaders, both inside the ruling Liberal Democratic Party (LDP) and outside it, had found objectionable in the 1952 version. The US forces based in Japan lost the capacity (which had never been tested in practice) to carry out military operations against domestic opponents of the Japanese government; the new Treaty committed the United States unequivocally to defend Japan in the case of any external attack by any other country; and it provided for mutual consultation from time to time on how the Treaty was being implemented. It also made some concession to the growing sensitivity of the Japanese public to the issue of having nuclear weapons on Japanese territory. While it still allowed the United States the option of keeping such weapons in Japan, it provided for the holding of consultations if any major changes in the deployment of US forces or equipment were contemplated. Finally, the new Treaty would be allowed to expire if either party decided to end it – mutual agreement was no longer necessary – and in any case was to be in force for only 10 years, unless formally renewed.

In spite of these concessions to Japanese public opinion, Prime Minister Kishi encountered organized opposition to the 1960 Treaty, not only from the JSP and other elements on the political left, but also from more moderate groups (see Packard). The opposition feared that Japan's close alliance with the United States enhanced the risk that their country could be drawn into a conflict with either China or North Korea, and they also objected to the persistent pressure from the United States for Japan to strengthen its defenses, which they interpreted as pressure to revive Japanese militarism. Kishi's employment of strong-arm tactics to get the Treaty ratified by the Japanese Parliament, in May 1960, only made matters worse. The unprecedented demonstrations by opposition groups on the streets of Tokyo, which were joined by thousands of radical students, forced the Japanese government to cancel a visit to Japan by US President Dwight D. Eisenhower, and eventually resulted in the resignation of Kishi himself from office.

Security Cooperation Today

The 1960 Treaty was renewed in 1970 amid further protests, which included the largest demonstration held in Japan since the Occupation years. The Treaty remains in force to this day, while public opposition to it has declined, as symbolized by the events of 1994, when the JSP, now renamed the Social Democratic Party of Japan, dropped its longstanding hostility to the Treaty upon entering office in coalition with the LDP. In recent years, against the background of the ending of the Cold War, the numbers of US personnel and bases have been considerably reduced, thus removing some of the targets of the remaining opposition groups.

Japan's security policy is clearly still largely based on the Treaty with the United States, and is likely to continue to center on the Treaty well into the coming century. That much is clear from the most recent formal statements of the security relationship: the Joint Declaration issued by US President Bill Clinton and Prime Minister Hashimoto Ryutaro following their summit meeting on April 17, 1996; and the revised guidelines on security cooperation, issued in September that year.

The Declaration not only reaffirmed the status of the Security Treaty but also effectively restructured the bilateral relationship with regard to East Asia (see Chin), by emphasizing that the security relationship between Japan and the United States offers a basis for stability in the whole region as it enters the 21st century. In particular, the United States committed itself to maintaining around 100,000 military personnel in East Asia. The Declaration also initiated a review of the guidelines on security cooperation (1978); affirmed closer logistic support and collaboration between the Japanese SDF and the US Armed Forces; and committed the two countries to continue sharing advanced technology, equipment and information, and to consult with each other on military policies and postures. The scope of the relationship has thus been somewhat broadened, from protecting Japan alone, to proposing joint efforts to maintain the security and peace of East Asia, and from being based on a bilateral pact to initiating a partnership capable of facing global security concerns, such as arms control, disarmament, nuclear test bans, and UN peacekeeping operations (see also Chapter 19).

Following the Declaration, the two governments proceeded to reconsider and revise the detailed guidelines that govern the implementation of their relationship under the Security Treaty. Since September 1996, these guidelines have made explicit provision for cooperation during emergencies in the region around Japan, including such activities as evacuating non-combatants from battle zones, sweeping for mines, search and rescue activities, rear area support, and inspection of ships on the high seas to ensure compliance with economic sanctions (see Mochizuki). It is sufficiently clear from past statements and agreements that the guidelines would allow Japan to take part in military and ancillary activities as far away from its home islands as the Taiwan Strait or the Korean peninsula: in that sense, the guidelines have simply been adjusted to reflect longstanding commitments more clearly. Whether China was right to interpret them as expressing a new policy of containing Chinese ambitions remains to be seen (see also Chapter 17).

Meanwhile, although the security relationship appears likely to survive pressures and criticisms on both sides, the continuing presence of US military personnel irritates many Japanese people, whose views must be taken into account in the formulation of security policy into the 21st century. Occasional friction still arises, largely as a result of the major presence of US personnel and bases in Okinawa long after the official reversion of the islands to Japan in 1972. These facilities include the US Air Force's base at Kadena, its largest outside the United States itself. In 1995, for example, the rape of a teenage Okinawan girl by three US servicemen led to considerable protest and diplomatic activity at very high levels in both countries. Again, in April 1997 the government decided it was necessary to obtain additional legal powers to make compulsory purchases of land in Okinawa for allocation to US forces, over protests from local landowners who had refused to renew leases on existing US bases.

Defense Spending, Arms Exports and Nuclear Weapons

Nevertheless, Japan has been able to implement certain defense-related measures that distinguish it from other elements in the worldwide network of US security relationships. In particular, Japan has insisted on pursuing distinctive policies on arms spending and exports, and on nuclear weapons. There have been adjustments in these policies in recent years – reflecting domestic political pressures at least as much as relations with the United States – but their continuation, even after these adjustments, suggests that "sovereignty without autonomy" (John W. Dower's term, cited above) may no longer be entirely appropriate as a description of Japan's position in international relations. Indeed, the pursuit of these distinctive policies, alongside the commitment to the security relationship with the United States, makes it difficult to call Japan an "ally" of the United States in the same full sense as (for example) the European signatories of the North Atlantic Treaty.

In January 1968, Prime Minister Sato Eisaku saw to it that the Japanese Parliament introduced a total ban on exports of military equipment, and also announced the "three non-nuclear principles", promising that Japan would not produce nuclear weapons, possess them, or allow their introduction into Japanese territory. (Sato received the Nobel Peace Prize in recognition of this policy in 1974.) Both policies remained intact until the early 1980s. In January 1983, however, the government of Prime Minister Nakasone Yasuhiro partially lifted the ban on arms exports to permit exports of "defense-related components", which have proceeded since then under the strict terms of a US/Japan agreement on transfers of military technology, signed in November 1983. There have also been persistent reports, which the US authorities will neither confirm nor deny, that nuclear weapons have in fact technically entered Japanese territory on US vessels entering or leaving US naval bases, notably at Yokosuka, the headquarters of the US Seventh Fleet.

In addition, in November 1976 Prime Minister Miki Takeo formalized the longstanding policy of keeping government spending on defense below 1% of GNP, but in December 1986 Nakasone formally reversed this policy. Nevertheless, defense spending had risen only to 1.004% of GNP by the time that

Nakasone left office the following year, and it has remained significantly lower than in other OECD countries, or in other countries allied to the United States, ever since, despite the criticisms of those US politicians who perceive Japan as gaining economic advantage from its "sheltering" under the US security umbrella.

Summary

The Security Treaty with the United States is not as unpopular in Japan as it once was, now that both the internationalist left and the ultra-nationalist right have lost much of their influence in domestic politics. The security relationship with the United States has come to be widely seen either as, at best, one aspect of Japan's commitment to multilateral efforts to maintain peace and stability in East Asia, or as, at worst, a necessary evil if Japan is to be taken seriously by the United States and other western countries. Nowadays, the serious problems in relations between the United States and Japan tend to arise, not from security issues, but from mutual disagreement and misunderstanding over trade.

Trade Policies: From Partnership to Rivalry

A pattern of tension between US and Japanese interests in international trade has been a feature of the relationship between the two countries since the end of World War II, As Japan has moved from being a junior partner in the US trading system to being a powerful rival in global markets, the United States has abandoned its initial policy of supporting Japan's industrialization and diversification, and now seeks to balance cooperation in securing economic goals with defensive measures against allegedly "unfair" Japanese trading practices.

A Partnership Against Communism

Initially, there was serious disagreement between US and Japanese policy-makers over trade with China, which has always been very important for Japan. After the victory of the Chinese Communist forces in 1949, the United States extended its policy of containing Communism from the Soviet bloc to China, and exerted pressure to prevent Japan from gaining access to the China market. Instead, US officials, collaborating with sections of the Japanese bureaucracy, tried to help Japan redirect its trade to Southeast Asia, over the objections of Britain, France and the Netherlands, all of which had their own interests in the region (see also Chapters 17 and 18). The US plan was apparently to make Japan an integral part of the regional economy of East Asia, if possible under US hegemony.

For many years the potential for problems over trade was greatly reduced, as Japan's economy benefited immensely from US policies, which constantly linked trade and security issues. Both the Korean War in the 1950s and the Vietnam War in the 1960s benefited Japan, as US military procurement funds flowed into its industries. The United States allowed Japanese exports into its markets, while Japan protected its own markets against any foreign competition. The United States also sponsored Japan's entry into GATT in 1955. Indeed, the impressive growth of the Japanese economy between the 1950s and the early 1970s can in part be explained by the "reverse course" initiated under the Occupation, and the many trade concessions that the United States granted to Japan.

However, as Chalmers Johnson has pointed out, the Japanese "developmental state" and its institutions are just as responsible for the "economic miracle" and the consequent ability of Japan to challenge US hegemony in the 1980s. As a developmental state, Japan committed itself firmly to economic growth, through a system of cooperation between government and industry. Japan's elite developed the necessary strategies to take advantage of the boom and bust cycles of the international economy, in order to lift the domestic economy out of the stagnation that had characterized the immediate postwar period. The elite, recruited from the best universities, made a decision to adopt a directed capitalist path for development, and (to borrow Johnson's terminology once again) generally succeeded in devising and implementing "market-conforming" rather than "market-displacing" strategies (see also Chapter 3).

Eventually, despite the hopes of successive US administrations that Japan could be kept in a strictly subordinate position within the sphere of US influence, Japan overtook the United States itself, both in the extent of its penetration of world markets and in its relative levels of capital accumulation. Here, too, security policy had an indirect impact, since Japan, like West Germany, enjoyed all the advantages of US technology, combined with a relatively cheap labor force and, crucially, extremely low levels of spending on defense. Japan's growing economic power began to affect the material base of US productivity and, as the economic hegemony of the United States began to wane, its policy-makers and their advisers began to become less tolerant of Japan's protectionist trade policies. For example, in the spring of 1970 Stephen D. Cohen, then the chief economist for the US-Japan Trade Council, wrote a memorandum in which he asserted that Japan's trade relations with the United States were not reciprocal. Japan's idea of "liberal trade" seemed to him, as to others in the United States, to mean no more than assuring open markets for its own products. Cohen's assessment was that US officials had lost all faith in discussing problems with Japanese officials, and that some form of direct action was now warranted.

The "Nixon Shocks"

Nothing contributed more to the fundamental transformation of relations between Japan and the United States than the series of "shocks" inflicted by US President Richard M. Nixon in rapid succession, soon after Cohen's memorandum was published. Not content with abolishing the Bretton Woods system of fixed exchange rates (1971), opening relations with Communist China (1972) and withdrawing troops from Vietnam (1973) – and failing to consult the Japanese government in any of these cases – Nixon also set about imposing tariffs on virtually all imports into the United States. These events were enough in themselves to alter Japan's accustomed ways of doing business, but in addition the circumstances surrounding many of Nixon's decisions, and the feelings of mistrust that were aroused on both sides, severely tested the relationship. Nevertheless, Japan, being the junior partner, had to accept the unilateral policy initiatives emanating from Washington, and adjust its policies accordingly.

In particular, the textile negotiations between the Nixon administration and the government of Prime Minister Sato provide a classic example of the kind of foreign pressure (*gaiatsu*) that Japanese decision-makers always dread. Confronted with domestic problems affecting the strength of the US dollar, Nixon decided to take a tough approach with his country's trading partners, especially Japan, which he felt ought to contribute more to the cost of its own defense. Central to Nixon's disaffection with Japanese leaders was what he regarded as Prime Minister Sato's inability, or reluctance, to persuade Japan's textile producers to impose voluntary restrictions on their exports to the United States. The textile issue had been brewing from the time of President John F. Kennedy (1960–63) onwards, and US jobs were at stake. For Nixon, the textile problem was intimately tied up with domestic politics and his own prospects of re-election in 1972, for which he sought the support of the US textile lobby. The pressure on Nixon to get a settlement from Sato was increased by the efforts of Wilbur Mills, a Democrat who was Chairman of the powerful Ways and Means Committee of the House of Representatives. Mills wanted to take the credit away from Nixon and his fellow-Republicans, by threatening to impose quotas on Japanese imports through legislation which would then attract the US textile lobby to support the Democratic Party. While neither Nixon nor Mills wanted to go as far as taking legislative action to make the Japanese comply, they both hoped that the threat of such legislation would ease compliance. Thus, the textile issue festered through a series of misunderstandings about what was said and what was meant between Nixon and Sato.

Nixon also linked the textile issue to the question of the return of Okinawa to Japanese control, which increased the pressure on Sato to find ways to settle the issue. This deliberate linkage of trade with security issues became further complicated when Nixon, having opened relations with China without consulting Japan, threatened to impose trade sanctions on Japan under the Trading with the Enemy Act. By abandoning the fixed exchange rate between the US dollar and the yen, Nixon had also compelled Japan to value its currency upwards, causing an overall slump in Japanese exports to the United States. Finally, it became clear to the powerful employers' organization *Keidanren* (the Federation of Economic Organizations) that voluntary compliance on textile exports would be far better than the prospect of all Japanese exports becoming subject to US tariffs. After Tanaka Kakuei, then Minister of International Trade and Industry, promised to introduce a relief package to compensate for lost exports, the Japanese Textile Federation agreed to a deal.

Negotiating Change

The Nixon shocks and the textile negotiations mark a watershed in the trade relationship between Japan and the United States. Ever since the early 1970s, negotiations have become somewhat more sophisticated, largely because both sides have learned from the experiences of those years and now prefer not to take disagreements to the point of open confrontation.

Between 1970 and 1980, Japanese governments followed a strategy of responding to US demands by announcing trade liberalization packages. These announcements were usually made before high-level bilateral leadership meetings, so as to pre-empt crisis talks. However, Japanese governments implemented these packages only fitfully and incompletely, and the strategy was abandoned after it was discovered that first resisting and then acceding to at least some US demands was a better strategy. Over the same period, small, lower-level bilateral groups, whether governmental or private, began to work on specific issues, combining this work with well-publicized high-level summits to provide symbolic reassurance. For example, the Ad Hoc Group on Petrochemicals worked behind closed doors during the 1980s, while US President Ronald Reagan and Japanese Prime Minister Nakasone allowed their "Ron/Yasu" meetings to become the focus of media attention.

More recently, under both the Bush and Clinton administrations, negotiations have come to center on US efforts to make the structures of Japanese industry and business practices resemble more closely the market-oriented structures of the United States itself – a policy which, conveniently, can be presented as favorable to Japanese consumers as well as their counterparts in the United States. Trade relations between the two countries now move on a dual path. On one level, there are moves to make Japan open its markets to specific US products, and success in such market openings is measured by quantitative criteria. On another level, trade and diplomatic negotiations are aimed at the eventual removal of what US policy-makers regard as structural impediments to open markets in Japan (see Schoppa).

It is striking that, no matter how sophisticated the resolution of trade disputes has become in recent years, they seem to go through some predictable patterns. It is always the US administration that spots a problem, demands a solution, and then specifies what it will be willing to accept as a solution. It is the Japanese government that, again and again, finds itself reacting, often with a ritual defensiveness, to US complaints. Since the scope of trade between these two economic giants is huge, there are potentially any number of issues that can become problems from the US perspective. While the process by which any specific problem is incorporated in the US agenda for trade negotiations remains unclear, it is obvious that a combination of domestic political factors plays a significant role. After US officials have declared the existence of a trade problem, Japanese officials usually come out with an initial denial that there is any such problem, but eventually agree to discuss the matter. When negotiations start, rhetoric and counter-rhetoric spread through the regular channels. US spokespersons accuse Japan of "unfair" trade practices or, what is even worse, subscribing to the sinful doctrine of "neomercantilism". Japanese officials then complain that such statements amount to "Japan-bashing" or scapegoating, and try to delay any resolution of the issue. The US administration responds by threatening dire consequences, in the form of protectionist measures in the US

Congress. The Japanese side finally gives in and grudgingly accepts at least part of the US demands, all the while appeasing domestic political forces by claiming that it has to give in to *gaiatsu*. The dreaded *gaiatsu* originating from the United States thus becomes part of Japan's negotiating strategy, since it serves as a useful device to help the Japanese elite to sell a particular US demand to recalcitrant domestic interest groups.

Conclusion

In spite of all the troubles that the relationship between Japan and the United States has undergone, it remains remarkably intact. Between the end of World War II and the end of the Cold War, the relationship passed through three great traumatic moments (see Campbell). The first was the diplomatic crisis at the end of the Occupation, in 1951–52; the second was the crisis associated with the revision of the Security Treaty in 1960; and the third was the economic crisis pertaining to the shocks inflicted by President Nixon in the early 1970s. Throughout these crises the two sides learned valuable lessons about their respective modes of doing business and engaging in political negotiations (see Kimura). Bilateral trade relations, in particular, rest on national interests and economic goals, but those economic goals themselves are mediated by postwar international relations: as has been indicated above, the security relationship and the trade relationship are not as distinct from each other in practice as they are in theory. One option has been to build cooperative relations outside areas of conflict, for example, through the Japan Foundation and the grants that the Japanese government gives to various universities in the United States. Japanese decision-makers, in government and in the business world, have become well-versed in combining micro-level corporate philanthropy within the United States with macro-level global partnership to address global issues such as hunger and the environment.

Throughout the Cold War era, the various strands of the relationship – security, trade, cooperation in other areas – could all be seen as subordinate to the overriding goal of keeping Japan within the anti-Communist camp, while Japanese political leaders saw the Security Treaty and the related trade advantages as being decidedly in the Japanese national interest, whatever specific problems might arise from time to time. However, now that the Soviet Union has collapsed the familiar foundations for the relationship no longer exist (see Mochizuki). In particular, the security relationship is likely to come under renewed scrutiny. Japan's defense budget remains small in relation to its GNP, yet it is now the third largest in the world: it is no longer entirely clear what it is for. Japan still contributes little to the security of East Asia in general, and in the United States more and more voices are being raised to assert, once again, that Japan is free-riding on the security that the US military presence provides. The continuing relative decline of US hegemony, coupled with the revival of China as an economic power – and perhaps some day as a military rival – may eventually lead to a further revision of the Security Treaty. These and related trends in the global economy will also bring further gradual changes in the trade policies of both the United States and Japan.

Further Reading

Bienen, Henry, editor, *Power, Economics and Security: The United States and Japan in Focus*, Boulder, CO: Westview Press, 1992

Borthwick, M., editor, *Pacific Century: The Emergence of Modern Pacific Asia*, Boulder, CO: Westview Press, 1992

Buckley, Roger, *US-Japan Alliance Diplomacy, 1945–1990*, Cambridge and New York: Cambridge University Press, 1992

Campbell, John C., "Japan and the United States: Games that Work", in Gerald L. Curtis, editor, *Japan's Foreign Policy After the Cold War: Coping With Change*, Armonk, NY: M. E. Sharpe, 1993

Chin Chu-Kwang, "The U.S-Japan Joint Declaration", in *World Affairs*, Winter 1998

Cohen, Stephen D., *Uneasy Partnership: Competition and Conflict in US-Japan Relations*, Cambridge, MA: Ballinger, 1985

Cohen, Warren, editor, *Pacific Passage: The Study of American-East Asian Relations on the Eve of the Twenty-First Century*, New York: Columbia University Press, 1996

Destler, I. M., with Sato Hideo, Fukui Haruhiro, and Priscilla Clapp, *Managing an Alliance: The Politics of US-Japanese Relations,* Washington, DC: The Brookings Institution, 1976

Destler, I. M., with Sato Hideo and Fukui Haruhiro, *The Textile Wrangle: Conflict in Japan-American Relations, 1969–1971*, Ithaca, NY: Cornell University Press, 1979

Destler, I. M., with Sato Hideo, *Coping with US-Japan Economic Conflict*, Lexington, MA: Heath, 1982

Dower, John W., "Peace and Democracy in Two Systems", in Andrew Gordon, editor, *Postwar Japan as History*, Berkeley, Los Angeles: University of California Press, 1993

Dunn, Frederick S., *Peace Making and the Settlement with Japan*, Princeton, NJ: Princeton University Press, 1963

Finn, Richard B., *Winners in Peace: MacArthur, Yoshida, and Postwar Japan*, Berkeley, Los Angeles: University of California Press, 1992

Frost, Ellen L., *For Richer, For Poorer: The New US-Japan Relationship*, New York: Council on Foreign Relations, 1987

Halliday, Jon, *A Political History of Japanese Capitalism*, New York: Pantheon, 1975

Johnson, Chalmers, *MITI and the Japanese Miracle: The Growth of Industrial Policy 1925–1975*, Stanford, CA: Stanford University Press, 1982; Rutland, VT, and Tokyo: Charles E. Tuttle, 1986.

Kades, Charles, "The American Role in Revising Japan's Imperial Constitution", in *Political Science Quarterly*, Summer 1989

Kimura Takayuki, "Japan-US Relations in the Asia-Pacific Region", in Richard L. Grant, editor, *The Process of Japanese Foreign Policy: Focus on Asia,* London: The Royal Institute of International Affairs, 1997

Kitamura Hiroshi, Murata Ryohei, and Okazaki Hisahiko, *Between Friends: Japanese Diplomats Look at Japan-US Relations*, translated by Daniel R. Zoll, Tokyo: Weatherhill, 1985

Krasner, Stephen D., "Japan and the United States: Prospects for Stability", in Inoguchi Takashi and Daniel I. Okimoto, editors, *The Political Economy of Japan*, Volume 2, *The Changing International Context*, Stanford, CA: Stanford University Press, 1988

LaFeber, Walter, *The Clash: A History of US-Japan Relations,* New York: W. W. Norton, 1997

Lauren, Paul G., and Wylie, Raymond F., *Destinies Shared: US-Japanese Relations,* Boulder, CO: Westview Press, 1989

Mochizuki, Mike M., editor, *Toward a New Alliance: Restructuring US-Japan Security Relations*, Washington, DC: The Brookings Institution, 1997

Packard, George R., *Protest in Tokyo: The Security Treaty Crisis of 1960*, Princeton, NJ: Princeton University Press, 1966

Schaller, Michael, *Altered States: The United States and Japan since the Occupation*, Oxford and New York: Oxford University Press, 1997

Schoppa, Leonard, *Bargaining With Japan: What American Power Can and Cannot Do,* New York: Columbia University Press, 1997

Yoshitsu, Michael, *Japan and the San Francisco Peace Settlement,* New York: Columbia University Press, 1963

Chitta R. Unni is Professor of Philosophy at Chaminade University, Honolulu, Hawaii. He has written extensively on the postwar Japanese polity, and on the philosophy of politics in Japan.

Chapter Sixteen

Japan and Europe

Julie Gilson

The beginning of Japan's relations with Europe was precipitated by the accidental arrival of Portuguese sailors on an island south of Kyushu in 1542. Bringing muskets and heralding the introduction of Christianity, these foreign intruders were welcome for a time, until the Shoguns, the military dictators of Japan, closed its doors to most of the outside world, in stages up to the 1630s. There followed more than 200 years during which interaction with Europeans was limited for the most part to trade with the (non-proselytizing) Dutch, through the island of Dejima near Nagasaki.

Even after re-establishing contact with the western world in the 19th century, Japan, unlike other countries in East Asia, was not destined to become a protectorate or possession of any of the western European powers. Indeed, the Pacific War saw Japan challenge their very presence in Asia. Nor did Japan's eventual defeat bring it under their yoke, since the Allied Occupation forces were mainly (but not exclusively) American, under the command of US Generals Douglas MacArthur (1945–51) and Matthew Ridgeway (1951–52). The Occupation years ensured that Japan's postwar relations with the rest of the world would be influenced to a large extent by US policy (see Chapter 15). The Security Treaty (1952, with subsequent revisions) forms the bedrock of this Japan/US relationship, and has ensured its durability and intensity, even in the face of growing trade friction, and the "Japan-bashing" of the 1980s and 1990s.

In contrast, Japan's relations with the countries of Europe for the past 50 years or more have been largely neglected by practitioners and observers alike, and, when conducted at all, have occurred on a predominantly ad hoc basis. For this reason, when Japan's relations with Europe are occasionally mentioned in academic literature or policy statements, they tend to be classified as the "weak side" in a Japan/US/Europe triangle; and the "Europe" that is mentioned often still means only the western part of the continent, or sometimes just those states that are members of the EU. The trilateral concept was developed, to include western Europe only, during the 1970s, in particular through the Trilateral Commission, and its use perpetuates the idea that little or no change has occurred in relations between Japan and Europe, despite the ending of the East/West division that underpinned it. Its application also ensures that attention remains focused upon the two "stronger" links, supported by the Japan/US Security Treaty, and by US and European involvement in the North Atlantic Treaty Organization (NATO), reinforcing the suggestion that cooperation between Japan and any part of Europe is negligible. The analogy is sustained by the fact that there are no joint security arrangements between Japan and Europe, and by the fact that their trade disputes, given that they are relatively minor when compared with Japan/US disagreements, are not covered widely by the world's media. In addition, the physical distance between Europe and East Asia, combined with linguistic barriers and other cultural divides, would seem to militate against more sustained forms of cooperation between Japan and Europe. Until recently, this mutual lack of interest has been compounded by an absence of institutional arrangements for regular meetings between Japanese officials and their European counterparts.

A further problem impeding the development of relations between Japan and Europe concerns the definition of the types of international actors involved. On the one hand, Japan represents for many observers a traditional nation-state, with a unique identity, a view reinforced by *Nihonjinron* ("Japaneseness") literature, and yet at the same time the Japanese are pressed more and more to clarify their global and regional roles in the post-Cold War world. On the other hand, several diverse cultures combine to make up "Europe", and the proliferation of different institutions, most notably but not exclusively the EU, makes it difficult for external partners to be sure about which "Europe" to address in which situation. In the face of such enduring problems and with enough concerns to occupy them in other areas of activity, there seems little reason to expect an imminent strengthening of this relationship.

Nevertheless, despite a dearth of political will and international compulsion, a number of developments between Japanese and European representatives in recent years have resulted in the establishment of several mechanisms for discussions covering a whole range of subjects. In addition, although the position of the United States continues to exert a significant influence on the foreign policies of the countries of Europe and Japan, the rationale that cemented these partners to the United States during the period of the Cold War no longer exists. For this reason, too, there have been some changes in Japan/Europe relations, particularly in the past 10 years. This is not, of course, to suggest that Japan/Europe relations are about to overtake either Japan/US or Europe/US relations in importance, but it does imply that it is worth trying to extricate the "third leg of the triangle" from its trilateral confines, in order to highlight the quantitative and qualitative changes that have taken place in Japan/Europe relations over the past 20 years or more, and to examine the potential importance of these relations at the start of the 21st century.

Cooperation Between Japan and Europe Today

When considered in quantitative terms, Japan/Europe relations are well-developed in a number of bilateral and multilateral forums. Representatives from Japan and various European countries, as well as those from EU institutions, meet frequently in multilateral forums, such as the World Trade Organization (WTO), the IMF, the Group of Seven (and now the Group of Eight), and as part of the "G24" process, to assist several countries in central and eastern Europe. Representatives also meet in various UN bodies and other regional groupings. These groupings include the Post-Ministerial Conference of the Association of Southeast Asian Nations (ASEAN) and its Regional Forum (the ARF); as well as the Asia-Europe Meeting (ASEM), held so far in Bangkok in 1996 and in London in 1998, which, despite its name, in fact brings together only the 15 member states of the EU and 10 countries in East Asia.

At a "bilateral", Europe/Japan level, most arrangements are formulated between Japan and the EU. At the highest level, annual summits have taken place since 1991 between the Japanese Prime Minister, the head of whichever national government is holding the EU presidency (which rotates every six months), and the President of the European Commission, the EU's appointed policy-making and executive body. At lower levels this structure is replicated in a number of forums. There are ministerial meetings between members of the Japanese government and their counterparts in the EU "Troika", representing the current, previous and subsequent EU presidencies. There are also meetings, again representing Japan and the Troika, between "political directors", a broad term that includes such officials as the Assistant Vice-Minister of Japan's Ministry of Foreign Affairs (MoFA), the Foreign Minister of the country holding the EU presidency, the head of the European Commission's Directorate-General in charge of External Relations (DG1), and members of EU embassies in Tokyo. Participation in such meetings varies in line with the level of representation for given meetings, and with the specific political issues under discussion. Both in political directors' meetings and in other forums, Japan, the European Commission and EU member states conduct high-level financial consultations, a dialogue on industrial policy and cooperation, a committee for trade cooperation, talks on competition policy, a science and technology forum, a social affairs conference, consumer policy talks, environment committees, talks on aid policy, a dialogue on standards and certificates, talks concerning deregulation; and committees to cooperate with regard to trade statistics. They also sponsor a league of friendship between Japan and Europe, as well as a business dialogue. While most of these forums do not enjoy media attention, they do illustrate that high-level officials from Japan and Europe – principally, the EU and its member states – participate in an increasingly broad and regularized dialogue. At other levels, too, exchanges take place with increasing frequency. For example, in 1993, 4,915 Japanese students went to study in the 12 countries that were then members of the EU, while 653 students from those countries were registered in Japan (according to the Asia Committee of the European Science Foundation).

Despite growing contacts, these disparate meetings lack an overarching sense of purpose by which to define Japan/Europe relations. The 1991 Hague Declaration (see below) was an attempt to address this concern at the level of Japan and the EU, but it too offered only general statements and provided a written compilation of mainly existing arrangements. As is shown below, however, the Hague Declaration was important in recognizing the set of relationships that had already developed over more than two decades of incremental activities.

Relations with the EU and its Predecessors

Contemporary relations between Japan and the EU now encompass a broad set of subjects, but their appearance on the agenda was not always planned (see Gilson). Many of the instances in which representatives from Japan and the EU, or its predecessor bodies, have come together derive from ad hoc approaches to global, regional or bilateral changes, and, as mentioned, the Hague Declaration itself represented in the main a codification of existing arrangements that had developed from occasional encounters. For this reason, the mechanisms now in place have assumed a permanent character not necessarily intended when they were formulated. By looking at the derivation of those mechanisms, it is possible to examine their potential consequences.

From the 1940s to the 1970s

At the end of World War II, Japan was preoccupied with the implementation of internal reforms needed to help rebuild a devastated economy and society. Similarly, western Europe was concerned with problems resulting from the devastation of the war. Both were directed by US control and financing, so that external relations were conducted principally through a US filter and by means of structures such as the Marshall Plan in western Europe and the Security Treaty between Japan and the United States. As the Cold War set in, the association of both Japan and western Europe with the United States was reinforced further.

That is not to say, however, that there were no contacts between Japan and the countries of western Europe during the early postwar period. Indeed, even while internal questions regarding the establishment of the European Economic Community (EEC), the first form of what is now the EU, were being addressed during the 1950s, attention in western Europe began to turn towards the growing success of the Japanese economy. Given the obvious concern of the United States regarding the influx of Japanese good onto its markets, several western European countries participated in blocking Japan's early entry into GATT. When Japan did eventually join GATT, in 1955, many western European countries insisted on maintaining unilateral "safeguard" clauses that would be eradicated only slowly over time, and through individual negotiations between Japan and the country in question. Similarly, in November 1962 the Council of the EEC, which then had six member states, decided that a safeguard clause should be included in all bilateral trade agreements with Japan, since Japan was seen to be benefiting unduly from GATT.

Despite such restrictions, and the accompanying emphasis on country-by-country negotiations with the Japanese, by the end of the

1960s decision-makers in what had become the European Community (EC) were beginning to entertain the prospect of developing bilateral trade deals with Japan. The Commission of the EC attempted, initially in vain, to represent the whole of its membership in a unified trade approach towards Japan. In 1974, the Commission's delegation was established in Tokyo, although the equivalent Japanese mission in Brussels was not created until five years later. However, given the ambiguous and limited role of the EC, particularly at that time, it was hard for their Japanese counterparts to accord legitimacy to its delegation representatives, a situation exacerbated by the thinly veiled disdain towards the delegation's role harbored by many Tokyo-based representatives of the individual member states. The 1970s, then, witnessed a division of approaches, between the search for EC-level agreement with the Japanese and the continuing need to deal on a country-to-country basis.

Although an overall agreement between Japan and the EC was not forthcoming during the 1970s, trade agreements were formulated and Japanese investment in Europe was increased (see Darby). Progress was stimulated in part by the entry of the United Kingdom, the Irish Republic and Denmark into the EC, in 1973, as well as by the need for Japan to diversify its economic and political activities, as a result of the three "Nixon shocks" of 1971 (see Chapter 15), and the "oil shocks" of 1973 and 1979 (see Chapter 1). The second half of the decade witnessed growing concern in western Europe at the intensification of Japanese export efforts into the region, and Japan became the particular target of anti-dumping practices.

The Japanese were to experience further western European discontent in 1976, during a trade mission organized by the Federation of Economic Organizations (*Keidanren*), headed by Doko Toshio. The criticisms leveled at the Doko mission were to bring into relief many of the growing trade-related problems between Japan and western Europe. The Keidanren representatives were attacked in particular over Japan's increasing penetration of western European markets, and over the obstacles to western European investment in Japan. As a result, Doko eventually returned to Japan with a list of complaints, but was to find division at home and criticism of the way that he had handled the entire mission (see Wilkinson). While the complaints continued, mechanisms for their resolution were not forthcoming.

Another of the criticisms directed at Japan during this period was that it allegedly operated a "divide and rule" policy among EC member states and other western European countries; in particular, it refused to deal with the EC Commission. Although this criticism may have had some validity, the EC itself was unable at the time to convince external partners to do business uniquely with EC bodies. The EC, perhaps especially in the eyes of the outside world, remained a hybrid body, with only limited power, and as a result external interlocutors such as the Japanese did not always know whom they were dealing with. As long as this problem persisted, Japan had little choice but to continue to focus upon individual states.

Key Changes in the 1980s and 1990s

For several reasons relations between Japan and the EC began to develop more quickly during the 1980s. In the first place, Japan found itself increasingly out of step with the US view of the world, not in the least because of the Iranian hostage crisis, and the Soviet invasion of Afghanistan. On those particular occasions, Japanese officials found themselves liaising more and more with their EC counterparts, with whose policies they more closely agreed. For example, in the case of the hostage crisis in Teheran, representatives of the Italian presidency of the EC suggested the possibility of common action with the Japanese, and as a result the EC's Ministerial Committee and the Japanese government decided that the Japanese ambassador in Teheran would take the same action as his EC counterparts (see Tanaka). In his memoirs, the former Foreign Minister Okita Saburo wrote that the meetings held between Japan and the EC during this period transformed their mutual relations into a more substantial dialogue that encompassed political questions and the possibility of joint action (see Okita). Although no major institutions were created to sustain these new relations, the need to broaden dialogue with western Europe became a more frequent theme in Japanese official statements, and in 1983 and 1984 regular meetings were established between foreign ministers, and then between political directors. Although these meetings were frequently neglected, representatives of the EC Commission were gradually welcomed in Tokyo, and accorded a status more closely resembling that of their member state counterparts.

It was also during the 1980s that Japan began a concerted effort to obtain a more influential role in the IMF, the World Bank and other multilateral institutions, in order to become a more proactive player in international economic affairs (see Rapkin et al.). As part of redefining its new international role in this way, and given the well-established involvement of western European countries in these key bodies, Japan began to include in its official statements a recognition of the need to view western Europe, principally as embodied in the EC, as an international ally.

The EC, too, was redefining its international role at this time. After an initial period of stagnation in its internal momentum, the mid-1980s saw it relaunch itself onto the world stage. With the promise/threat of the 1992 Single Market program, outside investors, especially those from the United States and Japan, began to issue a wave of concerned utterances about the prospect of a "fortress Europe". In the event, many of the fears subsided, but they did provide important press coverage for the EC in Japan and engendered a number of high-level exchanges during the decade. The 1992 project combined in the latter part of the decade with renewed interest in the process of European Political Cooperation (EPC), in which EC (and now EU) member states coordinate activities beyond the trade realm. A political dimension became associated with the EC when EPC was included in the 1986 Single European Act, and was integrated further – in the form of the Common Foreign and Security Policy (CFSP) – as a result of the Maastricht Treaty of 1992, under which the EC became the EU (for most purposes) in 1993. This new development provoked some discussion in Japan regarding the future possibility of a "Mr/Ms CFSP" for what is now the EU, although representatives of the EU themselves have been slow to take up the challenge of defining an external political role.

Changes during the 1980s also altered views of Japan in the EC and its member states. In particular, representatives of the EC began to change their approach toward Japan, and in a 1988 White Paper the EC pledged to develop a more complete relationship with Tokyo. This decision subsequently led to the Hague Declaration and to the creation of a new political dimension within the relationship. The EC Commission, in particular, adopted the view that a more balanced approach towards Japan offered one means of dealing with trade problems, and provided an approach distinct from what it viewed as the more confrontational position still held by the United States. At the same time, Japan's growing trade surplus continued to pose a problem. As Japan's share of EC, US and third markets grew, the EC and its members began to push for greater access to Japanese markets, and started to institute programs to facilitate market-opening measures, such as the EC/Japan Trade Expansion Committee. During this period, however, these measures were not translated into a comprehensive agreement between the two sides.

The 1990s witnessed changes in relations between Japan and the whole of Europe, for a number of domestic and international reasons. First, events after 1989 focused international media attention upon the European continent and revitalized the conception of Europe as including many countries outside the EC (see below). In July 1989, the Japanese Prime Minister, Uno Sosuke, attended the Arche summit, which entrusted the EC Commission with the coordination of the international aid effort to central and eastern Europe, on behalf of the Group of 24 (G24). Even at that time, however, Japanese officials remained uncomfortable about dealing with non-state actors over issues that concerned specific nation-states. For their part, the governments of western Europe did not embrace Japanese involvement in the G24 process, and continued to court the attention of the United States, despite Japan's much greater potential contribution.

Second, in the 1990s more regular meetings between representatives of Japan and the EC led to the signing of the Hague Declaration in July 1991. Following in the wake of the Transatlantic Declaration, signed by the United States and the EC, this Declaration in effect listed all the ad hoc meetings that were taking place already. In spite of its inauspicious origins, however, the Declaration consolidated a potentially useful alliance which is now able to cover a wide range of issues involving Japan and the EU. In the negotiations preceding the signing of the Declaration, Japanese officials stressed the need to address the political side of the agenda, while they suspected the EC of trying to impose managed trade as a price for political dialogue. This criticism notwithstanding, it seemed that the Japanese had come to acknowledge the importance of direct dealings with the EC Commission. Conversely, the EC apparently had come to terms with the need to engage Japan in a broader dialogue, to the extent that in June 1992 the conclusions of a meeting of its Council of Ministers placed political dialogue ahead of economic dialogue.

Changing EU Approaches to Japan

Changing conditions within Europe mean that, for example, even the United Kingdom, one of the more "Euroskeptic" members of the EU, cannot ignore its external role. In January 1998, during his first visit to Japan as British Prime Minister, Tony Blair was also acting in the role of head of the government then holding the EU presidency. In that capacity he commented to his Japanese audience that the Economic and Monetary Union (EMU), scheduled to start in 1999, is important, and that attempts to widen the EU to other countries must also be accompanied by "deepening" processes within it. He also sought to convince Japanese investors that the United Kingdom remained the "gateway" to Europe.

A large imbalance remains in trade relations: a record deficit of about US$33.1 billion was reached by the 12 members of the EC as against Japan in 1992, compared with about US$22 billion in 1994. Certain mechanisms and promotional projects have been strengthened or developed in order to address this deficit, including the "EXPROM" campaign, of which the Executive Training Program has been one successful outcome. The development of the Trade Assessment Mechanism (TAM) has provided Japan and the EU (now with 15 members) with the means to compare trade performance on the basis of objectively defined trade statistics, and to indicate those goods of which the export performance suffers in Japanese or EU markets when compared with other markets for the same products. This mechanism provides a permanent process to which either side can appeal without having to set the agenda first. Other projects such as that to assess Japanese deregulation have met with similar success.

In March 1995 the European Commission published *Europe and Japan, the Next Steps*, following an assessment of developments and changes made in bilateral relations over the previous few years. On May 29, 1995, the EU's Council of Foreign Ministers adopted conclusions on the basis of this communication, which now serve as guidelines for relations with Japan. Despite such promises of progress, however, a unified approach to foreign policy action has yet to be developed by "Europe" – whether that word is defined to mean only the 15 members of the EU, or a larger group of western European states, or the continent as a whole.

Relations with Central and Eastern Europe

As long as the countries of central and eastern Europe were controlled by Communist regimes, barriers imposed by the Coordinating Committee for Multilateral Export Controls (COCOM) inhibited Japan and other countries allied to the United States from exporting to them. Nevertheless, Japan began to pay some attention to the region during the early 1980s, when, in recognition of some of the changes taking place there, and in the light of worsening relations with the Soviet Union, the Prime Minister's Office in Japan developed an "East European Doctrine", which encouraged all western countries to support democratic reforms in the region. Four branches of the Japan External Trade Organization (JETRO) had been opened in central and eastern Europe during the period of *détente* and, despite the fact that trade with the region was still marginal from Japan's point of view, Prime Minister Nakasone Yasuhiro made the political move of visiting Poland, East Germany and Yugoslavia in January 1987, during a particularly low point in Japan's relations with the Soviet Union.

In 1989–90, while Japanese policy-makers were beginning to ponder further the possible role of the EC, events in central and eastern Europe swept the wider continent into the international spotlight, and Japan started to pay closer attention to the countries in the region. Their political instability, currency and budget problems, and relatively poor infrastructure have combined with Japan's lack of linguistic and cultural know-how to make Japanese policy-makers and businesses cautious – even after 1989–90 – and the news coming out of them during that year of upheaval was treated principally as a series of "human interest" stories, at least to begin with. However, in Brussels in late November 1989 Japan joined other western countries, through its involvement in the G24 process, to set up the European Bank for Reconstruction and Development (EBRD), in order to help Poland and Hungary. The Japanese government extended US$200 million-worth of food aid and low-interest credits to them, and since then Japanese participation in the EBRD has grown to equal that of the most important European contributors. Other high-level visits to the region and financial contributions, aid credits and technical assistance from Japan were soon forthcoming in 1990. In the spring of that year the Japanese Foreign Minister announced a "New East European Doctrine", which included the recognition that aid for the region had become a common concern for the United States, western Europe and Japan, and acknowledged support for the role of the EC in the process of aid disbursement.

The MoFA document setting out the new policy also contained Japan's first public bid for observer status at what was then the Conference on Security and Cooperation in Europe (CSCE). This series of summit meetings, starting in Helsinki in 1975, was unique in the Cold War period, in that it brought together almost all states in all parts of Europe – whether NATO members, Warsaw Pact members or neutral countries – as well as the United States and Canada. France, in particular, was initially opposed even to observer status for Japan, while some in Japan itself were concerned about the effect that it might have on the country's dispute with the Soviet Union, and now with Russia, over the Kuriles, the islands which the Japanese call "the Northern Territories", occupied by Russian troops and settlers since 1945. While the Kuriles issue remains unresolved, Japan eventually achieved observer status at the CSCE in 1992, and retained it after the Conference became the Organization for Security and Cooperation in Europe (OSCE), in 1995. The CSCE/OSCE question illustrates a change in Japan's approach to Europe, and suggests that the events of 1989–90, rather than the 1992 project, pushed Japan towards a broader dialogue with the EC and now the EU. However, while the Japanese government's reasons for acknowledging Europe may apply to the entire continent, it is the EU, not the OSCE, that has in place effective mechanisms capable of sustaining relations. As these develop further, the Japanese view of Europe may become more centrally focused on the EU itself, and as processes such as ASEM reinforce the "EU" definition of Europe, it may become more difficult for apparently peripheral areas to attract Japanese interest.

There are reasons for optimism, however. In the first place, those countries with long-term plans to join the EU have found themselves the recipients of official visits by Japanese delegations, and those, such as Hungary or Poland, which have been major beneficiaries of EU aid have, as a result, garnered Japanese contributions towards their economic growth. Other factors have also encouraged Japanese interest in the region in the 1990s. Some of the countries of central and eastern Europe can offer both relatively cheap labor and access to EU markets; they can also provide plentiful natural resources, and workforces willing to learn new skills.

What is more, the ending of the Cold War has led to the eradication of the sense in Japan that "Europe", along with Africa and the Middle East, forms an exclusive domain of the Europeans, and that Japan's interests center only around the Pacific Basin, North America and Oceania. More and more policy papers and academic discussions in Japan note that events occurring in Europe may affect Japan's own well-being. However, a comprehensive formula for relations remains to be established, and without a deeper understanding of the nature of events in other continents, and the resolution of questions regarding the definition of Europe, these factors will continue to pull and push Japan's attention towards different interpretations of Europe.

Japan, Europe, and the United States

Relations between Japan and Europe continue to be regarded, on both sides, as poor relations to the bilateral structures in place between Japan and the United States, on the one hand, and between Europe (mainly western Europe) and the United States, on the other. Despite Japanese attempts to play a greater role in international burden-sharing, in the 1990s in particular, and in spite of Japan's observer status in NATO, granted in 1992, relations with Europe in general, and with individual European states, remain predominantly in the realm of "soft" issues. Despite the security dialogue now in place in the ARF, to which both Japan and the EU belong, these issues have still to be fully addressed, and the roles to be played by Japan and European institutions remain to be formulated.

More fundamentally, as long as Article 9 of Japan's "Peace Constitution" limits its ability to play a military role even within its immediate geographical region (see Chapter 2), Japan will continue to depend on the United States for its defense needs, and any military cooperation with Europe will be of secondary importance. Europe, too, retaining a US military presence on the soil of many countries, is far from formulating a bilateral military alliance with Japan. For this reason, any security arrangements between Japan and Europe are likely to proceed through multilateral channels that involve the United States.

Lingering Problems

Although relations between Japan and Europe have certainly developed over the past 20 years, and particularly since the middle of the 1980s, there remain several fundamental problems which hinder Japan and Europe from developing a more substantial partnership.

The first problem concerns the very nature of the actors involved. In contrast to Japan, Europe does not have a relatively fixed identity. Although something of a "European" identity is developing – particularly within the EU, which now issues common passports to its citizens, and is actively planning further expansion in the number of member states – the boundaries of where that identity begins and ends are more fluid than in the case of Japan. What is more, since the EU is the only European body to have organized institutional contacts with Japan, and the Japanese are more used to dealing with western European countries, much of the development of this identity rests upon changes within structures internal to the EU. For the present, the EU seems largely unaware of the external repercussions of any internal changes, and is slow to change structures, such as the rotating presidency and the Troika associated with it, that make it difficult for non-EU partners to know whom they are dealing with in their discussions with "Europe". This situation is rendered more complicated by the fact that, despite having augmented their role in recent years, the foreign delegations of the EU have not been given real leadership positions in the external representation of the EU.

Unless and until the main bodies of the EU are accorded a more important role in the development of a single EU foreign policy, it will continue to be acceptable for Japan to divide official responsibility for the European region between departments in the MoFA. The separation between the West European I and II divisions, and the coordination of economic responses in an entirely different location, hamper Japan's ability to develop an overarching policy towards Europe or even towards the EU. At the same time, private institutions and businesses make representations to individual governments in Europe or lobby the

Commission, depending on their needs, and their uncoordinated actions further reinforce the dichotomy between the national and "European" levels of negotiation. Popular interest in Europe, except as a set of historically rich tourist destinations, remains low in Japan, and no public movement is inclined at present to encourage a coherent approach towards Europe as a whole.

The fits and starts in the integration of the EU have nevertheless ensured that Japan pays attention to Europe on some issues. In particular, the movement towards EMU, and the prospect of the introduction of the euro as a single "European" currency, from 1999, have focused the attention of Japanese businesses on the need to make adequate preparations for the changes. The Amsterdam Treaty of 1997 was important in reassuring the Japanese that the single currency will go ahead, even if, initially, without the participation of all EU member states. These developments have been reinforced by the creation of structural bodies within the EU to deal with specific issues, so that the Japanese delegations who come over increasingly know whom to contact regarding this issue. The European Commission, which is becoming more important, both within the EU itself and in Tokyo, has played no small role in disseminating information in this regard. The Japanese government has also recognized the need to cooperate with the arrangements for the euro, which has the potential to challenge the US dollar. Difficulties remain, however, in formulating a coherent approach to issues that derive from EMU but which are not limited to the trade realm.

The broader dialogue now undertaken between Japan and Europe – meaning, once again, predominantly the EU – has, to date, covered issues such as landmines, although not without disagreement, and the UN arms register, but most of the projects undertaken never reach the headlines (see Gilson). In addition, most of the bilateral interaction comes on the periphery of a growing number of multilateral regional and international forums, such as the ASEM, the ARF, the Korean Peninsula Energy Development Organization (KEDO), the UN, and the WTO. Since "Europe" is represented in different guises in these various forums, there remains the problem of defining what it actually is.

A second problem concerns, not the quantitative developments in relations between Japan and Europe, but their qualitative nature. For most practitioners involved, the existing dialogue between Japan and "Europe" (however defined) does not have a high priority on their busy agendas. The Hague Declaration failed to elevate that relationship, and the *pot pourri* of activities that it covers, based on a variety of meetings devised for different purposes, prevents the formulation of a clear rationale upon which Japan and "Europe" could found their cooperation. This lack of purpose is illustrated more starkly by the fact that such cooperation does not seem to have benefited to date from changes in the position of the United States. It remains, of course, a key partner both for Japan, for example, with regard to the issue of instability on the Korean peninsula, and for the EU, most notably, with regard to the continuing conflicts within and among the countries that used to form the Yugoslav federation. Yet alliances with the United States are no longer as sharply defined as they were during the four postwar decades dominated by the Cold War. Japan and the countries of Europe have not taken up the potential challenge that such a blurring of divisions could offer, with the result that, as long as both sides fail to devise a framework that gives greater meaning to their mutual relations, they will be unable to escape from enduring caricatures of them as inferior partners and weak links.

The Future

The growing importance of relations between Japan and the EU was emphasized at their fourth summit meeting in Paris, on June 19, 1995, and every year the number of state visits to Europe by Japanese officials gradually increases. The currency crisis in East Asia in 1997–98 has provided evidence of the interdependence of global financial markets, and the need for the EU to coordinate a response in conjunction with the Japanese. On the Japanese side, the crisis has shown that Europe still offers some relatively stable and desirable destinations for commerce and finance. Both sides have been slow to take up the challenge, however, and continue to rely on ad hoc cooperation as circumstances dictate.

It is possible to offer only a speculative glimpse into the future. What has already become clear is that the 21st century will witness further digressions away from traditional state-to-state diplomacy, in favor of a whole range of subnational and supranational relationships, and will bring to the fore official and unofficial actors whose actions will influence future international relations. Reflecting this multi-level engagement, Japan and the countries of Europe, particularly the EU, have come to realize that they provide potentially stable allies for one another, especially with regard to those issues which have traditionally been regarded as belonging to the domain of "low" or "soft" politics, but which have begun to assume greater prominence on the international stage. These issues include environmental protection, the fight against drugs trafficking, conflict prevention, global macroeconomic coordination, aid to Russia; redefining relations with the United States, and problems of development as between the North and the South. In this regard, Japan and the countries of Europe have the potential to cooperate on a whole range of issues that do not require military intervention. Two major problems remain, however: defining partners and creating a permanent dialogue.

First, even if an agenda for bilateral exchange could be expressed more explicitly, different representatives of the regions continue to take part in different types of meeting, so that no single "Japan/Europe" relationship has been defined as yet. The definition of "Europe" depends to a large extent on the success of the EMU project, as well as on the continuing widening of EU membership, while Japan, too, is now reconsidering its role in the post-Cold War world. If the EU is to develop into a more legitimate international actor, capable of taking responsibility for decision-making in foreign policy, a clear and increasingly strong focus on the EU – whether with 15 members, as now, or with up to 26 members over the next two decades, as planned – may act to the detriment of those European countries left on its periphery, particularly as fewer and fewer remain outside. In any case, even if government officials and business elites clarify their understanding of "Europe," mixed perceptions in the popular realm are likely to lag far behind, and a large political project to inform the wider public will be required to underpin any lasting relationship between Japan and Europe. For now, both popular interest and political will are in short supply. The apparent lack of interest among Europeans in the crisis in the financial markets of East Asia serves to illustrate that even globally resonant events cannot convince mass publics that "their" problems are of concern to "us" in the West.

Second, once the actors have been decided upon, permanent and enduring institutions must be established to sustain and facilitate their relations. To date, Japan and the countries of Europe have accommodated their dialogue within a complex structure that houses different participants for different sets of discussions. Unless Japan and Europe – at the very least, the EU – can create for themselves a body of permanent mechanisms with an overarching rationale to justify an internationally recognized alliance, the likelihood is that they will continue to come together for brief encounters of little import.

There are several areas of common interest on which Japan and Europe could usefully strengthen their partnership as the world enters a new century. However, unless these areas can be addressed in a concerted manner, by clearly defined representatives, a stronger partnership is unlikely to develop.

Further Reading

Asia Committee of the European Science Foundation, editors, *Asia and Europe towards the 21st Century*, Strasbourg: European Science Foundation, 1997

A general account of contemporary activities as part of the ASEM process.

Bridges, Brian, *EC-Japanese Relations: In Search of a Partnership*, London, The Royal Institute of International Affairs, 1992

A brief but clear introduction to the development of relations between the EC and Japan up to the early 1990s, with useful appendices.

Bridges, Brian, "Japan and Europe: Rebalancing a Relationship", in *Asian Survey*, Volume 32, number 3, March 1992

An overview of contemporary relations, including business and political issues.

Daniels, Gordon, and Reinhard Drifte, editors, *Europe and Japan: Changing Relationships since 1945*, Ashford: Paul Norbury, 1986

A short edited work, based on the results of a symposium, this book provides a comprehensive examination of relations between Japan and western Europe up until the early 1980s.

Darby, James, editor, *Japan and the European Periphery*, London: Macmillan, and New York: St Martin's Press, 1996

This edited work provides useful data and a contemporary examination of economic relations between Japan and Europe, and includes many references to non-EC areas.

Gilson, Julie, *The Institutionalisation of Political Dialogue between Japan and the EU*, unpublished Ph. D. thesis, University of Sheffield, 1997

This provides a detailed overview of non-economic relations between Japan and the EU up to the mid-1990s.

Maull, Hanns W., editor, *Japan and Europe in an Interdependent World*, Tokyo: JCIE, 1988

Based on the proceedings of the ninth Europe-Japan Conference, this edited work covers political and trade relations between Japan and the EC before the ending of the Cold War.

Nakanishi Terumasa, Tanaka Toshiro, Nakai Yasuro and Kaneko Yuzuru, *Naze Yo-roppa to Te wo Musubu no ka* [Why Join Hands with Europe?], Tokyo: Mita Publishing, 1996

A comprehensive account of Europe's relations with Japan in a post-Cold War context.

Nester, William, *European Power and the Japanese Challenge*, London: Macmillan, and New York: New York University Press, 1993

This work provides a useful examination of internal structures which affect the development of relations between Japan and Europe.

Nuttall, Simon, "Japan and the European Union: Reluctant Partners", in *Survival*, Volume 38, number 2, Summer 1996

This article offers a practitioner's perspective on the difficulties involved in realizing comprehensive relations between Japan and the EU.

Okita Saburo, "Waga Gaiko 250 Nichi" [250 Days of Our Diplomacy], in *Bungei Shunju*, October 1980

This work offers an historical overview of Japan's foreign policy dilemmas at the beginning of the 1980s.

Rapkin, David P., Joseph U. Elston and Jonathan R. Strand, "Institutional Adjustment to Changed Power Distributions: Japan and the United States in the IMF", in *Global Governance*, Volume 3, number 2, May-August 1997

This article assesses, among other topics, the ways in which Japan's diplomacy was affected by its growing involvement in international financial institutions during the 1980s and 1990s.

Rothacher, Albrecht, *Economic Diplomacy between the European Community and Japan: 1959–1981*, Aldershot: Gower, 1983

This book, the most-often cited work on relations between Japan and the EC, provides an ideal introduction to many issues up to the early 1980s.

Tanaka Toshiro, "Euro-Japanese Political Cooperation: In Search for [sic] New Roles in International Relations", in *Keio Journal of Politics*, number 5, 1984

Written by the leading Japanese authority on Japan/EC relations, this short article covers their principal characteristics up to the early 1980s.

Tsoukalis, Loukas, and Maureen White, editors, *Japan and Western Europe: Conflict and Cooperation*, London and New York: Pinter, 1982

This edited work offers insights into various aspects of relations up to the early 1980s.

Wilkinson, Endymion, *Misunderstanding: Europe versus Japan*, Tokyo: Chuokoronsha, 1981; updated and revised as *Japan versus the West: Image and Reality*, London: Penguin, 1991

Wilkinson's original book provides a useful introduction to the question of identifying Japan and "Europe" in an increasingly interdependent world; the later version draws further on his experience as an official of the EC, and places the relationship in the wider context of Japan's global alliances and commitments.

Dr Julie Gilson is a Lecturer in the Japan Centre at the University of Birmingham.

Chapter Seventeen

Japan and China

Quansheng Zhao

It is widely recognized that the most important bilateral relationship in East Asia is that between Japan and China, and, indeed, Japanese leaders have always attached great importance to the relationship. Nevertheless, for more than a century Japanese foreign policy in general has been guided by the concept of *datsua nyuo* – "leaving Asia and joining Europe" – which has helped in enabling Japan to become the first industrialized country in Asia. Today, Japan is the only Asian country in the Group of Seven, alongside the United States, Germany, Britain, Italy, France and Canada. Understandably, therefore, many Japanese believe that their country's foreign policy should continue to run along these lines in the next century. On the other hand, the recent rise of China and the development of the economic "miracle" in East Asia have provided new challenges for Japanese foreign policy, despite the financial crisis that began in 1997. Increasing numbers of people in Japan believe that the country now needs to "return" to Asia. One may therefore expect to see two parallel trends in Japanese foreign policy, especially as expressed through the newly emerging triangular relationship among Japan, China and the United States.

Historical Background

For centuries, Japan was a student of Chinese civilization, acquiring and adapting Confucian and Buddhist ideas, its writing system, and much else from its larger neighbor, which was widely regarded in Japan as an "older brother". By the mid-19th century, however, both Japan and China faced extraordinary challenges from the West, led by Britain, the world's first industrialized country, which looked to what was then (tellingly) known as the "Far East" for new fortune and new adventure. Yet again, Japan learned lessons from China, notably from its defeat in the Opium War with Britain (1840–42). In 1868, Japan started its own process of modernization with the Meiji Restoration, and its new leaders, adopting the populist slogan "rich nation, strong army" (*fukoku kyohei*), began trying to catch up with the western powers by making Japan an industrialized country within a short period. The First Sino-Japanese War (1894–95) and the signing of the Treaty of Shimonoseki at its end symbolized the rise of Japan and the further humiliation of China.

Since then, many Japanese have had mixed feelings toward China. On the one hand, China continued to be regarded as a fundamental source of Japanese civilization. On the other hand, many Japanese became contemptuous of modern China's chaotic internal situation, its disintegration, its apparently endless civil wars, and its repeated defeats by the western powers. For those Japanese who shared the western powers' fear of Communism, the victory of the Chinese Communists in 1949 provided additional reasons to distrust China. Similarly, the Chinese have also had mixed feelings toward Japan. Many Chinese still regard Japan with great hatred because of the brutal behavior of the Japanese Imperial Army during its invasion of China in the 1930s and 1940s. At the same time, there has also been admiration among the Chinese for the rapid modernization of the Japanese economy.

These mixed feelings on both sides have lasted for more than a century as historical events unfolded. Japan's invasion of China, from 1931 onwards, the brutal occupation of most of China's territory, and the Rape of Nanjing (in 1937) produced unforgettable memories among the Chinese people. This wartime criminal behavior toward China also produced powerful feelings of guilt among many Japanese, especially the older generation.

Political Relations

Japan's relations with China in the postwar period have been heavily influenced by the changing dynamics of international relations, and the changing configurations of Japan's domestic politics. In particular, postwar Japan's policy toward China has always been deeply influenced by changes in US policy toward China. During the early years after the establishment of the People's Republic of China in 1949, Japan was still under Allied Occupation, and the main foundation of Japanese foreign policy was already its close security relationship with the United States (see Chapter 15). The world was entering the Cold War era, as most countries joined one or the other of the two camps headed by the United States and the Soviet Union. Despite the fact that there was some domestic demand to develop relations with Beijing, the Japanese government faithfully followed US guidance and stuck with its official recognition of the Republic of China, the Nationalist regime in Taiwan, as China's official representative in the international community.

The situation dramatically changed when the United States announced, in the summer of 1971, that its National Security Adviser, Henry Kissinger, had visited Beijing, and that President Richard M. Nixon would visit China during the following year. Another significant event took place in the fall of 1971: after more than two decades of debates in the UN General Assembly, the majority of UN member countries voted to recognize the People's Republic as the sole official representative of China, replacing Taiwan, despite strong opposition from the United States and a number of other western countries, as well as from Japan. This change in the international atmosphere corresponded with the increasing demands from all walks of life in Japan to normalize relations with Beijing.

In September 1972, Prime Minister Tanaka Kakuei visited China, where he met Mao Zedong and Zhou Enlai. This time, Japan moved ahead of the United States, since it took another seven years for Washington to normalize its relations with Beijing, in 1979. After the normalization of relations brought about by Tanaka, Japan moved quickly to develop a cordial relationship with China in all areas. In 1978, Deng Xiaoping, then China's paramount leader, visited Japan to join Prime Minister Fukuda Takeo in signing the Peace Treaty between the two countries. State leaders of the two countries have since made frequent mutual visits, the most recent being those of Prime Minister Hashimoto Ryutaro to China in September 1997, and Premier Li Peng to Japan in November that year.

After the Tiananmen tragedy of June 1989, western industrialized countries, led by the United States, imposed economic sanctions on China. Japan was the only Asian country that joined in applying these sanctions. Nevertheless, Japan also worked hard to avoid pushing China into a position of isolation, and on many occasions Japan has acted as a mediator between China and the West. Indeed, Japan was the first country to lift the sanctions on China, a year after the Tiananmen incident. Japan's policy during and after Tiananmen reflects the dilemma that Japan has faced. Japan has always tended to follow the lead of the United States in world affairs, but Japanese policy-makers are also acutely aware of the continuing rise of China and would like to maintain close ties with it. This dilemma lies behind Japan's dual position: to maintain a friendly and cooperative relationship with China, while demanding transparency from China, notably on its military expenditures, as well as following the United States in occasionally criticizing China's record on human rights.

Economic Interdependence

By 1970, informal, unofficial trade between Japan and the People's Republic represented around 2% of Japanese exports and as much as 20% of China's imports. Since 1972, however, increasing economic cooperation between Japan and China has become one of the most important factors in their own development and in promoting stability in the Asia-Pacific region. Bilateral structures have flourished in trade, investment, joint ventures, technology transfer, and personnel exchanges, and many observers see these as ushering in a major new period in the history of East Asia. It has become clear that Japan's major interest in China is in the economic dimension. During the 1980s, there were complaints and protests among many Chinese, particularly among Chinese youth, over what they saw as an economic invasion by Japan. They criticized Japan's lack of investment in China and its reluctance to increase the level of technology transfers, and students demonstrated in Beijing, Shanghai and other major cities in China. On the Japanese side, there was also increasing concern about business practices in China, arising, for example, from a series of cancellations of bilateral economic contracts, such as the agreement on the Boshan steel complex near Shanghai (see Lee). However, none of these incidents had much impact on the momentum of economic interdependence between the two countries.

Trade

Trade between the two countries has increased dramatically. In 1972, when it was worth around US$1.10 billion, it represented just 2.1% of Japan's total world trade, but it was worth around US$8.84 billion, or 3.3% in 1982; around US$28.94 billion, or 5.0% in 1992; and around US$62.23 billion, or 7.6% of Japanese trade, in 1996 (see IMF). China has been Japan's second largest trading partner since 1993, while Japan has been China's most important foreign trading partner for many years. In 1993, for example, the total value of Sino-Japanese trade was about one third larger than that of Chinese trade with the United States. Japan has a large share of China's markets in virtually every field except aircraft technology, which is dominated by US companies.

Since the late 1980s, Japan has allowed China to maintain a substantial surplus on its bilateral trade. As is well known, Japan has maintained a huge trade surplus in its overall trade, in particular with the United States, and in general, Japan has only had trade deficits with energy-rich countries, such as the oil-producing countries of the Middle East. To explain why Japan has allowed a surplus to develop in the China trade over the past decade, Hong Kong must be brought into consideration as a significant factor. Hong Kong has been used by many Japanese firms as a toehold in China, and it is believed that Hong Kong's role as a management base for developing businesses in China is likely to expand. Accordingly, it is not surprising that Japan has enjoyed sizable surpluses on the balance of its trade with Hong Kong, rising from a net surplus of around US$9.27 billion in 1989 to around US$25.04 billion in 1995 (see IMF). Here we can see an interesting phenomenon. Although both China and Hong Kong are crucial to Japan's foreign trade, they have come to serve different functions: China is more important for Japanese imports, while Hong Kong's significance lies mainly in relation to Japanese exports. The overall result is that Japan has enjoyed consistently favorable balances in its trade with Hong Kong and China combined, the net surplus having risen from around US$6.66 billion in 1989 to nearly US$11.06 billion in 1995 (see IMF). In effect, Japan allows China to run a large trade surplus, so that the bilateral relations can remain smooth and stable, but at the same time the Hong Kong trade is kept in surplus so that the gap can be filled. Thus, Hong Kong's importance to Japan is by no means purely economic, but also has a political dimension within the framework of general Sino-Japanese relations.

Investment

Japan's direct investment in China has picked up additional momentum during the 1990s. Most of the investment is concentrated in the North and Northeast. For example, the port city of Dalian in Manchuria has become a favorite destination for many Japanese corporate investors. The rapid increases both in the actual value of Japanese investments, and in the numbers of projects, have been noteworthy: while around US$201 million was invested in 94 projects in 1986, nearly US$3.11 billion was invested in 2,935 projects in 1995 (see Harwit). During the first half of the financial year 1994–95 (that is, April to September 1994), China was the second largest destination for Japan's overseas direct investment, which reached US$1.14 billion, a 63.5% increase over the first half of 1993–94, and second only to Japanese direct investment in the United States, which amounted to US$6.60 billion (according to a report in the weekly edition of *The Japan Times*, December 12–18, 1994).

Development Assistance

One other important form of economic cooperation between the two countries is Japanese official development assistance to China – including government loans, grants, and technical aid – which started in 1979. Japan has remained the largest donor to China among the 21 countries which are members of the OECD's Development Assistance Committee, comprising all the western European donors along with Japan, Australia, Canada, and the United States. China has also consistently ranked among the leading recipients of Japanese assistance.

According to figures released by the Japanese Ministry of Foreign Affairs, the total value of official assistance to China rose from around US$2.6 million in 1979 to a peak of nearly US$1.48 billion in 1994, before declining to US$861.7 million in 1996. Since 1982, loans have formed the largest of the three components of this assistance, accounting for nearly 90% of the total in 1982 but falling to nearly 62% by 1996. Most of these loans have been provided as "soft loans" through four major packages. The first government-to-government loan, worth ¥350 billion (around US$1.5 billion) was granted in support of China's Five Year Plan for 1979–84, following a pledge made by Prime Minister Ohira Masayoshi on a visit to China in December 1979. This was followed by a package worth ¥470 billion (around US$2.1 billion), agreed by Prime Minister Nakasone Yasuhiro in March 1984 for the five-year period 1985–90. The third package was worth ¥810 billion yen (around US$5.4 billion), covering 1990–95, which was promised by Prime Minister Takeshita Noboru during a visit to Beijing in August 1988. The most recent package, arranged under an agreement between the two countries in December 1994, was worth ¥580 billion yen (around US$5.8 billion), covering 1996–98 (according to a report by Charles Smith in the *Far Eastern Economic Review*, January 26, 1995).

Cultural and Educational Exchanges

Ever since the normalization of bilateral relations, there have been tremendous increases in the bilateral cultural and educational exchanges between the two countries. For instance, in 1992 there were around 50,000 Chinese students in Japan. While many of them were pursuing advanced studies, around one half of their number were language students, working and earning money in Japan while studying Japanese (according to a report in the Chinese newspaper *Renmin Ribao*, November 27, 1992). Thousands of Japanese students travel to China to study Chinese culture. Such bilateral exchanges will no doubt have a direct impact on the future development of relations between the two countries.

Strategic Relations

In the 1970s and early 1980s, China was actively seeking strategic partners to deter what it called "Soviet hegemony". The rapprochement between the United States and the People's Republic of China was partly prompted by their shared concern over the apparent expansionist tendencies of the Soviet Union. Japan's move towards China at that time was also influenced by this concern, and in 1978 Japan agreed, although somewhat reluctantly, to include an "anti-hegemony" clause in the Peace Treaty with China. That treaty effectively completed the formation of an informal "united front" of China, Japan and the United States against the Soviet bloc.

Since the end of the Cold War, however, Japan's strategic considerations have changed considerably. Russia is no longer regarded as a security threat, while China has been on the rise both economically and militarily. Many Japanese therefore have increased concerns about a potential threat from China, notably in relation to the territorial dispute over the Diaoyu or Senkaku islands (discussed below). Meanwhile, Chinese concerns are focused mainly on a potential revival of militarism in Japan, and on developments in the Korean peninsula. In this connection, the revision of the guidelines on US/Japan security cooperation, announced in September 1996 (and detailed in Chapter 15), aroused Chinese suspicions, since the new guidelines appeared to envisage an extension of Japanese military activity, at least in emergencies, well beyond Japan itself. The guidelines accordingly became a focal point for bilateral discussions between China and Japan in 1997, notably during Prime Minister Hashimoto's visit to China in September and Premier Li's visit to Japan in November. Even though Hashimoto assured the Chinese leaders that the new guidelines did not target any third party, he refused to clarify whether the Japan/US defense zones included the Taiwan Strait or not. At the same time, he called for a strengthening of the security dialogue among China, Japan, Russia and the United States, in order to ensure peace and stability in the Asia-Pacific region.

Japan has also increasingly demanded that China increase the transparency of its defense policy and military expenditures. Japan has also expressed concern about the territorial disputes between China and several Southeast Asian countries over islands in the South China Sea. Japan regards the South China Sea as a crucial link with the Middle East, a major source of Japan's oil and other energy supplies. At the same time, Japan has worked hard to strengthen bilateral security cooperation with China.

Problems

Although there has been, in general, a friendly relationship between Japan and China since the early 1970s, a number of problems have appeared, or been revived. These range from disputes over economic relations and human rights policy, to disagreements over territory, history and symbolism.

Economic Policies, Nuclear Tests and Human Rights

In general, economic relations between Japan and China are interdependent and mutually beneficial, because of the complementary nature of their different levels of economic development. Understandably, however, the increase in economic relations has also brought some problems. For much of the 1980s, for example, China criticized Japanese protectionism against Chinese goods, while Japan claimed that China should exert better control over its imports of consumer goods, such as cars. In the area of technology transfer, China pressed Japan to increase its exports of high technology, while Japan was constrained by its membership of the Coordinating Committee for Multilateral Export Controls (COCOM), which policed trade with Communist regimes until it was disbanded in 1994. The Japanese also criticized the "red tape" of Chinese bureaucracy and the absence of the rule of law in many business transactions. In the 1990s, however, Japan has greatly increased its investment in China and expanded its technology transfers, while China has significantly improved its business environment, especially regarding those involved in foreign investment and joint ventures. China has also slowly moved towards more law-based practices in business transactions with foreign countries, although there is much room for further improvement.

A new problem that has arisen in the 1990s reflects differing perceptions of the proper relationship between economics and politics. One noteworthy example relates to Japan's guidelines for its official development assistance to China. Significant changes to these guidelines were announced by Prime Minister Kaifu Toshiki in April 1991, and confirmed by his successor, Miyazawa Ki'ichi, in January 1992. The guidelines now emphasize promotion of the market economy and human rights, non-military use of aid funds, defense transparency, and environmental protection. Accordingly, after China's underground nuclear test in June

1994 the Japanese government responded quickly and with displeasure, and then, when China conducted another test in October, Japan protested by deferring sending a team to Beijing for talks on projects to be covered by the fourth "soft" loan (see above). After China's next two rounds of underground nuclear tests, in May and August 1995, Japan announced that its aid to China for that year would be cut by 93%, from ¥7.8 billion to around ¥500 million, although the loan would not be affected. Premier Li Peng criticized this action as "economic blackmail". The disagreement continues. Similarly, as the issue of human rights increasingly becomes a focus of international relations, one may expect that the Japanese government, like any other government, will continue to use its economic power to support its foreign policy priorities. At the same time, however, it is also likely that Japan will go on maintaining a relatively low profile and avoiding confrontation in its dealing with China.

Territorial Disputes

One serious problem is the dispute over the group of islands which the Chinese call Diaoyu but which the Japanese know as the Senkakus, lying in the East China Sea around 110 nautical miles northeast of Taiwan and 240 nautical miles southwest of Okinawa, on the Chinese side of the Okinawa shelf. The group consists of eight small uninhabited volcanic islets, the largest of which is two miles long and less than one mile wide. They have long been a point of dispute between Beijing and Tokyo, at least partly because they are reported to lie in an area possessing significant oil deposits.

In the 1970s, the dispute triggered anti-Japanese demonstrations in Taiwan and Hong Kong. The controversy further intensified in April 1978, when more than 100 fishing boats from the People's Republic sailed around the islands. At that time it appeared that the dispute, if not rapidly resolved, could jeopardize negotiations on the Sino-Japanese Peace Treaty. Before the disagreement worsened, a Chinese Deputy Foreign Minister stated that both sides had agreed to handle the problem by separating it from the negotiations on the Treaty. During a visit to Japan the same year, Deng Xiaoping deflected queries about the islands by suggesting that the issue could be "handled better by the next generation" (Whiting pp. 68–9). The dispute was then shelved, but China was careful not to abandon its claim to the islands. In 1992, the Chinese government explicitly revived its claim by arranging for the National People's Congress to pass a Law on Territorial Waters and Contiguous Areas, which stipulated that China's territory included the Diaoyu Islands, among other areas (Kim pp. 150–1). In addition, China has repeatedly called for Sino-Japanese joint development of the disputed area. The issue was to remain a source of tension between the two countries.

That tension flared up once again in the fall of 1996, when a group of Japanese right-wingers set up a lighthouse on one of the islands. While both the People's Republic and Taiwan maintained low profiles in their protests to Japan, activists in Hong Kong took the lead by organizing large anti-Japanese demonstrations, and groups from Hong Kong and Taiwan even organized a joint protest trip to the islands in fishing boats. Despite a blockade by Japan's Maritime Self-Defense Force, these fishing boats reached the islands on October 7, 1996, and several people climbed onto them, waving (incongruously) the flags of both the People's Republic and Taiwan. A leading figure in this expedition, Chen Yuxiang, was drowned near the islands, and instantly became a hero for certain sections of Chinese communities all around the world. These extraordinary events drew broad attention to the territorial dispute in the international community, and Japan's image in Hong Kong reached a new low. In response, Japan quickly established a crisis management system, directly under the Prime Minister's Office, to deal with the dispute, while China's Ministry of Foreign Affairs warned Japan that relations between the two countries would suffer serious damage if the government did not stop Japanese right-wingers from setting foot on the disputed islands. After extensive consultations, the dispute was put on hold once again, but it is certain to re-emerge in the future.

"Japanese Militarism"

One other difficult issue in the development of bilateral relations concerns what Chinese official spokespersons habitually call "Japanese militarism", in reference to the enduring bitter memories of the Japanese invasion. The fact that some Japanese politicians have continued to downplay the significance of Japan's past militarism has only deepened suspicions among the Chinese, and the Beijing government has occasionally found it useful to reignite those memories.

In 1982, for example, the issue of "militarism" reappeared as a result of what is now known as the textbook controversy. Japan's Ministry of Education was sharply criticized by liberal and left-wing groups in Japan itself, as well as by governments and pressure groups in China, Thailand, Hong Kong, and North and South Korea, over its revisions of accounts of Japan's military past in school textbooks. The Ministry insisted, for example, that the textbooks should refer to Japan "entering", rather than "invading", China and other parts of East Asia. The Chinese government organized a campaign of opposition to "Japanese militarism", which was wound down only when the Japanese government undertook to review the issue in preparation for a visit to Beijing by Prime Minister Suzuki Zenko, which was to mark the 10th anniversary of the normalization of relations between the two countries.

The controversy was revived in 1985–86, when new editions of Japanese school textbooks were shown to contain the same wording as in 1982, and suspicions were further aroused by Prime Minister Nakasone Yasuhiro's visit to the Yasukuni Shrine in Tokyo, which honors Japan's war dead, including General Tojo Hideki and other convicted war criminals. The Chinese government ensured that the mass media under its control were filled with renewed attacks on "Japanese militarism", while students were permitted to organize demonstrations in major Chinese cities.

Since the textbook controversy died down for the second time, Chinese allegations about "militarism" in Japan have resurfaced from time to time, both in the mass media of the People's Republic and in bilateral meetings between high-level officials of the two countries. Some Japanese policy-makers have occasionally responded by expressing their own concerns about what they see as Chinese "militarism", notably in relation to the ever increasing defense budget of the People's Republic and the Chinese government's statements on territorial issues (discussed above). Nevertheless, the steady growth in the economic interdependence of the two countries makes it likely that Sino-Japanese relations will remain close.

The Issue of Taiwan

China formally ceded Taiwan to Japan in 1895, after China was defeated in the First Sino-Japanese War. The Japanese occupation lasted 50 years, until Japan's defeat in World War II. Adding to the complexities facing the former colonial power, there has been a separation between the Chinese mainland and Taiwan since 1949, caused by the rivalry between the Communists and the Nationalists, who have competed fiercely for recognition in the international arena.

Even though Japan issued a number of official statements in 1972 declaring Taiwan to be Chinese territory, the status of the island continues to be a potentially volatile issue between China and Japan, not least because Japan ruled Taiwan as a colony for so long. Some Japanese are keen to pursue a special relationship with Taiwan, and would therefore have preferred to maintain the status quo, viewing the People's Republic and Taiwan as separate countries. Such opinions irritate the government in Beijing, which views any suggestion of "two Chinas" or "one China, one Taiwan" – however trivial it may appear – as an impermissible assault on China's territorial integrity. For example, there has been a continuing controversy surrounding the Guanghua Hostel ("Kokaruo" in Japanese), a student dormitory in Kyoto. Both the People's Republic and Taiwan claim ownership of the dormitory, and Japan has been caught in the middle, even though its own preference would be to have the dispute resolved in the courts, rather than become a matter for government intervention (Whiting pp. 152–7). Tensions rose again in 1994, when Hsu Li-teh, the Vice-Premier of Taiwan, was invited to attend the Asian Games in Hiroshima. This, coupled with a failed attempt by Taiwan's President, Lee Teng-hui, to visit Japan at the same time, brought strong protests from the Chinese leadership, which even considered boycotting the Games (according to a report in the weekly edition of *The Japan Times*, September 12–18, 1994). In November 1994, Prime Minister Murayama Tomi'ichi took the opportunity presented by a summit meeting of the Asian-Pacific Economic Cooperation (APEC) organization in Jakarta to assure President Jiang Zemin that Japan does not support "two Chinas".

Nevertheless, some close ties remain between Japan and Taiwan, notably because many older members of Taiwan's elites were educated in Japanese during the colonial period, and some went on to study in Japan. President Lee, for example, was an undergraduate at Kyoto University; his command of Japanese is almost up to the standard of native speakers; and he has frequently indicated in conversations with Japanese politicians, journalists, scholars and businessmen, that he still has close ties with their country. Indeed, after he visited Cornell University, where he took his second degree, in the summer of 1995, the Taiwanese government lobbied Japan to allow him to make a similar visit to his *alma mater* in Kyoto. This attempt has so far been unsuccessful because of strong opposition from Beijing.

All of these developments have made the issue of Taiwan a significant source of problems between Japan and China. Japanese policy in this regard has been cautious and constrained, but the issue is likely to remain a sore point for the government of the People's Republic as long as Taiwan remains separate from it.

Implications and Future Directions

As China continuously increases its economic, political and strategic influence in the Asia-Pacific region, its rise is certain to be a main focus of the foreign policies of the other major powers in the region – the United States, Japan and Russia – as well as other neighboring states, including South Korea and the countries of Southeast Asia. The United States remains Japan's leading partner in virtually all fields, whether political, economic, security or cultural, but many people in Japan recognize that China's importance to Japan runs a close second. In the post-Cold War era, there are many overlapping interests, such as in trade and investment, among Japan, China and the United States. There are also occasions that place Japan in the position of making choices between China and the United States, particularly when there are confrontations between these two countries. This pattern of repeated dilemmas may well continue to be an important feature of Japanese foreign policy as it enters the next century.

China is also crucial to Japan's aspiration to become a global political power. Japan has long wished to become a permanent member of the UN Security Council, which would enable Japan to exercise much greater influence in world affairs (see Chapter 19). Japan has been supported by virtually all western countries and some developing countries, but China has yet to commit itself fully to endorsing Japan's candidacy. China has been ambivalent because of its fear of Japan's future remilitarization. In order to realize its political aspirations, Japan may need to make more of an effort to heal the wounds of Japan's Asian neighbors, and of China in particular, caused by the invasion before and during World War II.

Japan's relations with China also have a significant influence on regional stability. Both Japan and China would be important in any settlement between North and South in the Korean peninsula; and Japan is a significant actor in Southeast Asia, where China also has a keen interest. In recent years, Japan has also strengthened its security ties with China. In February 1998, for example, General Chi Haotian became the first Chinese Defense Minister ever to visit Japan, and held extensive discussions on security issues with Prime Minister Hashimoto, Kyuma Fumio, the Director General of the Defense Agency, Obuchi Keizo, the Minister for Foreign Affairs, and Kato Ko'ichi, the Secretary General of the

Liberal Democratic Party . He also visited the Air Self-Defense Force base at Hyakuri and the Maritime Self-Defense Force base at Yokosuka.

There will continue to be problems and controversies between the two countries over territorial issues, the status of Taiwan, China's military transparency and human rights record, and "Japanese militarism". Nevertheless, Japan's relations with China will continue to develop toward greater economic interdependence and integration, and more consultations on political and security issues. Considering their cultural and geographical proximity, and, what is more important, their consensus on economic development, as well as the general changes in international relations in the post-Cold War era, one may expect that bilateral cooperation between Japan and China will move to an unprecedented level in the 21st century.

Further Reading

Harwit, Eric, "Japanese Investment In China: Strategies in the Electronics and Automobile Sectors", in *Asian Survey*, Volume 36, number 10, October 1996

Howe, Christopher, editor, *China and Japan: History, Trends, and Prospects*, Oxford: Clarendon Press, and New York: Oxford University Press, 1996

This multi-disciplinary study represents an ambitious attempt to summarize the historical background and modern development of the relationship between the two countries, and its meaning for the rest of the world.

International Monetary Fund (IMF), *Direction of Trade Statistics Yearbook*, Washington, DC: IMF, annual publication

This is an invaluable source of data on patterns of international trade.

Iriye, Akira, *China and Japan in the Global Setting*, Cambridge, MA: Harvard University Press, 1992

Placing the bilateral relationship within its global context, this book covers three distinct periods: from the 1880s to World War I; between the two World Wars; and postwar. While emphasizing the importance of economic and strategic factors, the author stresses that the future of China and Japan depends on the successful cultural interdependence of what may be the most significant pair of countries in the world today.

Katzenstein, Peter J., and Takashi Shiraishi, editors, *Network Power: Japan and Asia*, Ithaca, NY: Cornell University Press, 1997

This book examines regional dynamics in contemporary East Asia, scrutinizing the effects of Japanese dominance on the politics, economics, and cultures of the area. It assesses the competitive logic of continental and coastal primacy in China. In starkest form, the question addressed is whether Chinese or Japanese domination of the Asian region is more likely.

Kim, Samuel, editor, *China and the World*, third edition, Boulder, CO: Westview Press, 1994

This collection of essays on the foreign relations of the People's Republic includes the editor's own essay, "China and the Third World in the Changing World Order".

Lee, Chae-Jin, *China and Japan: New Economic Diplomacy*, Stanford, CA: Hoover Institute Press, 1994

This original in-depth analysis concentrates on a few salient cases of Sino-Japanese economic interaction: the steel complex at Baoshan, the joint offshore oil development in the Bohai Sea, and Japanese loans provided to fund China's important construction projects.

Mandelbaum, Michael, *The Strategic Quadrangle: Russia, China, Japan, and the United States in East Asia*, New York: Council on Foreign Relations, 1995

Five experts on East Asia explore the new shape of power among the major players in the region, examining the web of alliances, historical rivalries, and conflicting world views that define their relations among these four powers. David Lampton examines Beijing's policy of military detente and economic cooperation with the other three powers; Michael Mochizuki discusses Japan's role in the region by looking into its domestic politics and foreign policy.

Ross, Robert S., editor, *East Asia in Transition: Toward a New Regional Order*, Armonk, NY: M. E. Sharpe, 1995

By analyzing basic trends of development in East Asia, this collection provides clear background pictures of Japan's role in the region's political, strategic, and economic relations. In addition to

China, Japan, Russia, and the United States, the book also addresses the development of South Korea and Southeast Asian countries.

Taylor, Robert, *Greater China and Japan: Prospect for an Economic Partnership in East Asia*, London and New York: Routledge, 1996

This study explores the ambiguous economic and political relationship between "Greater China" and Japan. It analyzes their mutual suspicions: the Chinese fear of a Japanese military revival and the Japanese concern over Chinese territorial ambitions. The author believes that there is potential for both Sino-Japanese rivalry and cooperation.

Whiting, Allen, *China Eyes Japan*, Berkeley, Los Angeles: University of California Press, 1989

This book provides a comprehensive examination of the perceptions of China and of Chinese foreign policy towards Japan, covering key issues in this important bilateral relationship, such as the historical background, the textbook controversy, the anti-Japanese student demonstrations, and economic cooperation.

Zhao, Quansheng, *Japanese Policymaking: The Politics Behind Politics, Informal Mechanisms and the Making of China Policy*, New York and Hong Kong: Oxford University Press/Praeger, 1993

This study probes the politics behind the formal screen of Japanese diplomacy. The book offers new insights into Sino-Japanese relations, and draws broader lessons for Japan's international relations through four case studies of Japanese policies towards China. These include the watershed Sino-Japanese rapprochement in 1972, and Japan's foreign aid policy towards China before and after the Tiananmen tragedy of 1989.

Dr Quansheng Zhao is Associate Professor and Asia Coordinator at the School of International Service of American University; he is also Associate in Research at the Fairbank Center for East Asian Research at Harvard University. In addition to the book cited above, he has also written *Interpreting Chinese Foreign Policy* (Oxford: Oxford University Press, 1996). The author would like to thank Barry Press and Emily Williams for research assistance in the preparation of this chapter.

Chapter Eighteen
Japan and East Asia

Christopher W. Hughes

This chapter provides an overview of Japan's economic, political, and security relations with East Asia since the end of World War II. East Asia is taken to comprise, in addition to China and Japan itself, those countries that are members of the Association of Southeast Asian Nations (ASEAN) – Singapore, Indonesia, the Philippines, Thailand, Malaysia, Brunei, Vietnam, Laos, Burma and Cambodia; the newly industrialized economies (NIEs) of Southeast and Northeast Asia (Singapore again, South Korea, Taiwan, and Hong Kong); and the Communist state of North Korea. It should be noted, however, that Japan's relations with China are not discussed here, since they are covered in Chapter 17; and that some data on "East Asia", including Japanese government statistics, in fact refer only to eight of these countries (Hong Kong, Indonesia, South Korea, Malaysia, the Philippines, Singapore, Taiwan, and Thailand), thus excluding the other eight (Japan itself, of course, but also Brunei, Burma, Cambodia, China, North Korea, Laos, and Vietnam).

The Historical Framework

Since its opening to international society by the western imperial powers and its subsequent drive to modernize, from the mid-19th century onwards, Japan has faced a conflict of identity that has influenced its relations with the rest of Asia. Japan's leaders were aware that their country formed part of Asia geographically, ethnically, culturally, and, to some extent, politically, due to its historic ties to the previous Sinocentric world order (Hamashita pp. 113–35). Hence, during the modernization era, and as expressed in such sentiments as "Pan-Asianism", Japanese leaders and people continued to feel a sense of solidarity with other Asian countries, and believed that Japan, as the first modern state in the region, should take a leading role in protecting Asia from the encroachments of western imperialism. At the same time, however, this vision of Japan's role in Asia was matched by an awareness that, in order to avoid the fate of China and dismemberment by the western powers, the priority for Japan was to ensure its national survival by acquiring the economic and military resources to rival those of the West. This meant, inevitably, acquiring its own colonies and becoming an imperial power along western lines. As a result of this policy of *fukoku kyohei* ("rich nation, strong army"), there arose in opposition to the identity of Japan as an Asian nation another conception, which stressed Japan's military, political, and economic ties to the West. This was typified in the Meiji period (1868–1912) by the political thinker Fukuzawa Yukichi's espousal of *datsua nyuo* (abandonment of Asia, and joining with the West). This westernizing and imperialist impulse proved to be dominant, and led after a series of military victories to Japan's annexation of Korea (1910), and its participation in the dismemberment of China by the acquisition of Taiwan (1895) and Manchuria (1931). However, "Asianist" ideology continued to exercise an influence upon Japan's Asia policy until the outbreak of the Pacific War in 1941. The conflict of identity could only be reconciled unhappily with the attempt to extend Japanese imperial and economic power southwards, to include Southeast Asia in the proposed Greater East Asian

Co-Prosperity Sphere, and the claimed liberation of the region from western rule.

The Legacy of Colonialism

Following Japan's defeat in 1945 the identity conflict remained unresolved, and relations with Asia were further complicated by the legacy of colonialism and Japan's integration into the bipolar Cold War structure under US hegemony. In particular, memories of Japanese colonial rule and its attempts to extinguish Korean ethnic identity have generated powerful anti-Japanese sentiment in Korea, which persists today. In addition, despite Japan's wartime portrayal of itself in Southeast Asia as the liberator from western imperialism and the destroyer of the myth of white supremacy, a deep suspicion of Japanese militarism has also persisted across the region until the present day.

Mistrust of Japan was compounded by its limited participation in the decolonization process in Asia. Although (as will be discussed in the next section) Japan did provide reparations to many of its former colonies under the 1952 Peace Treaty, the Allied Powers had quickly stripped Japan of its colonies in 1945, and so relieved it also of direct responsibility for dealing with the colonial past and rebuilding relations with the newly independent states of East Asia. Thus, in contrast to the experience of other former colonial powers, much of the Japanese leadership and people became divorced physically and psychologically from the problems of decolonization, and this perhaps helps to explain the reluctance of some conservative politicians even now to address fully the history of Japanese aggression in East Asia. However, the only outcome of this denial has been to hamper the reconstruction of Japan's relations with East Asia, and, as will be seen later, to exclude it from a deeper political and security role in the region.

The Cold War Relationship with the United States

Planning had already begun during the Occupation years (1945–52) for Japan's return to international society under US sponsorship, and this process was accelerated by the outbreak of the Korean War in 1950, which confirmed that, in order to assist in the containment of Communism, Japan had to be converted into a strong US ally and a bastion of capitalism in the "Far East". On the Japanese side, this thinking was matched by that of Yoshida Shigeru, Minister of Foreign Affairs (1945–46) and then Prime Minister (1946–47 and 1948–54). Although he did not share completely the fears of US policy-makers about the Chinese Communist threat, and was to meet domestic opposition from both conservative and leftist politicians, Yoshida did accept that Japan had to rely on the United States for military security while it concentrated on the goal of economic reconstruction (Welfield pp. 21–59). US containment strategy and the "Yoshida Doctrine" were given concrete form by the simultaneous signing in 1951, and operation from 1952, of the multilateral Peace Treaty with Japan, and the bilateral Security Treaty between Japan and the United States, which together served to condition Japan's relations with Asia.

The two treaties further distanced Japan from its former colonies. Communist China's position as a non-signatory to the 1952 Peace Treaty, and Japan's signing of a separate peace treaty with Nationalist Taiwan in 1952, meant that Japan was to lack diplomatic relations with mainland China until the normalization of relations in 1972 and the Sino-Japanese Peace Treaty of 1978. Japan's ideological association with the US and western camps represented by the treaties of 1951–52 also discouraged Japan from normalizing relations with that part of its former colony which had become North Korea. The bilateral relationship with the United States and the necessity to conform to US interests in the region therefore limited Japan's scope for interaction with East Asia, and accentuated the "de-Asianization" of Japanese identity. This was perhaps best demonstrated by Prime Minister Nakasone Yasuhiro's confident assertion in the 1980s that Japan was a "member of the West" (*nishigawa no ichi'in*). Although Japan's Constitution precluded any direct role in East Asian security, Japan did continue to play an indirect role by its provision to the United States, under the Security Treaty, of the exclusive right to use bases in Japan to contribute to the security of Japan and East Asia, so underpinning the US military presence in South Korea and the war effort in Vietnam. The bilateral relationship with the United States also ruled out consideration of Japan's participation in any type of multilateral security cooperation or dialogue in the region, and indeed many East Asian states came to see the most important function of the Security Treaty as constraining Japan's military strength and, thus, any regional role that its leaders might seek to play.

Japan's integration into US strategy also took place in the economic sphere. The United States launched Japan's economic recovery and rose to become its largest individual trading partner, notably by providing Japanese industry with massive military procurement contracts during the Korean and Vietnam Wars; by agreeing to the pegging of the yen at the highly favorable rate of 360 yen to the US dollar; and by opening its markets to Japan, along with the other export-oriented East Asian economies of South Korea and Taiwan (Stubbs pp. 367–72, Hook p. 173). In addition to shifting Japan's economic center of gravity towards the United States and away from East Asia, the United States also influenced the pattern of Japan's economic relations within East Asia itself. First, Japan's incorporation into the US political and economic sphere choked off its access to important markets in China and other newly Communist states, and forced Japan to conduct limited trade with them through a process of "separation of politics from economics" (*seikei bunri*). Second, to replace the lost markets in the Communist bloc and ensure Japan's economic survival, the United States worked to reopen Southeast Asian markets for Japanese corporations. US policy-makers planned to establish a triangular division of labor, with the United States as the core, producing high technology and capital goods; Japan as the semi-periphery, producing intermediate and consumer goods; and Southeast Asia as the periphery, providing energy and raw materials for Japan (Schaller pp. 178–211). As will be seen in more detail below, this triangular pattern of trade linkages, has persisted, with modifications, throughout much of the postwar period.

Thus, the colonial legacy and the advent of the Cold War bipolar structure created a framework of relations which suppressed Japan's Asian identity, and sealed it off from deeper political and security relations with East Asia. This framework initially left only economic avenues for interaction.

Economic Relations

Aid

The initial point for Japan's economic re-entry into East Asia was its provision of aid, or "economic cooperation", as part of the reparations agreement under Article 14 of the 1952 Peace Treaty, and under similar provisions in separate peace treaties with Burma (1954) and Indonesia (1958). Japan provided economic aid to Burma (1955), the Philippines (1956), Indonesia (1958), South Vietnam (1959), and, following the normalization of diplomatic relations, South Korea (1965). The start of what can be regarded as Japan's official development assistance (ODA) came in 1954 with the provision of technical assistance as part of the Colombo Plan, which Japan joined in that year. In 1961, Japan became a founding member of the Development Assistance Committee of the OECD. From the mid-1970s onwards, Japanese ODA expanded rapidly under a series of mid-term plans, rising from US$1.42 billion in 1977 to US$13.8 billion in 1995, and allowing Japan to surpass the United States as the world's largest ODA donor in 1989. During the 1970s and 1980s, Japan's ODA distribution began to diversify as Japan increased its aid to the Middle East to help guarantee its oil supplies, and employed "strategic aid" in support of allies of the United States in the Persian Gulf, the Horn of Africa, the Caribbean, and the states bordering Afghanistan (Yasutomo pp. 3–33). Yet despite these changes, 54.5% of Japanese ODA was still directed to East Asia in 1995, with the largest proportion (around 30%) allocated to Indonesia and China. Japan ranks as the main aid donor to these states, as well as to the Philippines, Malaysia, Thailand, Vietnam, and Laos. Japan's position as the principal donor in the region has been consolidated by its participation in multilateral assistance via the Asian Development Bank (ADB), founded in 1966, which relies on Japan for up to 50% of its funds (Ming p. 519; and see Chapter 19).

The provision of Japanese aid to East Asia has been motivated primarily by economic interests, but it also has served a political function. Reparations aid came mainly in the form of the export of outdated technology and plant, which allowed Japanese companies to re-enter Southeast Asian markets, and to create technological and product linkages between those countries and Japan. Japanese economic interests have also been furthered by the "tying" of aid to the purchase of Japanese goods and services, and especially infrastructure projects. In 1972, only 28.1% of Japanese aid was untied at the commitment stage. By 1989 the Japanese government could claim that 100% of Japanese aid was untied, but its allocation of ODA on the basis of a "principle of request" (*yosei-shugi*) means that, in practice, much of the aid is still tied, as in many cases it is Japanese companies which prepare and are awarded aid contracts on behalf of the governments making the "requests" (Soderberg pp. 72–88). Hence, Japanese ODA has supported the penetration of East Asian markets by Japanese corporations, and aid plans such as the New Asian Industries Development Plan (1987) or the ASEAN/Japan Development Fund (1988) reveal that Japanese government officials have conceived of ODA as a means to secure the vertical integration of the economies of East Asia into Japan's own, and thus establish a regional division of labor (Arase p. 203, Shiraishi pp. 189–90).

Japan has also deployed its ODA as a non-military form of power to further its political interests in East Asia. Japan has supported US strategy and its allies in the region, with the provision of aid to Taiwan in 1965 and to South Korea in 1967, and increased its bilateral aid to Thailand following Vietnam's invasion of Cambodia in 1978. Also (as will be discussed below), Japan used ODA to alleviate economic and political tensions with the member states of ASEAN in response to anti-Japanese rioting in 1974.

Investment

Japanese direct investment has played a role similar role to that of ODA in rebuilding the links between the Japanese and other East Asian economies. Japanese investments in resource extraction in Southeast Asia began in the 1950s and 1960s, but were limited by restrictions on the convertibility of the yen to preserve Japan's balance of payments. The first upsurge, or "wave", of Japanese investment came in the late 1960s and early 1970s. This was triggered by a combination of factors. The "Nixon shocks", which included attempts to restore the US economy by placing restrictions on Japanese imports and ending fixed exchange rates, led to the appreciation of the yen against the US dollar. Next, the first "oil shock" of 1973, which increased energy and production costs for corporations inside Japan, forced them to restructure and move away from reliance on heavy industry. In addition, there were significant rises in Japanese labor costs; companies were faced with domestic public pressure to move heavy polluting industries offshore; and the ASEAN countries imposed import restrictions on Japanese goods, as part of their policy of import substitution (Simone and Thompson Feraru pp. 340–2, Hook pp. 177–9, Selden p. 319). The consequent need for Japanese corporations to avoid import restrictions and to search for production sites offering lower costs, in conjunction with reduced barriers to the convertibility of the yen and the movement of Japanese capital, led to a nearly fourfold increase in the value of Japanese direct investment in the eight countries comprising "East Asia" in Japanese government statistics (Hong Kong, Indonesia, South Korea, Malaysia, the Philippines, Singapore, Taiwan, and Thailand), from US$96 million to US$358 million (see Tsusho Sangyosho 1995). A high proportion of this investment was concentrated in textiles (30–40%) and electronics (15–30%), with most production intended for export to third countries, and especially to the US market (Kanetsuna pp. 132–46).

A second upsurge in Japanese investment occurred in the late 1970s and early 1980s. This resulted from the continuing efforts of Japanese corporations to restructure and seek lower costs, and was characterized by an emphasis upon investment in chemical and metallurgical industries. The third and greatest upsurge in investment came in the mid- to late 1980s, in reaction to renewed US attempts, by the use of currency realignments and other measures, to curb its growing trade deficits and its absorption of imports from Japan and the NIEs (as defined above). The Plaza Accord of 1985 increased the value of the yen against the US dollar by up to 70%, and the Louvre Accord of 1987 increased also the value of the NIEs' currencies, and was followed by the removal of the NIEs from the General System of Preferences. The appreciation of the yen led to a threefold increase in Japanese direct investment around the world, with the greatest concentration in the United States and western Europe. Between 1985 and 1989, the share of the eight East Asian economies cited above in Japan's direct investment overseas remained stable, at around 20%, but it increased rapidly in value, from US$460 million to US$3,220 million (see Tsusho Sangyo-sho 1995), and was increasingly concentrated in electronics, automobiles, and manufacturing assembly. Further, the geographical concentration of Japanese direct investment began to switch from the four NIEs to four leading ASEAN members (Thailand, Malaysia, the Philippines, and Indonesia), reflecting growing wage costs in the NIEs, and increasing barriers to exports by Japanese corporations from NIEs' production bases to third countries.

Finally, as a result of the further appreciation of the yen to around ¥100 to the US dollar by 1992–93, a fourth upsurge of Japanese direct investment has taken place in the early and mid-1990s. By 1994, direct investment in the eight countries cited in Japanese government statistics had risen to US$5,181 billion, or 40% of the Japanese world total, and there have been ever greater concentrations in the four ASEAN countries mentioned above, compared to the four NIEs (see Tsusho Sangyo-sho 1995).

Trade

Concentrations of Japanese ODA and direct investment in East Asia have also given a distinct imprint to trade relations between Japan and the region, and have helped to bring about Japan's full economic re-entry into East Asia in the postwar period. While the United States has continued to be Japan's largest individual trading partner, its share of Japan's total international trade has varied significantly over the past three decades, accounting for 30.2% of the total in 1970, 21.0% in 1980, 27.4% in 1990, and 25.4% in 1996 (see JETRO). Meanwhile, Japan's combined trade with the NIEs, Thailand, Malaysia, the Philippines and Indonesia also varied from year to year but was generally on a rising curve, accounting for 17% of the total in 1970, 21% in 1980, and 25% in 1990. By 1991, when it stood at 27.1% of the total, it had come to exceed that with the United States; and by 1996 Japan's trade with these eight countries had reached 31.2% of its world total (see JETRO). In 1996 Japan also became the largest individual trading partner for Thailand, Indonesia, Malaysia, and Vietnam, and the second largest trading partner, after the United States, for South Korea and the Philippines. The share of Japan's total exports to these eight countries in East Asia increased rapidly, from 20% to 37.1%, between 1984 and 1996, and the share of their imports into Japan rose from 21% to 24.1% over the same period. In addition, an increasing proportion of these trade flows consist of manufactured goods, rising from 24.5% of total Japanese imports from East Asia in 1984 to 63.2% in 1996 (see Tsusho Sangyo-sho 1984 and 1996). Thus, it is possible to observe strengthening trade links between Japan and the East Asian economies, and to conclude that Japan, by absorbing a greater proportion of its manufactured imports from the region, is to some extent playing a role as the engine of growth for the export-oriented economies of East Asia.

However, it is also clear that the pattern of trade relations between Japan and East Asia remains asymmetrical in the late 1990s. The eight countries cited above have jointly reduced the share of their total exports to Japan, while Japan's share of their total imports remains high, at around 20%. These economies have therefore been far less successful in penetrating Japan's markets than the latter has been in penetrating theirs, and ran a combined trade deficit with Japan of around US$70 billion in 1996. The four NIEs have run the largest deficit with Japan, at US$61 billion in 1996, but the once favorable balance of trade with Japan enjoyed by Thailand, Malaysia, the Philippines, and Indonesia has also moved into deficit. Most of this deficit can be accounted for by the imbalance in the export and import of manufactures. In 1996, nearly 100% of Japan's exports to East Asia consisted of manufactured goods and technology, such as electronics, transport and precision machinery, and Japan ran a surplus in manufactures of close to US$100 billion (see Tsusho Sangyo-sho 1996). It is apparent also that, even though Japan has become dominant in many sectors of the East Asian economies, a triangular pattern of trade relations still exists between the United States, Japan, and East Asia. Despite East Asia's reduction of the share of its total exports to the United States from 31% in 1984 to 20% in 1996, this US share is still significantly greater than that of Japan. Further, according to other studies, in 1994 the United States accounted for between 23% and 35% of the manufacturing exports of each East Asian country, whereas Japan only accounted for between 4% and 16% (Bernard and Ravenhill p. 205). This pattern of trade suggests that Japan's economic activity in East Asia has been characterized more by its role as an exporter of technology than by its role as an absorber of imports, and that the continuing importance of the United States, as the principal external engine of growth for the export-oriented East Asian states, should not be overlooked.

Japan and the Economic Development of East Asia

These patterns of ODA, direct investment, and trade raise questions concerning the nature of Japan's past and future contribution to East Asia's economic growth. Two widely accepted models cast Japan in the role of leader of economic development in East Asia: the "flying geese" model of economic growth, first put forward by the Japanese economist Akamatsu Kaname (see Chapter 9); and its later refinement, the model of "production cycles" (Cumings pp. 1–40). Both models are based on the argument that, as Japanese industrial technology becomes more sophisticated, Japan moves up the "production cycle", there are shifts in comparative advantage, and older industries and technology are transferred from Japan to countries in East Asia. These countries then use these technologies to produce for export to Japan, and in turn move up the production cycle in Japan's wake.

Some evidence to support the "flying geese" model can be seen in Japan's constant progress up the ladder of production, from textiles, ships, steel, and chemicals in the 1950s and 1960s, to electronics and automobiles in the 1970s and 1980s, and then to high-technology computer and information industries in the 1990s. In turn, it would appear that geographical shifts in Japanese direct investment have led to the transfer of older industries, first to the NIEs, as shown by South Korea's overtaking of Japan as the world's leading shipbuilder in the 1970s; and then to the ASEAN countries in the 1980s and 1990s, as countries such as Malaysia become major electronics producers, and fears grow of the "hollowing out" and transplant of Japanese manufacturing industry to low-cost production sites in East Asia.

However, other commentators have criticized the "flying geese" model as too simplistic to explain the rapidly changing nature of the East Asian economy. They argue that, even though technology transfers may take place from Japan to East Asia, the costs of industries starting up and mastering the new technologies are so great that these countries ultimately remain dependent on Japanese technology, and so cannot close the production cycle to create their own fully fledged industries. Instead, they contend, Japan's economic relations with East Asia can be characterized as a system of complex production links, which are connected vertically backwards to Japan because of the dependence on exports of Japanese technology, and vertically forward to the United States because of its position as the main external export market for East Asian manufactures (Bernard and Ravenhill pp. 171–209). This model is certainly a useful refinement of the "flying geese" model and reflects more accurately the patterns of trade in East Asia already observed.

However, it may also be possible to synthesize from these two models a third model of complex production links on a horizontal plain. It is arguable that the costs of modern production technologies mean that East Asian states will remain dependent on Japanese technology, but that production links may become less hierarchical over time as the recipient Asian countries begin to take advantage of these technologies. This trend may be encouraged by Japanese parent companies, which see the need to devolve technology and decision-making power to their subsidiaries in East Asia, in order to respond more flexibly to rapidly changing market conditions in each country (Kanetsuna pp. 147–60, Mochizuki p. 126). For example, surveys have shown that Japanese companies are increasing, to varying degrees, both the level of their local procurement of parts, research and development, and their employment of local managers in East Asia (Yamashita pp. 65–76). All of these changes lead to increased integration between the Japanese and East Asia economies, but also to a corresponding flattening of production relations.

At present, the evidence is too contested to make a firm judgment on the validity of any of the models. What is certain is that, although Japan's economic presence in the region has produced an asymmetric pattern of trade and relations, and the United States is still a vital player, Japan has contributed significantly to the overall economic development of the region. Even if Japan is not the principal absorber of East Asian manufacturing exports, its inputs of ODA and direct investment have functioned to promote growth by linking the East Asian economies in chains of production. For example, the Japanese electronics firm Sharp provides electronic calculator technology and components to a Taiwanese company, which then manufactures the end product in Thailand (Hook p. 181). Much of the output from these types of production links between Japan and East Asia is still destined for export to third countries outside the region, such as the United States, but these production links crisscrossing the region, pioneered by Japanese corporations, have helped to stimulate the intraregional trade which is increasingly driving growth in East Asia (Lincoln pp. 172–5).

Finally, it can also be seen that Japan has served as a model of economic development for many East Asian countries. For example, South Korea has modeled some of its government economic institutions on Japan's Ministry of International Trade and Industry (MITI). Japan has also served as the inspiration for the "Look East" economic policy pursued by the Prime Minister of Malaysia, Mahathir Mohamad.

Political and Security Relations

Southeast Asia

In the postwar period, as during World War II, Southeast Asia has remained a region of vital strategic interest to Japan, not only as a source of raw materials and markets, but also because of its position astride the vital sea lines of communication which carry up to 80% of Japan's oil imports from the Middle East. Hence, Japanese policy-makers have been conscious of the need to promote stability in the region and to cultivate strong links with each individual Southeast Asian state. Yet (as discussed above) for much of the postwar period Japan also adopted a *seikei bunri* approach to relations with Southeast Asia, emphasizing bilateral economic relations over political contacts.

The *seikei bunri* approach was initially demonstrated by Yoshida Shigeru's first overseas trip in 1954, during which he visited only western countries and neglected to visit any of Japan's Asian neighbors. The extent of Japan's alienation from Southeast Asia was also shown by its role at the Bandung Conference in Indonesia in 1954. The progenitor of the Non-Aligned Movement, the conference was regarded as an important symbol of the solidarity of developing states in Asia and Africa. In what was regarded as something of a snub to other Asian states, the Japanese government, fearing the reaction of United States to Communist China's participation at Bandung, sent only its Minister of Foreign Affairs to the conference, and ordered its delegation to take a low political profile.

In the late 1950s, however, and on into the 1960s, Japan made greater efforts to improve bilateral relations with all Southeast Asian states, as indicated by the visits to Southeast Asia of Prime Ministers Kishi Nobusuke (1957) and Ikeda Hayato (1961 and 1963), during which particular attention was paid to establishing good personal contacts with Southeast Asian leadership elites. Japan also made progress in its relations with Vietnam and Burma. These two states were unusual in the region, in that their leaders looked favorably upon Japan as a liberator from French and British colonial rule during the war.

Vietnam, not yet formally divided between a Communist North and a US-backed South, signed the Peace Treaty with Japan in 1951. Two years later, however, Japan established diplomatic relations with the regime in South Vietnam, based in Saigon (now part of Ho Chi Minh City), recognizing it as the sole legitimate government of the country. Japan signed a reparations agreement with that regime in 1959. During the Vietnam War, Japan felt obligated, as an ally of the United States, to continue supporting South Vietnam, by providing economic aid, and Sato Eisaku, Prime Minister of Japan from 1964 to 1972, gave unreserved public support for the US bombing of North Vietnam, much of which was conducted from bases in Okinawa (administered by the United States until its reversion to Japan in 1972). At the same time, however, the private doubts of Japanese leaders about the wisdom of the US military strategy in Indochina, coupled with domestic opposition to the war, ensured that Japan avoided further entanglement and was even able to maintain a small trading relationship with North Vietnam. After the ceasefire in 1973, Japan continued to recognize the government of South Vietnam and provided it with economic support, but it also moved to acknowledge the existence of the Communist government based in Hanoi, by establishing diplomatic relations in September of the same year. Following the fall of Saigon in 1975 and the reunification of Vietnam, Japan accepted the Hanoi government as the sole legitimate government and concentrated on building up its trading links with the country.

Japan began its history of relatively close relations with the isolated military socialist regime in Burma with the provision of aid in 1955 and Prime Minister Kishi's decision to make Rangoon the first stop on his Asian tour in 1957. By 1987 Japan provided close to 70% of Burma's ODA. Even though Japan suspended all grant aid to the military regime, in protest at human rights violations, in 1988, it resumed both aid and trade relations in 1989 (Mendl pp. 103–4 and 112–9).

Japan's policy until the 1970s was thus to maintain good trading relations with all the states in Southeast Asia, on a bilateral basis and by adopting a *seikei bunri* approach. Consequently, the Japanese government's reaction to the establishment of ASEAN in 1967 was somewhat lukewarm, given that it feared that ASEAN would form a multilateral economic bloc to exclude Japan. It also came to fear that the ASEAN proposal in 1971 for a "Zone of Peace and Security" was designed to oppose the US security presence in the region.

Despite initial Japanese wariness of ASEAN, the limitations of bilateralism and the *seikei bunri* approach soon became apparent in the early 1970s, and forced Japan to take a more active political stance in the region. The withdrawal of US forces from Vietnam in 1973 signaled the relative decline of US hegemony in Southeast Asia and the need for Japan to increase its political commitment to the stability of the region. Just as importantly, Japan's growing penetration of the Southeast Asian economy, marked by its exploitation of the region's raw materials and its exports of cheap manufactures, had given Japan the image of an "economic animal", which sparked boycotts of Japanese goods and anti-Japanese riots on the occasion of Prime Minister Tanaka Kakuei's visit to Thailand and Indonesia in 1974 (Nester 1992 p. 126). In response to this growing anti-Japanese sentiment, Prime Minister Fukuda Takeo announced during his visit to Southeast Asia in 1977 what became known as the "Fukuda Doctrine". This stated that, from then on, Japan would seek to promote relations with the region based on the principle of "heart to heart" understanding on political, economic, social, and cultural issues; that Japan would eschew a military role in the region; and that Japan would seek to cooperate with the member states of ASEAN and with Indochina to contribute to the region's peace and prosperity. The Fukuda Doctrine was followed by promises of increased Japanese aid to the region, such as Prime Minister Takeshita Noboru's pledge of US$2 billion under his initiative of 1987, known as Japan and ASEAN: A New Partnership Towards Peace and Prosperity.

Japan's increased support for ASEAN has been criticized as long on words but short on action, with some of the aid promises remaining unfulfilled (Nester 1990 pp. 74–5). However, it is clear that in the 1970s and 1980s Japan did begin to move away from its *seikei bunri* approach and step up political support to the region. Hence, after Vietnam's invasion of Cambodia, beginning in December 1978, Japan acceded to pressure from the United States and ASEAN to cut off aid to Vietnam, and then increased its diplomatic and economic assistance to ASEAN in an attempt to promote stabilization. Japan did continue to maintain a trading relationship with Vietnam, but also made some earnest attempts to act as an intermediary between ASEAN and Vietnam, succeeding for example in having the Cambodia problem included in the Group of Seven's summit statement in 1981. Real progress on the issue only came with the end of the Cold War and the willingness of the superpowers to draw back from their interests in Cambodia. Japan then made the largest single financial contribution to the UN Transitional Authority in Cambodia (UNTAC), and despatched military personnel to Cambodia in a non-combat role (Morrison 1988 pp. 421–7, Mendl p. 117; see also Chapter 19). Thus, by the late 1980s and early 1990s, Japan had recovered a limited political role in Southeast Asia, and established a foundation for regional cooperation.

The Korean Peninsula

The geographical proximity of Korea to Japan has meant that throughout history its leaders have maintained keen interest in the stability of Korea. As with Southeast Asia, Japan's relations with North and South Korea have been conditioned by the problems of colonialism and Japan's bilateral links with the United States. Although there have been attempts by politicians, such as Nakasone Yasuhiro or President Park Chung Hee of South Korea, to shift the focus of relations away from the problems of the past, a number of issues connected with Japan's colonial rule have continued to fuel mutual distrust between the populations of both countries. These issues include periodic tensions over the reluctance of certain conservative Japanese politicians to apologize fully for the injustices of colonial rule; controversy over the correct historical reporting of the colonial period in Japanese school textbooks; the status of Korean citizens resident in Japan (see Chapter 11); and demands for compensation from Japan by Korean women forced into prostitution for the Japanese Imperial Army, known euphemistically as "comfort women" (*jugun i'anfu*). These issues of the colonial past also affect relations with North Korea.

Japan's relations with both Koreas have been complicated further by Japan's security ties with the United States. During the Cold War, the Korean peninsula formed the point of interaction for the strategic interests of the Soviet Union, China, and the United States, and also the point for "hot war" between the Communist and US-led camps following the outbreak of the Korean War in 1950. The bilateral security relationship with the United States obligated Japan to support the latter's containment strategy, which meant Japan's isolation from North Korea, the general prioritization of relations with South Korea, and an indirect commitment to South Korea's security, by providing bases in Japan for the US military to support its presence in the South.

Initially, Japan's bilateral relations with South Korea followed the pattern of those with Southeast Asia, because of the emphasis on economic relations over political contacts. Japan and South Korea did not normalize diplomatic relations until the 1965 Treaty on Basic Relations, which was concluded only after 13 years of hard negotiations and heavy US pressure on both sides. In particular, the leaders of South Korea had balked at normalization because of domestic anti-Japanese sentiment. Under the Treaty and other separate agreements signed at the same time, Japan recognized South Korea, in accordance with UN resolutions, as the only legitimate (although not necessarily the only) government on the Korean peninsula, and provided US$500 million in the form of "economic cooperation", which was to contribute significantly to the restructuring and economic growth of South Korea in the 1960s (Mendl pp. 61–4). In a communiqué issued by Prime Minister Sato and US President Richard M. Nixon in 1969, Japan confirmed that the security of South Korea was essential to its own, and ever since then the United States has encouraged Japan and South Korea to strengthen their bilateral dialogue on political cooperation and security. Relations are still hampered by memories of colonialism; problems of fishing rights and territorial claims to the island of Takeshima (known as "Tok-do" in Korean); and South Korea's fears that Japan has at times attempted to trade off North Korea against South Korea to achieve its desire for strategic balance and stability on the peninsula. Political cooperation between Japan and South Korea seems likely to deepen as a result of the election of Kim Dae Jung, regarded as friendly towards Japan, to the presidency of South Korea in 1997, and the recognition of shared security interests with regard to North Korea.

North Korea itself remains the only state in the world with which Japan has not normalized diplomatic relations since 1945. The Cold War division and the legacy of colonialism ensured that Japan and North Korea could only conduct economic relations on a *seikei bunri* basis, and a small trading relationship developed by the 1960s. In the 1970s the Japanese government showed signs of moving towards a policy of equidistance between the two Koreas by its efforts to improve relations with North, and this period also saw an increase in trade (Kim pp. 305–30). However, relations again deteriorated in the late 1970s and 1980s, because of North Korea's failure to repay debts to Japanese companies and its suspected involvement in international terrorism. The end of the Cold War brought another brief improvement in bilateral relations, as North Korea sought normalization of relations with Japan in order to break out of its increasing diplomatic and economic isolation (Ahn pp. 263–73). However, normalization talks held between 1990 and 1992 broke down over issues of compensation for colonial rule; alleged abductions of Japanese citizens by North Korea; permission for Japanese-born wives of North Korean citizens to visit Japan; and, in particular, North Korea's refusal to consider Japanese requests for inspections by the International Atomic Energy Agency of the facilities where it was believed to be engaged in a nuclear weapons program.

After a long impasse, Japan and North Korea agreed in August 1997 to restart normalization talks, but at the time of writing there has still been no resumption. There is unlikely to be further progress in relations until there is also corresponding progress in relations between the two Koreas, allowing Japan to avoid accusations from South Korea of trying to play North and South off against each other, as well as on issues such as the visits of wives to Japan and investigations of abductions of Japanese citizens. However, Japan provided food aid, both directly and through the UN, to alleviate the famine in North Korea between 1995 and 1997. It has also taken steps to integrate North Korea into international society and stabilize its failing economy by establishing, along with the United States and South Korea, the multilateral Korean Peninsula Energy Development Organization (KEDO). KEDO will construct light-water nuclear reactors to replace North Korea's existing reactors, which are capable of reprocessing weapons-grade plutonium, and so lead to the cessation of the North's suspected nuclear weapons program.

Conclusion: Japan and the Future of East Asian Regionalism

For almost the whole of the postwar period the twin problems of the colonial past and the Cold War imposed a particular pattern on Japan's relations with East Asia, characterized by the suppression of Japan's "Asian" identity, limitations on political and security relations, and concentration on economic relations. Japanese and East Asian leaders have had to struggle to overcome these problems in their attempts to improve bilateral relations, and, as already indicated, have achieved some progress, as increasing economic integration has necessitated the abandonment of *seikei bunri* and deeper political cooperation. Arguably, however, it is Japan's participation in regional multilateral organizations, such as KEDO, which will make for its fuller economic, political and security reintegration into East Asia.

Concepts of regional economic cooperation are not new for Japan. The academic Kojima Kiyoshi proposed a Pacific Free Trade Area as long ago as 1967, and Prime Minister Ohira Masayoshi's proposal for a Pacific Community in 1978 materialized as the Pacific Economic Cooperation Council in 1980 (Deng 1997 pp. 28–38). The ending of the Cold War has given a new fluidity to relations in East Asia, spurred the growth of regional trade and cooperation, and weakened the traditional bipolar obstructions to Japan's participation in East Asian political and security affairs. In particular, the relative decline of US dominance in the region has raised questions for some observers as to whether, for the first time since the end of World War II, Japan could convert its economic power in the region into political and security leadership, or even, perhaps, a new Japanese hegemony (Vogel pp. 159–83).

Japanese Leadership and Regional Organizations

Evidence for Japan's potential leadership role has been seen in the promotion of shared "Asian values" in both Japan and East Asia. Despite Japan's position as a developed country and a member of the "democratic West", many Asian leaders view it as sympathetic to the region's emphasis on economic developmentalism, and less willing than the United States to impose upon authoritarian Asian governments what it is argued are western human rights. For example, in their book *The Asia That Can Say No* the Japanese nationalist politician Ishihara Shintaro and the Malaysian Prime Minister Mahathir Mohamad defined a new form of "Asianism", in opposition to the values of the West. In recent years, a rise in Japan's "soft power" in the region, as a function of its economic power, has also been detected. The growth in readership of Japanese comic books (*manga*) across East Asia, and the diffusion of Japanese food, music, and other forms of high and low culture, is regarded by some as a possible Japanese challenge to US cultural leadership in the region (Drifte pp. 144–61). The outcome of increasing economic integration and affinity in cultural values between Japan and East Asia, at least according to some commentators, has been that Japanese identity is being "re-Asianized", and that Asian leaders can now countenance a greater leadership role for Japan. This is best illustrated by Mahathir's proposal, put forward in 1991, for an East Asian Economic Caucus (EAEC). This regional bloc would exclude such non-Asian countries as the United States, Canada, Australia, and New Zealand, and, it seems, give a leadership role to Japan as the largest economy in the region.

Japan has not committed itself to the EAEC but has given support to the establishment of other multilateral bodies in the region, some of which are concerned most significantly with political and security issues. The creation of the ASEAN Regional Forum as the first region-wide body to discuss security issues originated partly from a proposal by the Japanese Foreign Minister, Nakayama Taro, at the ASEAN summit in 1991 (Hughes pp. 232–3). In addition, Japan has worked to promote regional projects via the Asia-Pacific Economic Cooperation (APEC) organization, established in 1989. Japan has supported the establishment of the APEC Secretariat, proposed greater cooperation on energy at the APEC summit in 1994, and staged an APEC summit in Osaka in 1995.

Limits to Japanese Leadership

Nevertheless, despite the significant upgrading of Japan's economic, political, and security role in East Asia, this does not equate with a full leadership role, and far less with hegemony. Japan still has a "legitimacy deficit" in the region, because of the history of colonialism and fears of renewed militarism. US military power in the region remains supreme, while Japan's Constitution and its bilateral security links with the United States constrain it from playing a direct military role in East Asian security. The lack of depth to Japanese "soft power" is shown by the growing dominance of English as the region's language of commerce and culture. Japan also, arguably, lacks many of the domestic policy-making institutions that it would need to act as an effective leader of the region, such as a strong prime ministership (Calder pp. 1–24). The "reactive" nature of Japan's diplomacy in East Asia has been shown most recently by its taking a backseat to the United States in dealing with the currency crises of late 1997 and early 1998. Finally, Japanese leaders are aware that, even though the relative importance of the US market is declining, the United States is still Japan's largest trading partner and that Japan's participation in any regional economic project is dependent upon the inclusion of the United States. Thus, Japan has eschewed support for the EAEC in favor of APEC, entrusted the leadership of the latter to its ally the United States, and concentrated on playing a supporting role. Indeed, it is clear that APEC, with its mixed membership of Asian and "western" states on the Pacific Rim, represents for Japan a means to avoid a choice between Asian and western identities. Japan can now locate itself with less tension in the midst of the Asia-Pacific region, geographically, economically, and politically, and perhaps play a bridging role between Asia and the West.

Further Reading

Ahn Byung-Joon, "Japanese Policy Towards Korea", in Gerald L. Curtis, editor, *Japan's Foreign Policy After the Cold War: Coping with Change*, Armonk, NY: M. E. Sharpe, 1993

Japanese relations with North and South Korea after the Cold War.

Arase, David, *Buying Power: The Political Economy of Japan's Foreign Aid*, Boulder, CO: Lynne Rienner, 1995

Japan's aid policy in the 1990s and its business aspects.

Bernard, Mitchell, and John Ravenhill, "Beyond Product Cycles and Flying Geese: Regionalization, Hierarchy, and the Industrialization of East Asia", in *World Politics*, Volume 47, number 2, January 1995

This article challenges the theory of "flying geese" and posits instead the theory of complex production links.

Calder, Kent E., "The Institutions of Japanese Foreign Policy", in Richard L. Grant, editor, *The Process of Japanese Foreign Policy: Focus on Asia*, London: The Royal Institute of International Affairs, 1997

Calder examines Japan's internal policy-making institutions and the restrictions of reactive diplomacy.

Cumings, Bruce, "The Origins of the Northeast Asian Political Economy: Industrial Sectors, Product Cycles, and Political Consequences", in *International Organization*, Volume 38, number 1, Winter 1984

Cumings puts forward the theory of production cycles and Japan's role as the leader of industrialization in East Asia.

Curtis, Gerald L., editor, *The United States, Japan, and Asia: Challenges for US Policy*, New York: W. W. Norton, 1994

Articles by leading scholars on the triangular economic, political, and security relationships among Japan, East Asia, and the United States after the Cold War.

Deng Yong, *Promoting Asia-Pacific Economic Cooperation: Perspectives from East Asia*, London: Macmillan, and New York: St Martin's Press, 1997

Deng examines the history of Japanese leadership in promoting regional cooperation in East Asia.

Drifte, Reinhard, *Japan's Foreign Policy in the 1990s: From Economic Superpower to What Power?*, London: Macmillan, and New York: St Martin's Press, 1996

This book examines Japan's "soft power" and "leadership by stealth" in Asia and the world after the Cold War.

Grant, Richard L., editor, *The Process of Japanese Foreign Policy: Focus on Asia*, London: The Royal Institute of International Affairs, 1997

A useful and concise treatment of Japan's approach to Asian foreign policy.

Hamashita Takeshi, "The Intra-regional System in East Asia in Modern Times", in Katzenstein and Shiraishi, cited below

Hamashita explains the Sinocentric world order that prevailed in Japan before modernization began.

Hook, Glenn D., "Japan and the Construction of the Asia-Pacific", in Andrew Gamble and Anthony Payne, editors, *Regionalism and World Order*, London: Macmillan, and New York: St Martin's Press, 1996

An analysis of Japan's historical and contemporary economic, political, and security role in building the Asia-Pacific region.

Hughes, Christopher W., "Japan's Subregional Security and Defence Linkages with ASEAN, South Korea and China in the 1990s", in *The Pacific Review*, Volume 9, number 2, 1996

This article examines Japan's growing security dialogue with East Asia since the Cold War.

JETRO (Japan External Trade Organization), *Sekai no Boeki to Nihon* [World Trade and Japan], Tokyo: JETRO, various years

A useful source of statistical information on Japan's trading patterns.

Kanetsuna Motoyuki, "Nihon Kigyo no Gurobaru Netwakuka" [The Global Networking of Japanese Corporations], in Yokoyama Masaki and Wakui Hideyuki, editors, *Posuto Reisen to Ajia: Ajia no Kaihatsushugi to Kankyo, Heiwa* [Post Cold-War Asia: Asian Developmentalism, Environment and Peace], Tokyo: Chuo Keizaisha, 1996

A useful summary of Japan's economic relations with East Asia since the Cold War.

Katzenstein, Peter J. and Takashi Shiraishi, editors, *Network Power: Japan and Asia*, Ithaca, NY: Cornell University Press, 1997

Katzenstein and Shiraishi have brought together a wealth of articles on Japan's economic, political, security, and "soft power" relations with East Asia.

Kim Hong. N., "Japan's Policy Toward the Korean Peninsula Since 1965", in Kwak Tae-Hwan, editor, *The Two Koreas in World Politics*, Seoul: The Institute for Far Eastern Studies, Kyungnam University, 1983

A concise account of Japan's relations with the two Koreas up until the 1980s.

Lincoln, Edward J., *Japan's New Global Role*, Washington, DC: The Brookings Institution, 1993

A good all-round analysis of Japan's relations with Asia and the world following the end of the Cold War.

Mendl, Wolf, *Japan's Asia Policy: Regional Security and Global Interests*, London and New York: Routledge, 1995

A concise treatment of Japan's political relations with Asia, both historical and contemporary.

Ming Wan, "Japan and the Asian Development Bank", in *Pacific Affairs*, Volume 68, number 4, Winter 1995–96

An up-to-date description of Japan's role in the multilateral ADB.

Mochizuki, Mike M., "Japan as an Asia-Pacific Power" in Robert S. Ross, editor, *East Asia in Transition: Toward a New Regional Order*, Armonk, NY: M. E. Sharpe, 1995

An examination of Japan's role in Asia following the end of the Cold War.

Morrison, Charles E., "Japan and the ASEAN Countries: the Evolution of Japan's Regional Role", in Inoguchi Takashi and Daniel I. Okimoto, editors, *The Political Economy of Japan*, Volume 2, *The Changing International Context*, Stanford, CA: Stanford University Press, 1988

Japan and ASEAN relations during the Cold War.

Nester, William, *Japan's Growing Power Over East Asia and the World Economy: Ends and Means*, London: Macmillan, 1990

A critical treatment of Japan's economic dominance in East Asia.

Nester, William, *Japan and the Third World: Patterns, Power, Prospects*, London: Macmillan, and New York: St Martin's Press, 1992

Another critique of Japan's relations with Asia, which contains some very useful narrative.

Orr, Robert M., Jr., and Bruce M. Koppel, *The Emergence of Japan's Foreign Aid Power*, New York: Columbia University Press, 1990

A useful examination of Japan's ODA relations with developing states in Asia and across the world.

Schaller, Michael, *The American Occupation of Japan: The Origins of the Cold War in Asia*, New York and Oxford: Oxford University Press, 1985

An explanation of Japan's incorporation into the US security, political, and economic sphere before the start of the Cold War.

Selden, Mark, "The Regional Political Economy of East Asia", in Katzenstein and Shiraishi, cited above

Selden examines Japan's place in the East Asian economy.

Shiraishi Takashi, "Japan and Southeast Asia", in Katzenstein and Shiraishi, cited above

An interesting recent analysis of relations between Japan and Southeast Asia.

Simone, Vera, and Anne Thompson Feraru, *The Asian Pacific: Political and Economic Development in a Global Context*, New York and London: Longman, 1995

An accessible introduction to the topic of the growth of regionalism in the Asia-Pacific region.

Soderberg, Marie, "Japanese ODA: the Business Perspective", in Marie Soderberg, editor, *The Business of Japanese Foreign Aid: Five Case Studies From Asia*, London and New York: Routledge, 1996

An excellent analysis of Japanese ODA through case studies.

Stubbs, Richard, "The Political Economy of the Asia-Pacific Region", in Richard Stubbs and Geoffrey R. D. Underhill, editors, *Political Economy and the Changing Global Order*, London: Macmillan, and New York: St Martin's Press, 1994

A lucid analysis, from the point of view of political economy, of the East Asian economy, and the roles played by the United States and Japan.

Tsusho Sangyo-sho [Ministry of International Trade and Industry], *Tsusan Hakusho* [White Paper on Trade], Tokyo: Okurasho Insatsukyoku [Ministry of Finance Printing Office], 1984 and 1996 editions of annual publication; and *Waga Kuni Kigyo no Kaigai Jigyo* [The Overseas Activities of Japanese Corporations], 1995

These are valuable sources of statistics on Japan's direct investment in, and trade with, East Asia, bearing in mind that the Ministry generally defines "East Asia" as a group of just eight countries. An English-language version of the White Paper is also available.

Vogel, Ezra, "Japan as Number One in Asia", in Curtis 1994, cited above

One view of Japan's rise to hegemony after the Cold War.

Welfield, John, *An Empire in Eclipse: Japan in the Postwar American Alliance System: A Study of the Interaction of Domestic Politics and Foreign Policy*, London: Athlone Press, 1988

A detailed empirical study of the relation between Japanese foreign policy and the US alliance

Yamashita Shoichi, "Japanese Investment Strategy and Investment Transfer in East Asia", in Hasegawa Harukiyo and Glenn D. Hook, editors, *Japanese Business Management: Restructuring for Low Growth and Globalization*, London and New York: Routledge, 1998

Yamashita examines the transfer of technology and management practices by Japanese corporations to East Asia.

Yasutomo, Dennis T., *The New Multilateralism in Japan's Foreign Policy*, London: Macmillan, and New York: St Martin's Press, 1995

A concise history and analysis of the functions of Japanese ODA.

Christopher W. Hughes is a Research Fellow of the Institute for Peace Science at the University of Hiroshima.

Chapter Nineteen

Japan and the World: From Bilateralism to Multilateralism

Hugo Dobson

> . . . We desire to occupy an honored place in an international society striving for the preservation of peace, and the banishment of tyranny and slavery, oppression and intolerance for all time from the earth. We recognize that all peoples of the world have the right to live in peace, free from fear and want.
>
> We believe that no nation is responsible to itself alone, but that the laws of political morality are universal; and that obedience to such laws is incumbent upon all nations who would sustain their own sovereignty and justify their sovereign relationship with other nations.
>
> We, the Japanese people, pledge our national honor to accomplish these high ideals and purposes with all our resources.

As this excerpt indicates, the preamble to Japan's 1947 Constitution placed a heavy emphasis upon international cooperation through various multilateral institutions. Indeed, since Japan joined the UN, on December 18, 1956, a "UN-centered foreign policy" has been an oft-repeated phrase in the Japanese government's annual Blue Papers on Foreign Policy. Slowly but surely, after an unfortunate beginning to its experience of multilateralism, Japan is showing signs of realizing this ideal. However, attention should also be paid to other international organizations, notably the Asian Development Bank (ADB) and the European Bank for Reconstruction and Development (EBRD), that are used as forums for the promotion of Japan's multilateral foreign policy. These bodies and Japan's role in them are increasingly important, because only they have the capacity to deal with the security, economic and environmental problems of the post-Cold War world.

Prewar Experience of Multilateralism

Japan's experience of multilateralism has historically been an uncomfortable one, and its governments' longstanding preference for bilateral relations can be seen in the central position occupied in their foreign policies during the 20th century by the Anglo-Japanese Alliance (1902–23), the Tripartite Pact with Germany and Italy (1940–45), and the Security Treaty with the United States (1952, with subsequent revisions). Japan's first experience of multilateralism came with its participation as a victorious power in the creation of the League of Nations at the Versailles Peace Conference of 1919. However, Japan's first impressions of the League were tarnished by its failure to have a clause on racial equality written into the League's Covenant. Thus, from the beginning Japan was clearly regarded by the western powers as a racially inferior parvenu. The following period did see Japan's participation in various schemes for multilateral cooperation, including international conventions such as the Washington Conference (1921–22), the Kellogg-Briand Pact (1928), and the London Naval Treaty (1930). For a time it appeared that Japan was behaving as a responsible member of an evolving international society, cherishing what Nitobe Inazo,

Japan's representative to the League (1920–26), called "the maintenance of peaceful relations with the rest of the world" (quoted in Howes). Yet, despite the presence of a number of influential internationalists, Nitobe among them, occupying positions of responsibility and influence within the League and actively working for its promotion, Japan's participation was unproductive. It came to a sudden conclusion in March 1933, when Japan withdrew from the multilateral body following international chastisement based on the findings of the Lytton Commission, which had been dispatched by the League to report on the establishment of the Japanese puppet state of Manchukuo, in the aftermath of the engineered Manchurian Incident (1931). The concepts of internationalism and civilian control in Japan began to erode, as the country turned away from its World War I allies and found convenient bedfellows in the revisionist powers of Germany and Italy, even though they often pursued diametrically opposed objectives (see Nish).

Japan and the UN: An Overview

Following Japan's defeat in World War II and the installation of the Allied Occupation, Yoshida Shigeru, Minister of Foreign Affairs (1945–46) and then Prime Minister (1946–47 and 1948–54), surmised that, given the central role of the United States in the new international system, Japan's recovery of great power status would be best accomplished by concentrating on its own economic reconstruction, while playing a submissive and reactive role in relation to the United States. To ensure recovery, the "Yoshida Doctrine" which he formulated stressed the coercion of Socialist and Communist groups within Japan; minimum military spending; the transference of responsibility for Japan's security to the United States; and the avoidance of other international commitments, such as through the UN.

Against the background of continuing adherence to the Yoshida Doctrine by Japan's leading politicians, the country's next experience of multilateralism came with admission to the UN in 1956, at the height of the Cold War. A number of its previous attempts had been vetoed by the Soviet Union, which was fearful of adding another "lackey" of the United States to this adolescent international organization. Admission was regarded in Japan as a return to the international community and a symbol of Japan's regained legitimacy and status, and was greeted with an overwhelming display of public support. Throughout the postwar period the UN and its work have continued to command high levels of support from the Japanese public.

Upon admission the Japanese government enthusiastically advocated "UN-centered diplomacy". Prime Minister Kishi Nobusuke stated in the House of Representatives in January 1957 that Japan's policy would center on three pillars: cooperation with other democracies, maintaining a position as a "member" of Asia, and "UN-centrism". However, beyond the rhetoric there was no sense of what these ideas would concretely involve in a divided world. Rather than living up to its rhetoric and placing its faith in the UN, itself already crippled by this East/West confrontation, the Japanese government firmly placed itself in the western camp, displaying an overriding preference for the substantial economic benefits and minimum security commitments that accompanied cooperation with the United States. Thus, for the first few decades of Japan's membership the concept of UN-centered diplomacy was a hollow one, which amounted to little more than meeting budget contributions to the UN, joining various organs of the UN system, serving as a non-permanent member of the UN Security Council (UNSC), and voting in line with the United States on issues such as the Korean War or the representation of China in the UN.

During the 1970s, the nature of the UN changed drastically, as the process of decolonization created new, non-western power bases within its General Assembly. As a consequence, the United States began to lose influence and, later, interest in the UN. In this atmosphere, the Japanese government was given the opportunity to realize its concept of UN-centrism, as Japanese participation became more independent of the United States. For example, on the issue of Palestine, Japan abstained on the "Zionism equals Racism" Resolution, passed by the General Assembly in November 1975, and, despite opposition from the United States, endorsed resolutions recognizing the right of self-determination for the Palestinian people in 1976.

In the 1980s, the attitude of the United States towards the UN was compounded by the uncooperative stance of the Reagan administration and its domestic supporters, encapsulated in the Kassenbaum Amendment, adopted by the US Congress in August 1986, which mandated the withholding of US contributions to the UN's regular budget. In this political environment, Japan found it increasingly difficult to associate itself with the anti-UN line of the United States. Thus, a degree of independence and proactivism continued to develop on the Japanese side, with the aim of promoting conciliation and reforming the UN. This was exemplified by the proposal, tabled by the Japanese Foreign Minister Abe Shintaro at the 40th Session of the General Assembly in 1985, to create a Group of High-level Intergovernmental Experts, which became known as the Group of 18. This committee reviewed the UN's administrative and financial procedures, with the aim of regaining the trust, participation, and funds of the United States and other major contributing nations, and its 71 recommendations were adopted by the General Assembly in 1986.

The extent to which Japan branched out on its own is also reflected in voting patterns. During the 1960s there was generally a voting coincidence of 80% between the United States and Japan, but during the 1980s this rate dropped to 37% (see Urano). Only in the 1980s, following changes in international political structures, did the oft-repeated espousal of UN-centrism become imbued with proactive meaning.

With the Iraqi invasion of Kuwait in 1990, the issues of Japan's general participation in the UN and, specifically, its peacekeeping contributions were catapulted from the ivory towers of academia and government circles onto the front pages of the national and international press. One salient effect of this was the Tokyo Declaration of January 1992, issued during the honeymoon period of the UN's post-Cold War activism, in which US President George Bush and Japanese Prime Minister Miyazawa Ki'ichi committed "their resources and the talents of their peoples to the purposes of the United Nations Charter [in order to] invigorate the UN organizations" (see Immerman). The United States soon lost interest in the UN's post-Cold War role, largely because of the shortcomings of operations in Somalia and quarrels over budgetary contributions, but Japan's role still gives cause for optimism among those who support proactivism at the UN.

Peacekeeping

As the UN itself emerged from a long decade of financial and credibility problems, and began to raise its profile in the late 1980s, Japan began to display signs of a more proactive approach to participation in UN peacekeeping operations (UNPKO). Although the greatest change in Japan's peacekeeping policy came in the 1990s, the issue of Japan's participation in UNPKO had been a recurrent topic of debate since Japan's admission to the UN.

In 1958, Dag Hammarskjöld, then UN Secretary General, attempted unsuccessfully to solicit the dispatch of the newly created Japanese Self-Defense Forces (SDF) to the Lebanon. During the Cold War, Japan's support for UNPKO was purely financial, but with the relaxation of East/West tensions in the late 1980s, concrete contributions of personnel to UNPKO began with Japan's first participation, the dispatch of 27 civilians to Namibia in October 1989 as part of the United Nations Transition Assistance Group (UNTAG). Subsequently, six Japanese civilians joined UNPKO overseeing elections in Nicaragua and Haiti, in 1989 and 1990. Because of the legacy of World War II and perceived constitutional restrictions on Japan's right of belligerency, these were purely non-military operations, arousing little controversy over remilitarization within Japan or among its East Asian neighbors.

The Iraqi invasion of Kuwait and the consequent international effort during the Gulf War raised the issue of Japan's military role in UNPKO. The government of Prime Minister Kaifu Toshiki came under pressure, both domestically and internationally, to make more of a contribution to operations in the Gulf than the traditional minimum enshrined in the Yoshida Doctrine. The US$13 billion eventually agreed upon by the Japanese government was not forthcoming initially and failed to satisfy demands from the Allies in the Gulf for a human contribution on Japan's part to the Middle East, an oil-rich area which is of strategically vital concern to Japan. This not inconsiderable contribution was made with more regard for preserving bilateral relations with the United States than for promoting international philanthropy.

In response to outside pressure (*gaiatsu*), the government then presented a short-lived United Nations Peace Cooperation Bill to Parliament in October 1990. The bill failed, not only because of the lack of public and political support both inside and outside the government, but also due to the haste with which it was prepared. The debate over how best to contribute continued, and in September 1991 the government proposed the Law on Cooperation in UN Peacekeeping and Other Operations, known as the "PKO Bill" or, later, the "PKO Law", in the Japanese media. The government proposed that Japanese participation be permitted in the following 11 fields:

- monitoring ceasefires;
- stationing troops in, and patrolling, demilitarized zones;
- controlling the influx of weapons;
- collecting, storing and disposing of abandoned weapons;
- assisting disputants in settling borders;
- assisting with the exchange of prisoners of war;
- observing and supervising elections, and ensuring fair balloting;
- providing bureaucratic advice and guidance, for example, on police administration;
- medical care;
- transportation, communications and construction work; and
- humanitarian work, including the assistance, rescue and repatriation of war refugees.

However, in order to conform to the Constitution, the PKO Law, enacted in 1992, limits Japanese participation to traditional PKO, those which recognize state boundaries, do not use force except in self-defense, and maintain strict impartiality. It therefore became necessary to place a temporary freeze on the first six of the activities listed above, which could only be lifted after a government review.

It was possible for the Japanese Parliament to pass the PKO Bill into law in 1991–92, unlike its predecessor in 1990, because the international climate and domestic public opinion had both changed drastically. Resistance at home and in East Asia relaxed when the Japanese government made clear the kinds of UNPKO in which the SDF could and could not participate. Opposition groups at home and abroad also came to accept the necessity for Japan to make a human contribution.

One case in point was the resolution of the conflict in Cambodia and the UN's commitment, for the first time in its history, to undertake the governance of a sovereign state, in order to promote UN-sponsored elections. The five conditions attached to the dispatch of the SDF were all satisfied in Cambodia. The parties in the conflict had reached agreement on a ceasefire; the parties involved had given their consent to the deployment of Japanese forces; the deployed forces would remain impartial; the use of weapons would be limited to self-defense; and, if any of these conditions were not met, the Japanese government would be free to remove its forces. Thus, the establishment of the UN Transitional Authority in Cambodia (UNTAC) saw the first dispatch overseas of Japanese military forces since 1945. Six hundred SDF personnel served as engineers, engaged chiefly in repairing roads and bridges in areas where the use of force would be a faint possibility. Despite these precautions, hostilities did continue, particularly as the Khmer Rouge faction refused to honor the terms of the Paris Peace Accords. As a result, Japan experienced its first UNPKO casualties. In April 1993, Nakata Atsuhito, a UN volunteer, was killed in the dangerous region of Kompong Thom; and in the following month Takata Haruyuki, a civilian police officer, was shot and killed in an attack in the region of Banteay Meanchey. These events caused an emotional outcry in Japan, and calls were made to bring back all personnel, as it was evident that the ceasefire was not effective. Nevertheless, the government remained firm and contended that the ceasefire was generally holding and that a certain number of casualties was to be expected from participation in any UNPKO.

The educational effect of the Cambodia experience on Japanese attitudes cannot be underestimated. Not only did it raise the profile of the SDF, but it also aroused the Japanese public's interest in UNPKO and their humanitarian aspects. In particular, interest in the work of UN volunteers peaked in the aftermath of the UNTAC operation. Public opinion polls also reflect this change in the Japanese public's perception of the peacekeeping work of the UN. Immediately before the passage of the PKO Law, 41.6% of respondents supported SDF participation in UNPKO, but 36.9% opposed it. In addition, 50.3% believed that dispatch would be constitutionally problematic, while 28.2% thought otherwise. A poll carried out for the public broadcasting organization NHK in May 1993 saw 17% of those polled rate Japan's role in UNTAC highly and 47.8% rate it fairly highly, making a total of 64.8% holding a positive view of UNPKO. In January 1994, a poll conducted by the Prime Minister's Office revealed that 48.4% supported Japan's participation in UNPKO, an increase from 45.5% as polled in 1991. Significantly, the proportion opposed to Japan's participation fell from 37.9% to 30.6% over these three years. Opposition abroad to Japan's dispatch of the SDF also dissipated, as certain East Asian nations, such as the Philippines and Thailand, developed excellent working relationships with Japan in the field of logistics during the UNTAC operation. The traditional criticisms by China and South Korea of what they perceive as Japanese remilitarization have also softened in relation to dispatches of the SDF within the framework of UNPKO, and even the idea of joint training exercises between South Korea and Japan has been given some consideration.

After UNTAC the Japanese government found it considerably less difficult to dispatch SDF contingents, as staff officers, movement control units and electoral observers, to the UN Operation in Mozambique (ONUMOZ) between May 1993 and January 1995, and in January 1996 SDF troops were dispatched to the UN Disengagement Observer Force (UNDOF) on the Golan Heights, to provide secondary support for staff and transportation. In between these two commitments Japan also dispatched its first SDF contingent on a humanitarian relief effort, to Goma in Zaire, in order to provide medical assistance, sanitation, water supplies, and transportation for refugees. These dispatches provided the SDF with opportunities to develop and refine their peacekeeping skills. Encouragingly for supporters of UNPKO, these dispatches have encountered little opposition at home and abroad, in comparison to operations in the Persian Gulf and Cambodia.

However, despite these advances in Japan's UNPKO role, certain obstacles need to be tackled before Japan can shoulder greater responsibilities. These include the legally required but drawn-out government review to remove the restrictions on the activities of the SDF under the PKO Law (monitoring disarmament, stationing and patrol, inspection and disposal of abandoned weapons); the use of force within UNPKO; the necessity of creating an organization to contribute to UNPKO separate from the SDF; and the connection drawn between Japan's UNPKO policy and other aspects of its work within the UN system.

Funding Multilateral Institutions

In the same way as Japan has sought to increase its commitment to the international community via UNPKO, it has also demonstrated a growing financial commitment, both to the UN and other multilateral institutions. At the time of its admission to the UN in 1956, Japan paid a modest assessed contribution of 2.19% of the UN budget. Since that time, and in line with growth in Japan's GNP, the assessment has increased rapidly to overtake not only that of its chief rival for a permanent UNSC seat, Germany, but also those of three permanent UNSC members, the United Kingdom, France, and the former Soviet Union, along with its successor state, the Russian Federation. Reflecting this increase in Japanese contributions, not only to the UN but to other multilateral institutions (see below and Table 19.1), the term "checkbook diplomacy" has been applied to Japan's reliance upon making financial, rather than personnel, contributions to solve international problems. The 1980s in particular were a period when Japan's contributions expanded rapidly and came close to rivaling those of the United States. Despite Japan's monetary donations, however, its political influence in the UN has remained minimal, and the privileges of UNSC membership have been denied. Japan is not only the second largest contributor to the general UN budget (16.75% in 1997 and 1998) but it is also the second largest contributor to the PKO budget. As mentioned above, Japan contributed US$13 billion to the international effort during the Gulf War, and it also made large donations to the UN interventions in Cambodia, Somalia and Bosnia.

Japan regularly meets its payment commitments, although the differing time scales employed by the UN and the Japanese government for short periods before the Japanese Parliament has approved the budget mean that the statistics may occasionally demonstrate a considerable but deceptive amount of arrears (see Kwon). Indeed, Japan has been eager in the past to use reserve funds to assist the UN when funding has been in short supply. For example its voluntary contributions to the UN Force in Cyprus (UNFICYP) are surpassed only by those of the United States and four European states. In many cases Japan has often had to make up shortfalls in US contributions and has regularly made sizable voluntary contributions to various UN agencies. In 1963 Japan purchased US$5 million in UN bonds. Special contributions amounted to US$2.5 million in 1966, and US$10 million in 1974. In 1987 Japan made an unsolicited donation of US$15 million to assist in the establishment of UNPKO in Afghanistan, and between Iran and Iraq. Japan became the second largest contributor to the UN High Commission for Refugees in 1979, the UN Environment Program in 1981, the UN Development Program in 1984, and the World Bank in 1985. Despite all these considerable increases in Japan's contribution to the UN budget and the increased burden for the Japanese taxpayer, no vociferous opposition has been voiced in Japan, largely because the work of the UN is generally held in high regard.

Japan's activism has also extended to taking up the role of promoting reform of the UN's controversial funding system. Japan was able successfully to soften US intransigence over meeting contributions in 1985–86 with the promotion of the Abe proposal (described above), which introduced the principle of consensus in the budget programming process. The promotion of this idea of consensus was enough to bring about the resumption of UN contributions by the United States.

The Japanese Ministry of Foreign Affairs (MoFA) now emphasizes three points in its proposals for future financial reform of the UN. First, a conceptual distinction should be made between current problems of cashflow and the wider issue of financial reforms. The problem of cashflow can be addressed simply by member states fulfilling their financial obligations under the Charter. Second, financial reform should proceed together with other reforms, such as the reform of the UNSC and the reform of economic and social activities, in order to achieve the reform of the UN as a whole in a balanced way. Third, regarding the scale of assessments, it is necessary to establish a clearer linkage between the responsibility of each member state and the share of the financial burden that is apportioned to it for support of the UN. The MoFA seeks to supplement the current principle of "capacity to pay" with "responsibility to pay", which takes into account the special responsibilities and privileges of the permanent members of the UNSC. Thus, again with the limited commitment of the United States in the UN since the 1980s and the restructuring of the Cold War global system, Japan has attempted to plug the gaps in the UN budget and keep it afloat as a viable international organization.

Permanent Membership of the UN Security Council

The issue of Japan's inclusion as a permanent member of the UNSC was first raised in 1969 at the 24th Session of the General Assembly, when Foreign Minister Aichi Ki'ichi expressed Japan's ability and readiness to shoulder the burdens of permanent membership. In 1973, as Japan's assessed contribution to the UN budget reached 7.15% and surpassed those of France and the United Kingdom, a Joint Communiqué issued by Prime Minister Tanaka Kakuei and US President Richard M. Nixon touched on the issue, declaring the belief that: "a way should be found to assure permanent representation in that council for Japan, whose resources and influence are of major importance in world affairs" (see Ogata). Japan has served as a non-permanent member of the UNSC on more occasions (eight) than any other state, most recently from 1997 to 1998 (the other occasions were 1958–59, 1966–67, 1971–72, 1975–76, 1981–82, 1987–88, 1992–93). However, Japan has no stable claim to this position, as shown in 1978, when Japan lost to Bangladesh in the UNSC election, causing outcry at home. Not needing to run for election, along with the influence in exercising the veto that naturally goes with a permanent seat, are the chief incentives for Japan to join the UNSC.

Of course, the inclusion of Japan in the UNSC would be part of a wider reform of the UN system, with the objective of reflecting realities in the post-Cold War world rather than those of the post-World War II world. But this is no simple task, as reform of the UN Charter requires a two-thirds majority of member states, including all the present five permanent UNSC members. However, there is a precedent in the increase in the number of the non-permanent UNSC seats that took place in 1965. There is a myriad of proposals for restructuring the UNSC, which for reasons of space and relevance cannot be detailed here, but Japanese opinions center upon the need for a far-reaching and long-term reform, rather than a "quick fix", in order to enable the UN to face an expanded security agenda in the coming century. To maintain the present UNSC ossified by the Cold War would do considerable harm to the status and reputation of the UN in general. However, there is also a fear in the MoFA that an expanded UNSC could become a second General Assembly, with a consequent dilution of the influence and prestige of membership. In addition, there is no desire in Japan to downplay the importance of established members such as the United Kingdom or France. Their international importance may have declined relatively since 1945, but they still occupy positions of influence and "soft" power, and in any case would not acquiesce in a reduction of their central role in the UN. Thus, a degree of conservatism must be balanced against an attempt to include the rising global and regional powers, and here lies the difficulty in reforming the UNSC.

Of all of the states vying for inclusion in the exclusive permanent club of the UNSC, Japan is the most likely and most widely promoted candidate, chiefly because of the financial contributions that it makes. In addition, there is the fact that it has begun to make a visible human contribution. As the largest donor in the world of official development assistance Japan is also in a position to assist with any expansion of the humanitarian and social work of a newly reformed UN. Prominent politicians such as the opposition party leader Ozawa Ichiro agree with the MoFA in regarding inclusion in the UNSC as a natural direction for Japan as it embarks on a process of becoming a "normal state" unhindered by overtly pacifist norms. Yet these pacifist opinions, embedded in Japanese society, are also audible, as fears are expressed within Japan that a permanent seat would result in an expanded peacekeeping role, possibly including peace enforcement duties. It is often argued that Japan's role in the UN should be primarily economic and social, not military. These fears are also expressed in East Asia, especially by South Korea and China, which regard any increase in Japan's peacekeeping role as creeping remilitarization.

Support for Japan's membership of the UNSC has been forthcoming from the United States and western Europe, and Japan's policies of promoting the idea of "no taxation without representation" and of eliciting as many endorsements of support from member states as possible has kept the issue of inclusion alive. However, a number of other regional powers have been more keen to promote their own chances of permanent representation, Brazil, Nigeria and India being among the favorites. Thus, this issue will not be dealt with unilaterally, since a comprehensive overhaul of the UNSC is the preferred solution of the UN and the international community.

Changing the text of the UN Charter would also be a necessary condition for allowing Japan into the family of permanent UNSC members. The choice of the name, "United Nations", itself is a relic from the World War II alliance against Germany, Italy and Japan, although, fortunately, it is translated into Japanese as "International Confederation" (*Kokusai Rengo*, or *Kokuren*), thus avoiding any association with the war. The "enemy clauses" within the UN Charter are more problematic. Article 53 allows regional security organizations to take peace enforcement actions, without the approval of the UNSC, in the event of the Allies' former enemies renewing their aggression. Article 107 states that "nothing in the present Charter shall invalidate or preclude action, in relation to any state which during the Second World War has been an enemy of any signatory to the present Charter, taken or authorized as a result of that war by the Governments having responsibility for such action". This was used by the former Soviet Union to justify the occupation of the Kuriles, the disputed islands which the Japanese call the Northern Territories, and which are still a source of friction between Japan and Russia. The chief importance of the two clauses is the psychological weight they carry in still branding Japan an enemy within an organization propped up financially by Japanese contributions and in which Japan is gaining an increasingly higher profile.

Japan has participated actively in other organs of the UN system, notably with almost continuous representation on the Economic and Social Council, and regular representation on the bench of the International Court of Justice. Thus, one recourse to obtain permanent UNSC membership left open to Japan seems to be to promote its own representation within the UNSC, and especially through the Secretariat as the "central processing unit" of the UN. Japanese representation has traditionally been low-key and there is a scarcity of qualified Japanese international public servants, because of language limitations, and the influence of life-time employment systems at home, so that in 1990 Japan had fewer staff at the UN than most other industrialized nations. However, through individuals such as Ogata Sadako at the UN High Commission for Refugees, Akashi Yasushi, the UN Secretary General's Special Representative in Cambodia and the former Yugoslavia, and Owada Hiroshi, Japan's Permanent Representative to the UN, Japan's stock has recently risen in value. Although a Japanese Secretary General may be a long way off, the idea of Ogata Sadako occupying the recently created post of Deputy Secretary General was strongly supported by the present Secretary General, Kofi Annan. Annan sought a female from a developed nation to complement his role, but Ogata herself declined the post (according to a report in *The Times*, January 12, 1998).

Other Multilateral Commitments

Within other international organizations, Japan has received greater representation in return for its not inconsiderable financial support (see Table 19.1), ensuring that the UN is not the only important outlet for Japan's commitment to the international community. At both the IMF and the World Bank, Japan has seen its representation through voting rights increase in line with its contributions, making it the second largest contributor to each organization. Yet, as at the UN, Japan's comprehensive participation in other multilateral organizations has suffered from a lack of personnel. In 1991, the World Bank was thought to have only 86 Japanese nationals employed within a total staff of 6,700, only 1.3% of the total. The IMF and the International Finance Corporation at the same time were thought to have Japanese as only 1.3% and 0.8% respectively of their total staff (see Lincoln).

The ADB is an exception to this trend. Japan has occupied its presidency since it was created in 1966 and, in line with the high level of contributions and voting rights, Japan's personnel representation has been consistently high. Thus, the ADB is a case worthy of study, as it demonstrates Japan not increasing in influence, as within the UN, but as the dominant power in an international institution. In the

ADB Japan was credited along with the United States with a leading number of votes, and initially used this influence to win more contracts for Japanese companies. The aim appears to have been to place Japanese nationals in leading positions and provide a solid financial basis for the new multilateral bank to assist Japan's economic growth, by promoting loans to its major trading partners in the region – Thailand, Malaysia, South Korea and the Philippines.

However, this self-serving policy gradually changed as Japan took up a role more characteristic of a regional hegemony providing public goods. Japan's already high profile was built upon and redefined in the 1980s and the 1990s with expanding contributions. The ADB became the central pillar of Japan's debt diplomacy with the creation of the Japan Special Fund in 1989 to assist surplus finance recycling. This activism spilled over into the political sphere as Japan attempted, ultimately unsuccessfully, to outstrip the United States in voting rights within the ADB. In addition, Japan pointed the ADB in the direction of reconstruction in Cambodia, the provision of loans to Vietnam, the admission of China (in 1986), and support for Corazon Aquino's government in the Philippines. At least one observer, Ming Wan, has praised Inoue Shiro, the President of the ADB, for rising "above narrow considerations of nationalism or national interest to assert the internationalism of the bank at a time when some observers were beginning to write it off as a Japanese organization".

Japan's participation in the ADB demonstrates a regional slant to its traditional style of foreign policy, but Japan's activism within the EBRD is an example of a broader, global outlook and sense of responsibility. Japan's interest in Europe can be seen in a number of multilateral channels, including the Organization for Security and Cooperation in Europe, the North Atlantic Treaty Organization, the EU and the EBRD. The EBRD opened in 1991 with Japan as a member contributing an amount similar to those of the chief European powers, the United Kingdom, Germany, Italy and France. In addition, Japan created the Japan/Europe Cooperation Fund, with the objective of channeling Japanese aid to central and eastern Europe, and the countries formerly within the Soviet Union, through multilateral agencies. Again, political motives can be seen in Japan's use of EBRD negotiations to promote the adoption of resolutions in Japan's favor concerning the dispute with the former Soviet Union, and now Russia, over the Kuriles. Further, the peaceful and lasting implementation of democratization, market economies, and human rights in central and eastern Europe was a chief reason for Japan's activism within the EBRD. Japan's participation in such an overtly political multinational institution demonstrates a new direction away from the apolitical and overtly economic Yoshida Doctrine of the immediate postwar period – a direction also in evidence at the ADB and the UN.

Conclusion and Prospects

Inoguchi Takashi has described Japan as having "the will, the need and the capacity to assume more global responsibilities. It is driven by a tenaciously held aspiration to occupy an honorable place in the world, increasingly dictated by the self-interested need to sustain international stability and economic prosperity". This chapter has taken the example of the UN, the leading global multilateral institution, to demonstrate this "tenaciously held aspiration" through the three areas of peacekeeping, funding, and representation. All three are entwined deeply with each other: the MoFA's goal of a permanent UNSC seat may well be fulfilled through increased peacekeeping and budgetary contributions, and increased budgetary contributions bring demands, both domestic and international, for Japan to be more fully represented within the UN.

More of the same can be expected in the future, as Japan's commitment to, and participation in, the UN continues on the steady upward curve that it has been tracking for the past few decades. In the field of UNPKO, rather than the legally required review of the PKO Law, the exigencies of the phenomenon of "mission creep" – namely, how SDF personnel in the field adjust to altering mission profiles day to day – will define the extent of Japan's participation. As peacekeeping develops, the range of Japan's participation will also develop and the activities restricted under the PKO Law (see above) will become new areas of possibility for Japan to define its post-Cold War security policy. As for UN reform, a stubborn insistence by Japan on pursuing the symbolic prestige associated with a permanent seat on the UNSC is likely to divert a great deal of energy from more constructive activities. The issue of representation needs to be addressed, however, not with reference to Japan's sense of national pride, but rather to address weaknesses within the UN, and to prepare it for the social, economic, security, and financial challenges of the 21st century.

As the cases of the UN and the ADB demonstrate, there has been a break with the traditional low-profile, low-risk, US-dependent foreign policy that served Japan so well for so long. Now, in a post-Cold War world, Japan is demonstrating signs of having emerged from the cocoon of the Yoshida Doctrine, and is exercising a degree of regional and international leadership. It is adjusting its previous emphasis on bilateral relations, especially with the United States, to focus on multilateral relations, and the development of a greater emphasis on making a human contribution. Both Japan's UNPKO experience and its participation in the EBRD demonstrate the kind of controversial, highly politicized fields into which Japan will move in a world where the kind of multilateral institutions mentioned in this chapter will play progressively more important roles, as the multilateral development banks deal with debt strategy and reconstruction, and UN peacekeeping contributes to the definition of security policy. Japan's increasingly active role within various multilateral institutions has meant that its commitment to the international community has become imbued with a greater sense of morality and participation, which goes beyond the immediate postwar aims of simply ratifying and implementing international treaties and laws.

Further Reading

Curtis, Gerald L., editor, *Japan's Foreign Policy after the Cold War: Coping with Change*, Armonk, NY: M. E. Sharpe, 1993

A solid textbook for a course on Japan's international relations, introducing various topics written by students of John William Morley in memory of their professor.

Dore, Ronald, *Japan, Internationalism and the UN*, London and New York: Routledge, 1997

A manifesto describing what Dore thinks Japan ought to be doing within the UN, translated from a Japanese-language text.

Howes, John F., editor, *Nitobe Inazo: Japan's Bridge Across the Pacific*, Boulder, CO: Westview Press, 1995

A collection of scholarly essays reviewing the life, work and legacy of the prominent prewar academic and diplomat, whose ideas on international cooperation have contributed to the postwar policy of "UN-centrism" (and earned him posthumous glory as the face on the ¥5,000 bill).

Immerman, Robert M., "Japan in the United Nations", in Craig Garby and Mary Brown Bullock, editors, *Japan: A New Kind of Superpower?*, Baltimore: Johns Hopkins University Press, 1994

A solid chapter, highlighting the domestic sources of proactivism in Japan's attitude towards the UN.

Inoguchi Takashi, *Japan's Foreign Policy in an Era of Global Change*, London and New York: Pinter, 1993

Another text which ought to feature on any reading list for students of Japan's international relations.

Kwon Gi-Heon, "The UN Burden Sharing in the Post-hegemonic Era: Declining Contributions of the Large Industrialized Nations", in *The Korean Journal of International Studies*, Volume 25, number 1, 1994

A highly statistical review of "collective action" theory with reference to UN regular budget contributions.

Lincoln, Edward J., *Japan's New Global Role*, Washington, DC: The Brookings Institution, 1993

A thorough analysis of the economic changes in the 1980s which raised Japan's international profile.

Ming Wan, "Japan and the Asian Development Bank", in *Pacific Affairs*, Volume 68, number 4, Winter 1995–96

An excellent article demonstrating Japan's developing altruism in the ADB.

Morrison, Alex, and James Krias, *UN Peace Operations and the Role of Japan*, Clementsport: The Canadian Peacekeeping Press, 1996

A comprehensive review of Japan's UNPKO participation taken from a variety of viewpoints on a nation-by-nation basis.

Nish, Ian, *Japan's Struggle with Internationalism: Japan, China and the League of Nations, 1931–1933*, London and New York: Kegan Paul, 1992

An historical text taking the Manchurian Crisis and Japanese disillusionment with the League of Nations as its case study.

Ogata Sadako, "Japan's Policy towards the United Nations", in Chadwick F. Alger, Gene M. Lyons and John E. Trent, editors, *The United Nations System: The Policies of Member States*, Tokyo: United Nations University, 1995

A representative article by a high-profile Japanese UN official and professor of international relations who has written widely on the topic.

Saito Shizuo, "The Evolution of Japan's United Nations Policy", in *Japan Review of International Affairs*, Volume 7, number 3, 1993

An account of growing multilateralism, rather than bilateralism, as the salient characteristic of Japan's foreign policy.

Simai, Mihály, "The Future Role of Japan within the UN", in *Global Governance*, Volume 1, number 3, 1995

A theoretical article providing a good introduction to the topic of Japan's role in the UN from a political science background.

Tadokoro Masayuki, "A Japanese View on Restructuring the Security Council", in Bruce Russett, editor, *The Once and Future Security Council*, London: Macmillan, and New York: St Martin's Press, 1997

A personal view on the necessity and feasibility of reforming the UNSC, examining what, in particular, Japan has to offer.

Urano Kio, *Kokusai Shakai no Henyo to Kokuren Tohyo Kodo, 1946–1985* [Changing International Society and UN Voting Behavior, 1946–1985], Tokyo: Kokusai Chiiko Shiryo Senta, 1989

A large number of statistics, based on General Assembly roll calls, demonstrating Japan's increased proactivism in the UN.

Hugo Dobson is a doctoral candidate at the University of Sheffield, and a Researcher at the International Center for Comparative Law and Politics in the Faculty of Law at the University of Tokyo.

Table 19.1 Financial Contributions to the Budgets of Selected Multilateral Organizations by Japan, the United States, the United Kingdom, and Germany, 1995 (%)

	Japan	*United States*	*United Kingdom*	*Germany*
World Bank	6.41	17.48	4.74	4.95
International Development Association	25.00	24.28	7.30	11.89
International Finance Corporation	7.53	22.43	5.33	5.68
Multilateral Investment Guarantee Agency	5.095	20.519	4.860	5.071
ADB	19.055	9.528	2.494	5.282
Asian Development Fund	37.83	17.61	4.35	6.67
African Development Bank	4.680	5.680	1.433	3.513
African Development Fund	14.50	13.29	3.91	9.39
EBRD	8.52	10.00	8.52	8.52
Inter-American Development Bank	2.09	34.10	0.98	1.24

Source: *Annual Report of the International Finance* [Kokusai Kinyu Nempo], Tokyo: Financial and Fiscal Research Group [Kinyu Zaisei Jijo Kenkyukai], 1996, pp. 470–487

Appendices

Appendix 1

Chronology

Dates include the Japanese numbering of years by named eras (*nengo*). In modern Japan, these are identical with the reigns of emperors: thus, for example, 1945 was the 20th year of Hirohito's reign, known as the Showa era (shown here as S20), and 1997 was the ninth year of Akihito's reign, known as the Heisei era (H9).

1945 (S20) **August** (14) World War II ends with Japan's unconditional surrender to the Allies, as implied (15) in Emperor Hirohito's first ever radio address to the Japanese people.

September (2) The instrument of surrender is signed aboard the USS *Missouri* in Tokyo Bay.

(8) General Douglas MacArthur arrives in Tokyo as head of the Occupation authorities (SCAP), under an 11-nation Far Eastern Commission, in Washington, DC, and a four-nation Allied Council for Japan, in Tokyo. In practice SCAP largely implements policies laid down by the State, War and Navy Departments of the United States.

October Workers take over the *Yomiuri Shimbun* (newspaper), initiating the nationwide "production control" movement, which lasts until it is banned on June 13, 1946.

December Parliament passes the Labor Union Law, the first of the "three fundamental labor laws" (but see July 1948), as well as the first Agricultural Land Reform Law, which is denounced by SCAP as inadequate (see October 1946).

1946 (S21) **January** (1) Emperor Hirohito issues his last ever "imperial rescript", a decree rejecting "the false conception that the emperor is divine" and the notion that the Japanese are superior to other peoples.

(4) A directive from SCAP initiates the "purging" of more than 200,000 individuals held responsible for the war from government agencies and private corporations, and bans 120 political organizations associated with militarism and nationalism (but see June 1951).

February Faced with inflation and a growing black market, the government contracts the money supply by replacing the old yen with the new yen, and temporarily freezing bank accounts.

May After the first postwar general election – in which women vote, and are elected, for the first time – Hatoyama Ichiro, the leader of the Liberal Party, is "purged" from public life, and replaced by his rival Yoshida Shigeru, who forms a center-right coalition government.

The International Military Tribunal for the Far East assembles in Tokyo to begin the trial of 28 "Class A" war crimes suspects – 19 military officers and nine civilians – which continues to November 1948.

		Numerous protests over food shortages include an "invasion" of the Imperial Palace grounds in Tokyo.
	September	The Labor Relations Adjustment Law is enacted, as the second "fundamental labor law", permitting strikes, promoting collective bargaining between management and labor, and mandating arbitration of labor disputes.
		SCAP and the government begin the year-long process of abolishing 83 companies which formerly dominated the economy as holding companies of *zaibatsu* (conglomerates) and/or as sectoral monopolies.
	October	Parliament enacts the second Agricultural Land Reform Law, based on proposals by the Australian diplomat William McMahon Ball, Commonwealth representative on the Allied Council for Japan, leading by August 1950 to the transfer of around 4.4 million acres to former tenants, and an increase in the proportion of owner-cultivators from 33% of farmers to around 62% (rising to 85% by 1985).
1947 (S22)	**February**	The public service workers' general strike planned for the first day of this month is banned by SCAP, initiating the "reverse course" in Occupation policy.
	March	Parliament passes the Basic Law on Education and the School Education Law, introducing the "6-3-3" system of elementary, junior high and senior high schools, making schooling compulsory to the age of 15, and abolishing segregation by gender.
	April	Parliament passes the third "fundamental labor law", the Labor Standards Law, which provides for an eight-hour working day and a 48-hour working week, and imposes controls on the employment of women and children; and also passes the Anti-Monopoly Law, which creates the Fair Trade Commission (but see September 1953).
	May	(3) The new Constitution of Japan comes into force.
		After the first general election held under the new Constitution, Yoshida's government gives way to a coalition government of the Socialists and Democrats, which lasts until October 1948 (the only government with Socialist members until June 1994).
	December	The revised Civil Code abolishes the legal privileges of heads of *ie* (households), of husbands, and of oldest sons in respect of marriage, divorce, property and inheritance, with significant effects on investment patterns and labor markets.
		The Unemployment Insurance Law and the Unemployment Allowance Law extend the scope of the welfare state.
		The powerful Home Ministry (*Naimusho*) is dissolved on orders from SCAP, and replaced by Ministries of Labor and Construction, and, later, the National Police Agency (1954) and the Ministry of Home Affairs (1960).
1948 (S23)	**April**	Parliament enacts the Laborers' Accident Insurance Law, supplementing existing welfare provision.
	July	SCAP orders the government to remove civil servants' and other public sector workers' rights to strike and take part in collective bargaining, through revisions to the Labor Union Law of 1945.

	October	Yoshida Shigeru returns to office as Prime Minister (up to December 1954), and continues to cooperate with SCAP in laying the foundations of the postwar political and economic system.
	November	The International Military Tribunal for the Far East adjourns after passing death sentences on seven war criminals (who are hanged in December) and jailing 18 others.
1949 (S24)	**March**	The government introduces the "Dodge Line" (for Joseph Dodge, an adviser to SCAP since December 1948), mandating a balanced budget; the closure of the Reconstruction Finance Bank; cuts in subsidies, price controls, and government personnel; a fixed exchange rate of ¥360 to the US dollar; and the limited reopening of international trade to the private sector.
	April	The Foreign Exchange and Foreign Trade Control Law imposes a system of strict controls by the Ministry of Finance over all transactions involving foreign currencies (liberalized in stages between 1980 and 1998).
	May	The Ministry of Commerce and Industry (founded in 1925) is relaunched, after being purged of several leading officials, as the Ministry of International Trade and Industry.
	August	The Rice Price Deliberation Council is created to determine the price which the state's rice distribution system pays to farmers.
1950 (S25)	**May**	The Comprehensive National Land Development Law permits prefectures to introduce land use planning, and makes the Ministry of Construction a powerful source of funds and decisions on public works and land development.
		The Foreign Capital Law introduces a regime of licensing and supervision for all activities of non-Japanese investors.
	June	The Korean War begins. Japan's position as a leading supplier to the UN forces defending South Korea helps to bring about the "Doran boom", lasting to June 1951.
	July	SCAP begins the "Red Purge" by suppressing the activities of the Communist Party leadership, banning more than 12,000 Communists from holding office in labor unions, and closing the party's newspaper, *Akahata* [Red Flag].
	August	SCAP orders the government to create a National Police Reserve of 75,000 personnel to replace US troops withdrawn from Japan to fight in Korea.
	December	Parliament enacts a law to establish the state-owned Export Bank of Japan, renamed the Export-Import Bank of Japan in April 1952.
1951 (S26)	**March**	Parliament enacts a law to establish the state-owned Japan Development Bank.
	April	General MacArthur is relieved of his command by US President Harry S Truman, and replaced as SCAP by General Matthew Ridgeway (up to April 1952).
	June	SCAP lifts the purge on about 69,000 individuals (and on another 14,000 in July), permitting them to resume political and business activities.

	November	The economy enters its second postwar boom, the "consumption boom", lasting to January 1954.
1952 (S27)	**April**	(28) Two treaties signed in September 1951 come into force: the Treaty of San Francisco, under which Japan makes peace with 48 other nations (notably excluding the Soviet bloc, China, and India), and regains full sovereignty; and the US/Japan Mutual Security Treaty (*Ampo*), under which the US armed forces maintain their bases on Japanese territory. On the same day the United States ends direct financial aid to Japan, which totaled around US$2 billion during the Occupation years.
		Japan signs a separate peace treaty with the Chinese Nationalist government of Taiwan (a Japanese colony until 1945), while continuing informal trade relations with the People's Republic of China.
	July	The government creates the Fiscal Investment and Loan Program, a "second budget" around two thirds the size of the regular government budget, which the Ministry of Finance still uses to help fund pensions, public works, and industrial investment.
1953 (S28)	**July**	The end of the Korean War brings an economic recession to Japan.
	September	Parliament amends the Anti-Monopoly Law to reduce the powers of the Fair Trade Commission and permit "cooperative behavior" (cartels) in cases of "rationalization" and "adverse economic conditions".
	December	The Chisso Corporation begins the secret dumping of methyl mercury waste into the sea off Minamata, its "company town" in Kumamoto Prefecture, which continues to 1960, contaminating fish and poisoning the people who eat them. After at least 300 deaths, thousands of cases of brain damage and deformities, and years of litigation, Chisso pays out ¥664 million in compensation from 1971.
1954 (S29)	**March**	Twenty-three fishermen and their catch aboard a tuna-fishing boat, the *Fukuryu-maru* (*Lucky Dragon*), are irradiated by fall-out from a US hydrogen bomb test on Bikini atoll in the South Pacific.
	May	The government introduces a system of "welfare pensions" (*kosei nenkin*), based on contributions from private sector employees.
		Parliament passes legislation to recentralize control of the education system in the hands of the Ministry of Education, instead of the elected school boards created during the Occupation.
	June	The creation of the National Police Agency, and the consequent reduction in the autonomy of prefectural police services, partially reverses SCAP's policy of decentralization.
	July	The National Police Reserve is reorganized as the Self-Defense Forces, with a mandate to defend Japan against direct or indirect aggression.
	August	A Mutual Security Agreement with the United States supplements the 1952 Security Treaty, providing for cooperation between US forces in Japan and the Self-Defense Forces.
	November	Hatoyama Ichiro, purged from office in 1946 but depurged in 1951, leaves the Liberals to take over the Democratic Party, and continue his rivalry with Prime Minister Yoshida, whom he replaces in December.

1955 (S30) **January** The economy enters the "Jimmu boom" (up to July 1957), during which GNP grows at an average of around 8% a year.

July The Japan Housing Corporation is established to lead public sector construction of housing and help in funding private land developments.

September Sony Corporation launches the world's first transistor radio, inaugurating a new era of cheap electrical and electronic appliances, initially based on technology first developed in the United States.

Gensuikyo, the first and, arguably, most influential Japanese movement against nuclear weapons, is founded.

December Prime Minister Hatoyama launches the "Five-Year Plan for Economic Self-Support", setting a target of 5% growth in GDP each year (which is easily exceeded).

1956 (S31) **October** Japan resumes diplomatic relations with the Soviet Union, and signs a trade pact, but leaves the dispute over the Kurile islands (near Hokkaido) unresolved.

Following the closure of the Suez Canal by the Egyptian government, the major oil companies began using supertankers to transport oil from the Middle East, stimulating the growth of the Japanese shipbuilding industry to become the world's largest by 1960.

1957 (S32) **May** After the Bank of Japan increases official interest rates, the stock market crashes and a recession begins.

1958 (S33) **June** The economy enters the "Iwato boom" (up to December 1961), during which GNP grows at an average of around 10% a year.

1959 (S34) **January** Japan formally adopts Systeme Internationale weights and measures (meters, liters, kilograms, and others) in place of its traditional system.

April Parliament enacts the National Pension Law, creating "national pensions" (*kokumin nenkin*) based on contributions from all working people, to supplement the *kosei nenkin* introduced in May 1954.

A national minimum wage is introduced.

November The state's retirement pension system is completed with the introduction of non-contributory pensions for non-workers. (The public pension schemes together still accounted for around 70% of all pension assets in Japan in 1996.)

1960 (S35) **January** A nationwide strike by coalminers spreads from Mitsui's Mi'ike mine, continues for 282 days and becomes the longest strike in Japanese history, but ends in defeat for the miners, and a setback for the labor federation Sohyo, which supported them.

June The revised US/Japan Mutual Security Treaty (*Ampo*) is ratified by Parliament, during a boycott by opposition members and a mass demonstration around the building. Protests against the Treaty culminate in a one-day general strike by around 6 million workers, and a demonstration in Tokyo, during which a student is killed in disputed circumstances. The government cancels a planned visit by US President Dwight D. Eisenhower.

	October	Asanuma Inejiro, leader of the Socialist Party, is killed by a right-wing activist, the son of an officer in the Self-Defense Forces, in the only successful political assassination in postwar Japan.
	December	Having succeeded Kishi Nobusuke as Prime Minister in July, Ikeda Hayato launches the "Income-Doubling Plan", drafted by Okita Saburo, emphasizing encouragement of chemicals, machinery and other heavy industries; promotion of exports; upgrading of advanced research and education; improvement of infrastructure; and equalization of living standards between regions. The doubling of Japan's GNP, targeted to take 10 years, is achieved in seven.
1961 (S36)	**April**	The government establishes a system of universal health insurance.
	June	Parliament passes the Agricultural Basic Law.
1963 (S38)	**November**	An explosion at the Mi'ike coalmine (see January 1960) kills 458 miners, in Japan's single worst industrial accident.
1964 (S39)	**April**	As Japan becomes the first non-western member of the OECD, the government announces that its GNP equals the GNPs of all other Asian nation-states combined.
	October	The 18th Olympic Games, the first to be held in Asia, take place in Tokyo.
		The first *Shinkansen* ("bullet train") service starts, on the New Tokaido Line between Tokyo and Osaka, in the same year that Japan National Railways (JNR) goes into the red for the first time (see also April 1987).
	November	Ikeda Hayato is succeeded as Prime Minister by Sato Eisaku, who continues his plans for economic growth and support for the US war effort in Vietnam.
1965 (S40)	**April**	Beheiren, the Japanese movement for peace in Vietnam, holds the first of many rallies.
		The issuance of seven-year construction bonds inaugurates the Japanese bond market, which by 1996 was the second largest in the world (with about 95% of trading being in government bonds).
	June	Japan signs a Treaty on Basic Relations with South Korea, recognizing its government as the sole legitimate government of the Korean peninsula (a Japanese colony until 1945).
	October	The economy enters the "Izanagi boom" (up to June 1970), during which GNP grows at an average of around 11% a year.
1966 (S41)	**February**	National Foundation Day is revived and celebrated for the first time since the war, marking for some observers the revival of nationalism in Japan.
	November	The Asian Development Bank is founded at a multilateral conference in Tokyo.
1967 (S42)	**April**	The government announces that Japan has overtaken West Germany to become the world's second largest non-Communist economy.
	June	The government launches a Fundamental Plan for Capital Liberalization, to be implemented in stages up to 1974, effectively reducing the role of state-owned financial institutions in the economy.

1968 (S43)		Mass student protests erupt throughout the year and into 1969, at 152 of Japan's 377 universities, on issues ranging from student fees and housing, to the government's support for the US war effort in Vietnam.
	January	Prime Minister Sato announces a total ban on arms exports, and the "three non-nuclear principles", promising that Japan will not produce nuclear weapons, possess them, or allow their introduction into Japanese territory (and receives the Nobel Peace Prize as a result, in 1974). (See also January 1983.)
	March	Protests by local farmers and radical groups begin at the site of the new international airport at Narita, near Tokyo.
	June	The Ogasawara (Bonin and Volcano) groups of islands are returned from US to Japanese control, becoming part of Tokyo Prefecture.
	October	Student demonstrators from several universities and different radical groups destroy parts of Shinjuku station in Tokyo, the world's largest rail terminal.
1969 (S44)	**January**	Student radicals occupying buildings on the campus of Tokyo University are dispersed by riot police.
1970 (S45)	**February**	Japan launches its first satellite into orbit around the Earth.
	June	The renewal of the Security Treaty with the United States leads to the largest demonstrations in Japan since 1947.
1971 (S46)	**July**	US President Nixon announces his intention to visit Communist China (in February 1972) – the first "Nixon shock".
	August	Nixon announces a 10% surcharge on imports and abandons the Bretton Woods system of fixed exchange rates – the second "Nixon shock", or the "dollar shock". Japan reacts by finally abandoning attempts to fix the exchange rate of the yen in February 1973.
	October	The United States introduces restrictions on imports of synthetic textiles following the breakdown of negotiations with Japan – the third "Nixon shock".
1972 (S47)	**January**	The government introduces national children's allowances (welfare payments for third and later children), partly in response to concern about the falling birth rate, and partly to replace allowances introduced by many left-dominated local governments since 1967.
	May	The Ryukyu Islands revert from US to Japanese control and are designated Okinawa Prefecture.
	June	Following several amendments to existing laws in 1970–71, the promulgation of a comprehensive Natural Environment Preservation Law marks the adoption by Japan of the strictest controls on pollution in any industrialized country.
	July	Sato Eisaku resigns as Prime Minister, following criticism over his handling of the three "Nixon shocks" in 1971, and is succeeded by Tanaka Kakuei.
	September	Japan opens diplomatic relations with Communist China and ceases to recognize Taiwan.

1973 (S48)	**January**	The government introduces free medical care for the aged throughout Japan, following its introduction by many left-dominated local governments from 1960 (but see April 1984).
	September	Japan formally opens diplomatic relations with the Democratic Republic of Vietnam, but maintains relations with the Saigon regime in the South until the reunification of the country in April 1975.
	November	Japan experiences its first "oil shock" after the main petroleum-exporting countries raise prices and cut output.
1974 (S49)	**January**	During a tour of Southeast Asia by Prime Minister Tanaka, major anti-Japanese demonstrations take place in Bangkok and Jakarta.
	November	Prime Minister Tanaka resigns from office after the magazine *Bungei Shunju* exposes his use of his private fortune and Liberal Democratic Party funds to bribe voters and party members (see also July 1976 and October 1983)
1975 (S50)	**May**	The first of several groups of refugees from Vietnam ("boat people") lands in Japan.
1976 (S51)	**July**	A group of Liberal Democrat politicians breaks away from the party to form the New Liberal Club, following the indictment of former Prime Minister Tanaka Kakuei on charges of receiving bribes from the US aircraft-maker Lockheed (see October 1979 and December 1983).
	November	Prime Minister Miki Takeo formalizes the longstanding policy of keeping government spending on defense below 1% of GNP (but see December 1986).
1977 (S52)	**May**	The government announces that more than half the population were born after the end of World War II.
1978 (S53)	**February**	Former Prime Minister Tanaka Kakuei is arrested and charged with accepting bribes from the US aircraft-maker Lockheed (see October 1983).
	May	Narita Airport is officially opened, seven years later than planned, in the face of continuing protests.
	August	Japan and China agree on a Treaty of Peace and Friendship.
1979 (S54)	**February**	Japan joins China and the United States in suspending aid to Vietnam in protest at its invasion of Cambodia and its removal of the genocidal Khmer Rouge regime.
	March	Japan experiences its second "oil shock" as the price of crude petroleum rises sharply following the Iranian revolution.
	July	Sony Corporation launches its "Walkman" portable cassette player, another example of Japanese technology based on adaptation and miniaturization of earlier inventions.
	October	Following a general election, the Liberal Democrats lose their majority in the House of Representatives for the first time since 1955, but form a government with support from the New Liberal Club.
1980 (S55)	**April**	Exchange controls are abolished and the holding of foreign securities by Japanese citizens is liberalized.

1981 (S56)	**March**	The establishment of the Administrative Reform Commission marks the beginning of an era of slow but significant changes in the scope and methods of government activities, including some privatization measures.
1983 (S58)	**January**	The government partially lifts the ban on arms exports, announced in January 1968, to permit exports of "defense-related components" (notably under the US/Japan agreement on transfers of military technology, signed in November 1983).
	October	The Tokyo District Court finds former Prime Minister Tanaka guilty of receiving bribes from the US aircraft-maker Lockheed. Tanaka's appeal against the verdict is abandoned after he suffers a stroke in 1985.
	December	Following a general election the Liberal Democrats form a coalition government with the New Liberal Club (which rejoins the Liberal Democratic Party in August 1986).
1984 (S59)	**April**	The Health Care for the Aged Law of February 1983 comes into effect, reimposing fees for medical treatment for the aged.
1985 (S60)	**March**	Trade disputes between Japan and the United States culminate in the US Senate's resolution that Japan is "an unfair trading partner".
	April	The reorganization and partial privatization of Nippon Telephone and Telegraph (NTT) marks the first attempt since 1949 to reduce the size and role of the public sector in the economy.
	May	Parliament enacts the Equal Employment Opportunity Law.
	August	Government ministers pay an "official" visit to Yasukuni Shrine in Tokyo, where the spirits of the military dead are honored, in apparent defiance of the separation of state and religion (but see September 1991), and in the midst of controversy with other countries in East Asia over the censorship of information on the former Imperial Army in Japanese school textbooks (but see August 1997).
	September	The signing of the Plaza Accord by the governments of the Group of Five leads to a series of rises in the value of the yen against the US dollar and an economic recession, which continues into 1987.
1986 (S61)	**December**	Prime Minister Nakasone formally reverses the policy (announced in November 1976) of limiting government spending on defense to 1% of GNP or less.
1987 (S62)	**April**	JNR, state-owned since 1906 and responsible for two thirds of the rail network, is partially privatized, creating six regional Japan Railway (JR) companies for passenger services, as well as a freight corporation, and an agency charged with recovering some of JNR's enormous debts through real estate sales and development.
1988 (S63)	**April**	The opening of the Seikan Tunnel between Hokkaido and Honshu (the world's longest tunnel), and of the Seto O-hashi (bridge) between Honshu and Shikoku, completes the road and rail transport links among the four home islands.
	July	Recruit Cosmos, a company involved in publishing, telecommunications and land development, is accused of bribing numerous LDP politicians and government officials.

1989 (H1) **January** Emperor Hirohito dies and is succeeded by his son Akihito.

April Prime Minister Takeshita Noboru resigns from office, and former Prime Minister Nakasone resigns from the LDP, following revelations of their involvement in the Recruit Cosmos scandal. Takeshita is succeeded by Uno Sosuke.

The government introduces an extremely unpopular consumption tax on most goods and services, at 3% initially, then at 5% from April 1997. This has come to be widely seen as having damaging effects on consumer demand, and thus on the whole economy.

June Following the Chinese government's suppression of students' and workers' protest movements in Beijing and elsewhere, the Japanese government imposes limited economic sanctions (but lifts them one year later).

August Prime Minister Uno resigns over his failure to provide financial support for his mistress.

December The Bank of Japan increases interest rates in an attempt to dampen the boom in speculative loans.

1990 (H2) **January** The "bubble economy" starts to collapse as the speculative booms in shares and landholdings come to an end.

April The Ministry of Education orders schools to enforce attendance by staff and students at ceremonies centered on the national flag and the song *Kimigayo*.

October A crash on the Tokyo Stock Exchange, indicated by a steep fall in the Nikkei share index, signals the start of the Heisei recession, the longest since the war, which lasts until late 1995.

November A resolution passed by both Houses of Parliament calls for the relocation of the capital city.

1991 (H3) **January** Japan begins to give financial support to the United Nations coalition against Iraq in the Gulf War, eventually amounting to US$13 billion.

September The Supreme Court rules that visits by the Emperor or by ministers to the Yasukuni shrine, honoring the war dead, violate the constitutional separation of state and religion.

1992 (H4) **June** Parliament passes the UN Peacekeeping Operations Law, permitting military personnel to be dispatched overseas, under strict conditions, for the first time since 1945.

August Kanemaru Shin, the "Godfather" of the Liberal Democratic Party, resigns the vice-presidency of the party after admitting receiving ¥500 million from a delivery company, Sagawa Kyubin.

1993 (H5) **April** The Financial System Reform Law of June 1992 comes into effect, permitting banks to enter the securities business, and securities firms to enter trust banking, through strictly regulated affiliated companies.

June Prime Minister Miyazawa Ki'ichi's government loses a vote of confidence when 39 Liberal Democrats vote with the opposition parties against it, and 16 others abstain. Ozawa Ichiro, Hata Tsutomu and 42 others then leave the Liberal Democratic

		Party to establish the New Reform Party; 10 others leave to form the Harbinger Party; and still others join Hosokawa Morihiro's Japan New Party, founded in 1992.
	July	In a general election for the House of Representatives, the Social Democrats and the Liberal Democrats both lose seats to the new parties and the centrist opposition parties.
	August	The new House of Representatives elects Hosokawa Morihiro, leader of the Japan New Party, as Prime Minister of a broadly based coalition government, including other defectors from the Liberal Democratic Party as well as the Social Democrats.
	November	Parliament enacts a new Basic Environmental Law, consolidating and amending previous legislation, and mandating the drafting of a Basic Environment Plan.
	December	Under the terms of GATT's Uruguay Round Agreement on Agriculture, Japan undertakes to allow limited imports of foreign rice for the first time.
1994 (H6)	**April**	Prime Minister Hosokawa resigns over allegations of corruption and is succeeded by Hata Tsutomu, who forms another coalition government but loses his parliamentary majority when the Social Democrats depart after just 12 hours.
	June	Hata Tsutomu's government is replaced by a new coalition government formed by the Liberal Democrats, the Social Democrats, and the Harbinger Party, and led by the veteran Social Democrat Murayama Tomi'ichi.
		Parliament decides to raise the retirement age from 60 to 64 in stages between 2001 and 2013.
	August	Prime Minister Murayama announces a government plan to spend ¥100 billion on projects in Southeast Asia and elsewhere, to "atone" for the Japanese Army's brutality in World War II.
	October	The Ministry of Finance completes a 15-year process of gradual deregulation of interest rates, freeing banks to compete for deposits.
	November	Parliament passes modified versions of the electoral reform laws first proposed by Hosokawa Morihiro's government in 1993, introducing a new voting system and imposing formal limits on election spending.
1995 (H7)	**January**	Kobe, Nishinomiya and other cities are badly damaged, and more than 5,300 people are killed, by the Hanshin Earthquake, the worst natural disaster in postwar Japan.
	May	Asahara Shoko, head of the Aum Supreme Truth sect, is arrested and charged in connection with nerve gas attacks in the Tokyo subway in March, which killed 12 people, as well as other murders carried out by his followers.
	December	Shinozawa Kyosuke resigns as Permanent Vice-Minister of the Ministry of Finance, following controversy over its failure to inform US authorities about Daiwa Bank's loss of US$1.1 billion in the US bond market; over its plans to spend ¥680 billion on sorting out the finances of seven failed housing loan finance companies (*jusen* – see also July 1996); and over its apparent inability to respond to the rapid appreciation of the yen against the US dollar during the year.
		Public confidence in Japan's nuclear power industry is shaken by news of a leakage of coolant at the Monju fast-breeder reactor, and then by officials' attempts to cover up details of the accident.

1996 (H8)	**January**	Hashimoto Ryutaro, leader of the Liberal Democratic Party, replaces the Social Democrat Murayama Tomi'ichi as Prime Minister, heading the continuing coalition between their two parties and the Harbinger Party.
	April	Two major Japanese banks merge to form the Bank of Tokyo-Mitsubishi, the largest financial insitution in the world (until the completion of the proposed merger between Citibank and the Travelers Group, announced in April 1998).
		Prime Minister Hashimoto and US President Bill Clinton issue a Joint Declaration reaffirming the security relationship between their countries (and see September 1997).
	June	Sumitomo Copper reveals that one of its traders, Hamanaka Yasuo, concealed losses of as much as US$4 billion on trading in New York and London.
	July	The Housing Loan Administration Corporation is established to recover around ¥3 trillion in non-performing loans by the *jusen* (housing loan finance companies), which collapsed in the early 1990s.
	October	In the first general election held under the new system introduced in November 1993, the Liberal Democrats increase their share of the vote, although on the lowest turnout of voters since 1946. Lacking an overall majority in the House of Representatives, Prime Minister Hashimoto forms an exclusively Liberal Democrat government, supported by the Social Democrats and the Harbinger Party.
		Deregulation of the insurance industry allows life and non-life companies to enter each other's area of business.
	December	The United States announces cuts in its military presence in Japan, resulting in the full or partial closure of 11 bases, and the return of 21% of the land occupied by US forces to use by Japanese civilians.
		Faced with worsening economic conditions, the government announces a five-year plan to relax controls on the economy and use public funds to create jobs.
1997 (H9)	**March**	Nomura Securities is revealed to have paid large sums, illegally, to Ko'ike Ryuichi, a *sokaiya* (corporate racketeer).
		The Mi'ike coal mine, the oldest and largest in Japan, closes after 124 years, leaving only two mines open.
		A fire and explosion at the Tokaimura nuclear fuel reprocessing center lead to more adverse publicity for the nuclear power industry.
	April	Nissan Mutual Life, Japan's 16th largest life insurance company, becomes the first since before World War II to cease operations, with a capital deficit of ¥200 billion.
		Parliament enacts a law giving the government powers to make compulsory purchases of land in Okinawa for allocation to US forces, over protests from local landowners who had refused to renew leases on existing US bases.
	June	Sakamaki Hideo, having resigned as President of Nomura Securities following the Ko'ike scandal, is also arrested in connection with the case; thousands of customers leave the firm; and the government bans it from dealing in government bonds.

Okuda Tadashi resigns as Chairman of Dai-Ichi Kangyo Bank following allegations that he too made illegal payments to Ko'ike Ryuichi. He and three other former officials of the bank are convicted in January 1998.

August After 32 years of litigation, the Supreme Court finds that the Ministry of Education unconstitutionally removed unfavorable references to the former Imperial Army from a senior high school history textbook written by Professor Ienaga Saburo, but upholds the Ministry's right to censor textbooks.

September The Liberal Democrats regain an absolute majority in the House of Representatives, for the first time since June 1993, after defections from opposition parties.

The US and Japanese governments agree new guidelines on security cooperation, allowing emergency use of Japanese military bases by US forces, and providing for Japanese involvement in keeping sea-lanes open.

November Sanyo Securities becomes the first brokerage to file for bankruptcy since before World War II; Yamaichi Securities, Japan's oldest and fourth largest brokerage, also ceases trading, in the country's largest ever bankruptcy, involving debts of at least ¥330 billion, of which around one half were linked to illegal transactions through the Cayman Islands.

Hokkaido Takushoku Bank and Tokuyo City Bank become the first banks to cease trading in postwar Japan.

December A new law on holding companies comes into effect, permitting the revival of this type of corporate structure for the first time since the *zaibatsu* conglomerates were broken up during the Occupation.

1998 (H10) **January** Mitsuzuka Hiroshi resigns as Minister of Finance, taking responsibility for the actions of officials who received bribes and gifts from six of Japan's 10 largest banks. In this and subsequent months, two officials are arrested, two others resign, and two more kill themselves.

March Investigations and arrests begin at the Bank of Japan in connection with allegations that some of its officials took bribes from, and passed secret information to, major private sector banks; the Bank's Governor, Matsushita Yasuo, resigns. Meanwhile, the Deposit Insurance Corporation announces an injection of ¥1.8 trillion of public money into the banking industry.

The Economic Planning Agency forecasts the first decline in Japan's GDP since 1974, and the number of bankruptcies reaches record levels.

April The Akashi Straits Bridge, the world's longest suspension bridge, is opened to traffic between Honshu and Shikoku.

Prime Minister Hashimoto announces the latest in a series of "rescue packages" for the economy, comprising tax cuts, extra spending on public works programs, an increase in aid to other countries in East Asia, and other measures, to a total value of nearly ¥16.7 trillion.

The Ministry of Finance disciplines 112 of its officials for accepting entertainment and gifts from representatives of major corporations.

Unemployment reaches 4.1% of the workforce, the highest level in the postwar period.

Appendix 2

Glossary

This glossary does not contain names of institutions or organizations included in Appendix 4. Readers who are not familiar with the Japanese language should note that there are normally no grammatical distinctions between singular and plural, or between masculine and feminine.

Ainu: the indigenous people of Hokkaido, the northernmost of Japan's four main islands. The Ainu, of whom there are now fewer than 20,000, are probably the remnant of an indigenous people that inhabited large parts of Honshu, Japan's largest island, before being driven out, up to about 1,100 years ago. Their culture, centered on hunting, fishing and gathering plants, probably originated in Siberia. Their language and traditions are still being studied and debated, while the Ainu themselves are more and more assimilated into mainstream society. Thus, for example, "Ainu sculptures" have become well-known tourist souvenirs of Hokkaido, even though Ainu culture forbade the making of images of living things until well into the 20th century.

amakudari: literally meaning "descending from Heaven", the practice whereby senior bureaucrats take retirement positions in the private sector or in specialized government agencies. Every year more than 200 leading bureaucrats join the business world as directors or senior advisors of private companies that they monitored during their bureaucratic careers. Since the retirement age for civil servants is around 55, they may serve in their new company for up to 20 years, ensuring smooth relations between government and industry. The National Civil Service Law stipulates that retiring officials must wait two years before accepting employment from a profit-making company that has close ties with their ministry or agency, unless the move is approved by the National Personnel Agency.

burakumin: literally, "[special] hamlet people", the official term for the hereditary outcasts, numbering about 3 million and living mostly in western Japan, whose long-established association with leather-tanning, butchery, and other "unclean" occupations still exposes them to prejudice and discrimination. The term itself derives from laws, formally repealed in the 1870s, which restricted such outcasts to living in designated hamlets. It is quite common even now for *burakumin* to register themselves at government offices away from their homes, to move house frequently, or to change their family names, in order to disguise their origins.

datsua nyuo: literally meaning "sloughing off Asia, entering Europe", this slogan was popularized by the westernizing thinker and educationist Fukuzawa Yukichi during the Meiji period (1868–1912). It has been used frequently since then to summarize, favorably or not, one approach to the modernization of Japan.

doken kokka: "construction state", a description of the Japanese polity that emphasizes the crucial role of the construction industry, and its sponsors in government ministries and agencies, in fostering economic growth and regional development through massive public works programs. The particular concerns of scholars and other observers investigating the system have included the environmental impact of these programs, and their links with the election campaigns of politicians.

endaka: literally, "yen high", referring to periods when the value of the yen against the US dollar rises sharply, with recessionary effects, as in 1985–87.

fukoku kyohei: a slogan, meaning "wealthy country, strong army", that was popularized before, during and after the Meiji Restoration (1868), and that has occasionally been revived since then, to summarize the goals of a modernizing Japan seeking parity with the western powers.

fukushi: "welfare", in the broadest sense of social or communal well-being, used in such phrases as *Nihon-gata fukushi shakai* ("Japanese-style welfare society") or *fukushi kokka* ("welfare state"). It is sometimes contrasted, and sometimes interchanged, with the older, more formal term *kosei*, which can be translated as "welfare" in the narrower sense of good works, usually by officials, on behalf of the disadvantaged. The "fuzziness" of both terms reflects the ambivalence and confusion of public attitudes to welfare, just as usage of their English equivalent does.

gaiatsu: "external pressure", a term commonly used to describe demands made on Japan from overseas, usually through intergovernmental channels.

gososendan hoshiki: the "convoy system", encouraged and supervised by the Ministry of Finance since the 1920s, whereby regulatory intervention hurries the pace of slower firms and slows the pace of faster firms providing financial services, while large banks or companies are expected to provide support for smaller institutions, originally through *zaibatsu* structures but, in the postwar period, through *keiretsu* relationships. This system showed signs of breaking down in 1997–98, when the Ministry and some *keiretsu* "main banks" allowed especially weak institutions to go bankrupt.

gyosei kaikaku: "administrative reform", a prominent political slogan of the 1980s and 1990s. Its adoption by all the major political parties except the Communists reflects the impact of British and North American ideas about reducing the scope of the state's activities, cutting the numbers of its employees, and privatizing state-owned enterprises.

gyosei shido: "administrative guidance" by a ministry or agency, usually delivered through informal channels.

habatsu: faction, notably within political parties, but also within government ministries and other institutions, such as university departments.

hakurai-hin: this term, meaning "goods that arrive by ship", was coined in the 19th century to refer to all goods imported from overseas, but it is now often used to describe foreign-made luxury goods sold at impressively high prices.

hoshu jin'ei: the "conservative camp" in Japanese politics, referring to the Liberal Democratic Party, its predecessors, and, arguably, at least some of the parties that broke away from the Liberal Democrats in 1993.

ijime: a word that covers the range from "teasing" through "bullying" to "tormenting", but used especially often in connection with mistreatment of schoolchildren, both by their fellow students and by their teachers.

jimu jikan: Permanent Vice-Minister, the leading civil servant in each government department.

jishu kanri: literally, "autonomous management", a term referring to the wide range of systems in place in Japanese manufacturing that foster consultation between workers and management, and permit workers to take part in formulating and implementing decisions about production. Opinions of *jishu kanri* systems differ sharply, both in Japan and among foreign observers. Should they be seen primarily as tools for management to exploit workers' ideas, as well as their labor power, or do they offer effective channels for workers' direct participation in management, in contrast to Japan's relatively weak labor unions?

jiyuka: "liberalization", one of the buzzwords of the 1980s and 1990s in Japanese debates about the financial system, education, trade and other matters. This is just one of many Japanese words ending in *-ka*, often representing English words with the ending "-ization", and meaning as much, or as little, as they do.

johoka: "informationalization", a buzzword much used by journalists, scholars, and other (actual or would-be) opinion-formers, to refer to the increasing economic and social importance of information flows, and associated technological changes. As in western countries, debates around this and other such buzzwords fascinate some audiences, bore others, and – not always by accident – omit to address or involve the majority of the population, as if the process in question is necessarily beyond their comprehension or control.

juken jigoku: "examination hell", a vivid summation of the anxieties and stresses associated with preparing for and taking examinations for entrance to schools, universities and government ministries, the results of which have a major impact on career paths and life chances.

juku: "cram school", privately owned institutions, operating in evenings and at weekends, supplementing the work of junior and senior high schools. The separate cram schools that help to prepare students for the entrance examinations of major universities (often after a first attempt has failed) are not *juku* but *yobiko* ("preparatory schools").

junanka: "flexibilization", a buzzword commonly used by journalists and other commentators to highlight the rigidities of the education system and of segmentation within the financial system.

jusen: the private sector housing loan corporations, sometimes compared to the savings and loan institutions in the United States. The *jusen* flourished during the "bubble" years of the 1980s, collapsed in the 1990s, and have left the Japanese financial system burdened with non-performing loans.

kakushin jin'ei: the "progressive camp" in Japanese politics, referring broadly to the Social Democrats and the Communists and, arguably, other parties which historically opposed the dominant conservative parties.

kamban: paper, metal or plastic notices attached to sets of components in factories, allowing each set to be tracked throughout the production process. These notices, apparently first used in Toyota car plants in 1948, form a crucial part of the "just in time" production system, which is therefore often referred to as the *kamban* (or *kanban*) system.

karoshi: "death from overwork", a topic of major controversy from the late 1980s, when it was first claimed that Japanese workers were dying from stress-related conditions in relatively greater numbers than in other developed countries. The debate continues, raising some serious questions, not only about the definition of the term and the reliability of the evidence, but also about why the issue has not been addressed so intensively in the West, even though such deaths presumably occur there too.

kaso chi'iki: "depopulated regions", as designated by the National Land Agency.

keiretsu: the loose business conglomerates, centered on "main banks" or large manufacturing companies, which have become characteristic of certain sectors of the Japanese economy in the postwar period.

koenkai: the local personal support organization of a parliamentary candidate, performing fund-raising and vote-gathering functions, originally in the Liberal Democratic Party but later in other parties as well.

kokka: a word with at least as broad a range of meanings and connotations as any of the English words – "state", "country", "nation" – used to translate it.

kokusaika: "internationalization", a buzzword that has been widely used in Japan in recent years, although – or, sometimes, because – it has at least as many vague meanings as has its equivalent in English.

koreika: "aging", as in the phrase *koreika shakai*, "the aging society".

kosei: see *fukushi*.

kudoka: "hollowing out", referring to the twin processes of rising Japanese direct investment overseas and declining investment in heavy industry in Japan itself, as the economy increasingly shifts from reliance on the manufacturing base characteristic of the era of high growth towards becoming an information and services economy.

kyoiku mama: "education mother", a middle-class woman who pushes her children (usually sons) through the education system into high-status careers which reflect well on her.

Nihonjinron: literally "Japanese people discussion", a genre of books, articles, television programs, and other outlets for assertions and theories about the meaning of "Japaneseness". These range from semi-scholarly research on physical anthropology, to claims that Japanese people's brains process sensory input in entirely different ways than the brains of other human beings. As with equivalent discourses in other countries, there is considerable overlap between the writers and readers of *Nihonjinron*, on the one hand, and conservative/nationalist politics, on the other. It is indicative of the state of some types of Japanese studies that, even now, such texts are still taken seriously by some scholars and commentators.

Omu Shinrikyo: the Aum Supreme Truth sect, which sprang into the headlines when members released nerve gas in the Tokyo subway in March 1995, killing 12 people. The sect's beliefs combine elements from Buddhism, Christianity, and the *Foundation* novels of the US science fiction writer Isaac Asimov. Its members, many of them from scientific or technical backgrounds, claim to be preparing to rescue civilization following the ecological disasters predicted by its leader, Asahara Shoko.

pachinko: a gaming machine, resembling pinball but with a vertical board and numerous balls, that has been extremely popular, even addictive, in Japan since the 1950s. Unlike pinball, *pachinko* requires little or no skill. Its name is said to derive from the sound – *pachin* - made by the balls as they pass endlessly across the board.

sairyo rodo: "discretionary labor", meaning that working hours and other conditions are no longer regulated for an expanding range of white-collar occupations.

sarariiman: an example of an "English" word made in Japan, this term, combining "salary" and "man", is widely used to refer to male office workers, whether perceived favorably, as "corporate warriors" upholding traditional virtues of loyalty and diligence, or unfavorably, as sycophantic and conformist transmitters of the prevailing business culture. Maeda Hajime (1895–1978), the leader of the employers' association Nikkeiren in the 1950s, is generally credited with coining the term.

seikei bunri: the "separation of politics and economics" that is said to characterize Japan's international relations, in which political considerations (and, some would argue, ethical principles) take second place to the needs of the economy.

shingikai: advisory or deliberative council, attached to a ministry or agency of the national government, and usually bringing together representatives of relevant interest groups with leading bureaucrats and specialists.

shorei: ministerial ordinance issued by a ministry or agency using powers conferred on it by law.

shunto: "spring offensive", the annual round of negotiations on wages and conditions in large firms, often involving ritualized protest actions by labor unions. The results of *shunto* provide unofficial guidelines for pay settlements in the rest of the economy.

sokaiya: corporate racketeers who extort money from corporations by buying up blocks of shares to gain access to general meetings (*sokai*) of shareholders, while threatening in advance to disrupt them by asking awkward questions unless their demands are met. To minimize the chances of being a victim of this crime, many listed corporations hold their shareholders' meetings on the same day, in the knowledge that each *sokaiya* can only be in one place at one time. Others, such as, most famously, Nomura Securities, have paid up.

tan'itsu minzoku: "single race" or "single nation". This pseudo-scientific term is still used by some within the mass media in Japan, where it is deployed to assert the monoethnic nature of the Japanese people, in the face of all the evidence – from language, physiognomy, material culture, and geography – that they are descended from numerous waves of immigrants from (at least) central Asia, China and Korea, Polynesia, and the Malay archipelago.

tanshin funin: "solitary transfer", referring to workers, virtually always men, who are required to take up posts far from their marital homes, often for years on end. This can arise because their spouses are reluctant to move, or because children are expected to remain at the schools they have begun attending, or, in some cases, because it is impossible to move the whole family at short notice, especially where there are financial constraints.

tokushu hojin: "special legal entity", an official term which the Japanese government itself translates as "public corporation". In fact, it has a much wider meaning, referring to a range of public sector entities, including business corporations, financial institutions, pension funds, and research institutes.

tsutatsu: notification issued by a ministry or agency using a generally recognized power to supervise economic or other activities; a major tool of *gyosei shido* (see above).

yakuza: member of one of the organized criminal groups, based mainly in western Japan, which have amassed large fortunes from racketeering, prostitution, gambling, drug-dealing, and related activities.

Yakuza have acquired some of the glamour that attaches to gangsters in other cultures, and they have occasionally even been portrayed as contributing to the maintenance of Japan's relatively high level of public order, on the grounds that *un*organized criminals would be worse. Some *yakuza* gangs have specialized as *sokaiya* (see above); others have assisted employers' groups and conservative politicians, notably in combating left-wing militants in the 1950s and 1960s, and financing election campaigns in the 1970s and 1980s.

zaibatsu: a business conglomerate, centered on one family's holdings, or on a bank that provided finance for other companies in the group. The *zaibatsu* dominated the economy until they were broken up under orders from the Occupation authorities. Some have since re-formed as looser conglomerates, known as *keiretsu*.

zaitekku: "financial engineering", a term coined in the 1980s (in imitation of the English term "high tech"), and referring to the use of financial instruments, such as futures, options or swaps, to hedge against cross-border currency and interest rate risks. Most multinational corporations now use such derivatives as a matter of course in managing their international activities, but the line between employing them to hedge risks and using them to speculate on future market movements is a fine one.

zaru ho: literally, "basket law", referring to a law with loopholes in it.

zoku: This term is widely used for "tribes" of various kinds, such as the *bosozoku* (gangs of youths who enjoy speeding in cars or on motorcycles). In politics it refers to "tribes" of parliamentarians concerned with particular policy areas.

Appendix 3
Personalities

Emperors (*Tenno*)

1926–89	Hirohito	referred to since his death as *Showa Tenno* (the Showa Emperor)
1989–	Akihito	now referred to as *Tenno Heika* (His Majesty the Emperor), but due to become known as *Heisei Tenno* (the Heisei Emperor) after his death

Prime Ministers (*Sori Daijin*)

1945	Higashikuni Naruhiko (Prince)	Retired general and member of the imperial family
1945–46	Shidehara Kijuro (Baron)	Retired admiral
1946–47	Yoshida Shigeru	Former diplomat, leader of the Liberal Party
1947–48	Katayama Tetsu	Leader of the Socialist Party, in coalition with the Democrats and the People's Cooperative Party
1948	Ashida Hitoshi	Leader of the Democrats, in coalition with the Socialist Party and the People's Cooperative Party
1948–54	Yoshida Shigeru [2nd term]	Leader of the Liberal Party
1954–56	Hatoyama Ichiro	Leader of the Democratic Party of Japan up to 1955, then leader of the Liberal Democratic Party (LDP) after merger with the Liberals
1956–57	Ishibashi Tanzan	LDP leader
1957–60	Kishi Nobusuke	LDP leader
1960–64	Ikeda Hayato	LDP leader
1964–72	Sato Eisaku	LDP leader
1972–74	Tanaka Kakuei	LDP leader
1974–76	Miki Takeo	LDP leader
1976–78	Fukuda Takeo	LDP leader
1978–80	Ohira Masayoshi	LDP leader
1980–82	Suzuki Zenko	LDP leader
1982–87	Nakasone Yasuhiro	LDP leader, in coalition with the New Liberal Club from 1983 to its re-entry into the LDP in 1986
1987–89	Takeshita Noboru	LDP leader
1989	Uno Sosuke	LDP leader
1989–1991	Kaifu Toshiki	LDP leader
1991–93	Miyazawa Ki'ichi	LDP leader
1993–94	Hosokawa Morihiro	Leader of the Japan New Party, in coalition with other anti-LDP parties
1994	Hata Tsutomu	Leader of the New Reform Party and head of a short-lived anti-LDP coalition
1994–96	Murayama Tomi'ichi	Leader of the Social Democratic Party, in coalition with the LDP and the Harbinger Party
1996–	Hashimoto Ryutaro	Leader of the LDP; in coalition with the Social Democrats and the Harbinger Party until October 1996, head of an exclusively LDP government since then

Aikawa Yoshisuke (1880–1967): Also known as Ayukawa Yoshisuke, founder of Nissan Motors and one of the pioneers of Japan's postwar industrial development. The son of a *samurai* (warrior), Aikawa founded the Tobata Casting Company in 1910, with loans from the Mitsui family and other wealthy relatives. This became the basis of his holding company, Nippon Sangyo, or Nissan, founded in 1928, which also had Nissan or Hitachi subsidiaries in many other fields. In 1933 it bought DAT Motors, a company which produced Datsun cars on a small scale. Car-making was banned in 1939, and Aikawa's businesses spent the war years supplying the Army and investing in Manchuria. Aikawa was jailed from 1945 to 1947, and then purged, but later became a member of the House of Councillors and an adviser to Prime Minister Kishi Nobusuke (a close relative and a collaborator with Aikawa, first in Manchuria, and later in formulating industrial policy). Aikawa was excluded from direct involvement in Nissan after 1945. However, he had appointed most of the managers who went on to build it up as one of the world's leading auto-makers; and his policy of selective, profitable collaboration with government agencies has continued to influence management, both at Nissan and at other large companies.

Akihito (1933–): Crown Prince from birth until he succeeded his father as Emperor in January 1989. He was enthroned at the old Imperial Palace in Kyoto in November 1990.

Hashimoto Ryutaro (1937–): President of the Liberal Democratic Party (LDP) since October 1995, and Prime Minister of Japan since January 1996. Hashimoto entered the House of Representatives in 1963, taking over the seat formerly held by his father, who had been a cabinet minister under Kishi Nobusuke. Initially, Hashimoto developed a reputation as a nationalist, and a campaigner on behalf of war veterans. Later, as a member of Tanaka Kakuei's faction, he made the rising Liberal Democrat's customary progress through various ministries: Health and Welfare (1978–80); Transport, where he was in charge of privatizing the railways (1986–87); Finance, where he arranged financial support for the UN coalition in the Gulf War (1989–93); and International Trade and Industry, negotiating an agreement with the United States on automobiles and auto parts (1994–96). After replacing the Social Democrat Murayama Tomi'ichi as Prime Minister in January 1996, he led the LDP into the House of Representatives election of October that year, and regained an absolute parliamentary majority by September 1997. He has become known as an advocate of administrative reform and economic liberalization, but has also been criticized for inconsistency and indecisiveness.

Hata Tsutomu (1935–): Former leader of the New Reform Party, briefly Prime Minister in 1994. After working as a bus conductor for 10 years, Hata took over his father's local branch of the LDP, and his seat in the House of Representatives, in 1969. He joined the party's largest faction, led by Tanaka Kakuei, but distanced himself from it after Tanaka's resignation and arrest. He briefly achieved international fame in 1987 when, as Agriculture Minister, he announced that the Japanese could not eat US beef because their intestines were longer than those of Caucasians. He allied himself with Ozawa Ichiro in calling for liberalization of the economy and reform of the electoral system, joining him in defecting in 1993 to create the New Reform Party. Hata succeeded Hosokawa Morihiro as Prime Minister, at the head of the anti-LDP coalition, in 1994, but the Social Democrats withdrew support 12 hours after he took office and he resigned two months later. Later that year, he helped Ozawa to form the New Frontier Party, a merger of nine groups, but in 1996 he resigned from it, citing dissatisfaction with Ozawa's leadership.

Hirohito (1901–89): Emperor from 1926 until his death. Grandson of Mutsuhito, the Meiji Emperor (reigned 1867–1912), and son of Yoshihito, the Taisho Emperor (reigned 1912–26), Hirohito had become regent when his father was declared insane in 1921. While successive governments took Japan into war, first with China (1931–45) and then with the West (1941–45), Hirohito generally avoided political decisions and did whatever governments told him to do, in line both with the 1889 Constitution and with his understanding of the British model of constitutional monarchy, which he admired. He spent much of his time pursuing his lifelong interest in marine biology, but is known to have expressed opposition to the Army's murder of the Chinese general Chang Tso-lin in 1928, a mutiny by sections of the Army in 1936, and the government's attempts to avoid the unconditional surrender in 1945. The Allied governments decided to let him stay on the throne, as a force for stability in Japanese life. In 1946, he renounced the claims to divine status which some politicians had made on his behalf. He made frequent tours of Japan and visits overseas, again on the model of the British monarchy, until his death from cancer.

Honda So'ichiro (1906–91): Founder of the Honda group of companies and a pioneer of Japan's postwar industrial development. Honda's first business venture was a motor repair shop, which he opened in 1928 and then expanded into a factory producing auto parts. Like others in the industry, it was compelled to support the war effort between 1931 and 1945. After the surrender Honda's company began producing motorcycles, among which the C-100 model, introduced in 1953, achieved record sales around the world. Honda then shocked the business world by entering the automobile industry, in defiance of government plans to "rationalize" it into three giant firms that would not include his. He retired in 1973, soon after the launch of the energy-efficient and very popular Honda Civic.

Hosokawa Morihiro (1938–): leading reformist politician of the 1990s, Prime Minister of a coalition government 1993–94. Hosokawa, a grandson of an earlier Prime Minister, Prince Konoe Fumimaro, entered national politics as a member of the LDP in the House of Councillors, but achieved greater prominence as an outspoken Governor of Kumamoto Prefecture in Kyushu. In 1992, he and others dissatisfied with the LDP's resistance to reform of the electoral system and party funding left the LDP to form the Japan New Party (*Nihon Shinto*, or JNP). As Prime Minister, he headed a fractious coalition of eight political parties, and succeeded in pushing his reform plans through the House of Representatives, only to see them defeated in the House of Councillors by the LDP and a large number of Socialists. He left office amid renewed allegations that he too had accepted illegal payments from business interests.

Iwai Akira (1922–1997): Secretary General of the left-wing labor federation Sohyo from 1955 to 1970. He was regarded as the real power within the federation, although his close colleague Ota Kaoru occupied the nominally superior position of Chairman. Iwai quietly supported the efforts of Communists to maintain influence within Sohyo, thus contributing to antagonisms with "moderate" leaders such as Miyata Yoshiji.

Kanemaru Shin (1914–96): The "Godfather" or "King-maker" of the LDP in the 1980s and early 1990s. After one year in the Army, Kanemaru returned to his native Yamanashi Prefecture in 1938 to manage his father's saké business. In the late 1940s he became prominent in local conservative politics, chiefly as a fund-raiser: in 1953 he famously ate the name-cards given to him by businessmen who had made illegal campaign donations, rather than allow the police to see them. Having joined the LDP when it was founded in 1955, he entered the House of Representatives in 1958, and came to prominence in 1960 when he carried the Speaker of the House to his seat as opposition parties mounted protests against the revised the Security Treaty with the United States. He served in government from 1972 to 1987, collaborating with Tanaka Kakuei and Takeshita Noboru in the leadership of the LDP's largest faction, and was also the patron of Ozawa Ichiro. After Tanaka suffered a stroke in 1985, Kanemaru took control of his faction, the largest in the LDP, in Takeshita's name, and was largely responsible for elevating both Takeshita and Miyazawa Ki'ichi to the prime ministership, in 1987 and in 1991 respectively. Kanemaru resigned from the vice-presidency of the LDP in August 1992, after admitting that, like other politicians, he had illegally received money (in his case ¥500 million) from Watanabe Hiroyasu, the President of Sagawa Kyubin, a trucking firm. He resigned from Parliament in October that year. In 1993 investigators found gold bars, stock certificates and other evidence of enormous undeclared wealth in Kanemaru's home.

Kishi Nobusuke (1896–1987): Prime Minister from 1957 to 1960 and a leading figure in formulating postwar industrial policy. Born Sato Nobusuke, he was adopted as the heir of the Kishi family. A prominent bureaucrat before World War II, he served as Minister of Commerce and Industry, and later as Minister of Munitions, in wartime governments. After the war he was briefly "purged" from office, but entered the House of Representatives in 1952. Swiftly rising back to Cabinet rank, he helped to secure a central role for the Ministry of International Trade and Industry in planning and directing leading industries. He was pressured to resign as Prime Minister in 1960, after forcing the revised Security Treaty with the United States through Parliament, in the face of a boycott by the opposition parties and mass demonstrations outside the Parliament building. In retirement he continued to advocate a strongly pro-American, anti-Communist foreign policy, was an active lobbyist for small and medium-sized enterprises, and maintained close relations with the political "fixer" Kodama Yoshio. Kishi was a relative and close collaborator of Aikawa Yoshisuke, the founder of Nissan; the older brother of Sato Eisaku, also Prime Minister; and the father-in-law of Abe Shintaro, a prominent LDP politician and Foreign Minister in the 1980s.

Kodama Yoshio (1911–84): A key figure in the networks of influence linking politicians, business leaders and gangsters. Kodama, a businessman himself, was active in ultra-nationalist political groups up to 1945. He was jailed as a minor war criminal, but after his release in 1948 he helped to revive and finance the political careers of two Prime Ministers, Hatoyama Ichiro and Kishi Nobusuke. At the same time, Kodama collaborated with leading underworld figures and a number of business leaders in suppressing militant labor unions and groups protesting against the Security Treaty with the United States. In the 1970s, he was an intermediary between the US aircraft-maker Lockheed and the Japanese politicians, including Prime Minister Tanaka Kakuei, that Lockheed paid bribes to. After Tanaka resigned from office, Kodama's activities were investigated by the police, and he was convicted of tax evasion in 1977. Kodama continues to be the subject of sensationalist media stories, which make it difficult to establish exactly what he did during his long career.

Miyata Yoshiji (1924–): Probably the most influential labor leader in postwar Japan, though less prominent than the left-wing labor leaders Ota Kaoru and Iwai Akira. Miyata advocated a philosophy of company-based cooperation between management and labor, consonant with enterprise unionism. This approach continues to influence the activities of labor unions today. Miyata gained control over the Federation of Steelworkers Unions (*Tekko Roren*), the industrial union of what was then the country's most important industry, in around 1960. He went on to play the leading role in the formation of the International Metalworkers Federation-Japan Council (IMF-JC) in 1964. Miyata also helped to transform wage-bargaining practices to cope with reduced rates of economic growth from 1975 onwards.

Morita Akio (1921–): Co-founder of Sony and leading figure in business lobby groups. After serving in the Imperial Navy during World War II, Morita and his close collaborator Ibuka Masaru founded what is now the Sony Corporation in 1946. They and their teams of young, innovative engineers and designers have done much to make electrical and electronic equipment into familiar items in homes around the world, starting with reel-to-reel tape-recorders (1950), and then developing transistor radios (1955), transistorized televisions (1959), desktop calculators (1964), and the world's first video-cassette recorder (1969), which was the basis of their Betamax system, a technical success but a rare commercial failure. Sony also pioneered miniaturized portable cassette-players, televisions and CD-players, with its "Walkman", "Watchman" and "Discman" products, and introduced the 3.5-inch floppy disk for computers. Morita has also become prominent as an advocate of economic liberalization and a supporter of close ties with the United States.

Murayama Tomi'ichi (1924–): Prime Minister from 1994 to 1996. A labor union official and then a member of the prefectural assembly in his native Oita Prefecture (in Kyushu), Murayama was first elected to the House of Representatives as a Socialist in 1972. He actively opposed what he saw as the militarism of the ruling LDP, helping to organize the Socialists' attempts, in 1991–92, to delay or defeat the LDP's proposals to allow Japanese personnel to join UN peacekeeping operations. In 1993, he supported the decision of the party leader, Yamahana Sadao, to join with other anti-LDP parties in the coalition government formed by Hosokawa Morihiro. In June 1994, however, having succeeded Yamahana as leader nine months earlier, he became Prime Minister himself, in coalition with the LDP and the conservative Harbinger Party. He then accepted the constitutionality of the Self-Defense Forces, permitted imports of foreign rice, agreed to the raising of the retirement age, and otherwise reversed the policies that his party had upheld since the 1940s. Most notably, perhaps, in 1995 he became the first postwar Prime Minister to arrange for the House of Representatives to pass a resolution regretting the actions of the Japanese Army in the 1930s and 1940s. Murayama was succeeded as Prime Minister by Hashimoto Ryutaro.

Nakasone Yasuhiro (1917–): Prime Minister from 1982 to 1987, one of the longest-serving and most "activist" in the postwar period. After wartime service in the former Imperial Navy, Nakasone entered Parliament in 1951 and attracted attention as an advocate of national revival and remilitarization. A long ministerial career, from 1967 onwards, culminated in the prime ministership, secured for him by the LDP's largest faction, led by Tanaka Kakuei, and by the party's elder statesman, Kishi Nobusuke. The subsequent "Tanakasone" government agreed in 1983 to permit transfers of military technology with the United States; made "official" visits to the Yasukuni shrine, dedicated to the war dead (which were later declared unconstitutional by the Supreme Court); and, in 1986, abandoned the policy of limiting defense spending to 1% of GNP or less, although it rose only to 1.004% of

GNP by the time that Nakasone left office. He also weathered international controversies over the censorship of information about Japanese war crimes in school textbooks, and over his own racism, typified by his claim that the multiracial composition of US society reduced its average IQ. His vocal commitment to far-reaching reform led to the creation of an influential inquiry into the education system, and the partial privatization of parts of NTT and Japan National Railways, but little more. In 1989, having admitted taking part in insider trading of shares in Recruit Cosmos, Nakasone resigned from the LDP, but remained in Parliament and was readmitted to the party in 1991.

Ono Tai'ichi (1912–): The Toyota Motors executive who revolutionized its system of production management. Ono worked for Toyota from 1943 to 1978, when he retired as Executive Vice-President. His importance to the development of the postwar economy lies in his innovative techniques: the elimination of stockpiles, the "de-specializing" of workers, and the "just-in-time" (*kamban*) system, in which components are passed from unit to unit as and when needed. Critics, concerned about the effects of Ono's demanding production norms on workers, have seized on his admission that most of his ideas were not tested before being applied. However, the *kamban* system has been widely adopted by managers and management theorists, not only in Japan but also in the West.

Ota Kaoru (1912–): Probably the most prominent left-wing labor leader in postwar Japan, Ota was Chairman of the left-wing labor federation Sohyo from 1958 to 1966. He is regarded as the father of the *shunto* ("spring offensive") system of collective bargaining on wages. Ota intended *shunto* to help mobilize workers to confront management more assertively, but, to his regret, it became an effective mechanism of wage restraint in the mid-1970s.

Ozawa Ichiro (1942–): Prominent politician of the 1980s and 1990s, who helped to split the LDP in 1993 and created the New Frontier Party in 1994. Ozawa took over his father's personal support organization (*koenkai*) upon his death in 1968 and entered the House of Representatives in 1969. He became a protégé, first of Tanaka Kakuei, and later of Kanemaru Shin, the leading figures in the LDP's largest faction. He was briefly a member of Nakasone Yasuhiro's government, from 1985 to 1986, but was most active in organizing election campaigns and in trade negotiations with US officials. In 1989 Kanemaru arranged for Ozawa to become Secretary-General of the LDP. After his patron's resignation and arrest, in 1992, Ozawa formed his own small faction, which in 1993 became the nucleus of the New Reform Party, formed by Ozawa, Hata Tsutomu, and other defectors from the LDP. Ozawa then served in the coalition governments of Hosokawa Morihiro and Hata, but after the LDP had returned to office he led the nine opposition parties in forming a single New Frontier Party, in December 1994. Partly because of resentment of Ozawa's apparent arrogance, this party fell apart three years later. Ozawa, whose direct manner, knowledge of English, and advocacy of free market policies have marked him out from the mainstream of LDP and ex-LDP politicians, remains in opposition, less likely than ever to fulfill his dream of becoming Prime Minister.

Sato Eisaku (1901–75): Japan's longest-serving postwar Prime Minister (1964–72). He was a leading official in the Ministry of Transport until 1948, when he joined the Liberal Party led by Yoshida Shigeru, who became his political patron. He entered the House of Representatives in 1949 and Yoshida's government in 1951. With Yoshida he stayed aloof from the new LDP up to 1957, when his older brother, Kishi Nobusuke, became Prime Minister. After serving as Finance Minister under Kishi (1958–61), and heading the Ministry of International Trade and Industry under Ikeda Hayato (1961–64), he succeeded to the prime ministership. His government opened diplomatic relations with South Korea, eventually faced down the student protests of 1968–69, renewed the Security Treaty with the United States in 1970, and arranged the reversion of Okinawa to Japan. He retired in 1972, partly in order to take responsibility for not foreseeing the "Nixon shocks" that disrupted the Japanese economy. It was Sato who, in 1968, announced a ban on arms exports and the three non-nuclear principles, for which he received the Nobel Peace Prize in 1974.

Takeshita Noboru (1924–): Prime Minister from 1987 to 1989. As a leading member of the faction dominated by Tanaka Kakuei and Kanemaru Shin, Takeshita served in several Cabinets in the 1970s and 1980s, notably at the Ministry of Finance, where he succeeded in delaying administrative reforms and, notoriously, saw to a sharp rise in government spending on projects in Shimane Prefecture, which he represented in Parliament. Following Tanaka's stroke and withdrawal from political activity in 1985, Kanemaru arranged for Takeshita to assume the formal leadership of the faction, and then helped him to become Prime Minister in succession to Nakasone Yasuhiro. Two years later, having further delayed reforms, he resigned amid scandal over his acceptance of secret shares in the Recruit Cosmos company.

Tanaka Kakuei (1918–93): Leader of the LDP's largest faction in the 1970s and 1980s, Prime Minister from 1972 to 1974, and then extremely influential up to 1985. Like his most powerful aide, Kanemaru Shin, Tanaka was one of the few modern Japanese politicians to have risen from a rural background without university education. He became rich through construction projects and marriage to an heiress, before entering politics in 1947, as a representative of his native Niigata Prefecture. He first entered the Cabinet under Kishi Nobusuke, and served in several posts before becoming Minister for International Trade and Industry (1970–72) and launching an ambitious plan to "restructure" Japan through massive public works and regional development. Having succeeded Sato Eisaku as Prime Minister, he resigned after the magazine *Bungei Shunju* exposed some of the corrupt dealings that underlay his private fortune and his electoral victories. In 1976, he was arrested in connection with his taking bribes from the US aircraft-maker Lockheed. From the fall of Prime Minister Fukuda Takeo in 1978, Tanaka dominated politics from behind the scenes, as Japan's *Yami Shogun* ("Shadow Shogun"). In 1983, however, he was at last convicted of receiving bribes; his lawyers were still appealing the verdict when a stroke ended his active life, in 1985. Tanaka's most lasting legacy is in Niigata, where he secured the construction of Niigata International University, two super-highways, the Joetsu Shinkansen line, and several auto parts factories around Nagaoka City. He was perhaps the most successful of postwar Japan's numerous "pork barrel" politicians.

Toyoda Ki'ichiro (1894–1952): Founder of Toyota Motors and pioneer of Japan's postwar industrial development. After working for many years in his father's automatic loom business, Toyoda established an auto plant in 1933 and launched it as a separate company, Toyota Motors, in 1937. After producing trucks for the Army during World War II, the company reverted to cars, launching the Toyopet model in 1947. Toyoda resigned as chief executive officer in 1950, and died after a stroke. However, Toyota continued his policies of emphasizing original research and development, and avoiding tie-ups with foreign auto-makers (unlike Nissan, which collaborated with the British company Austin throughout the 1950s). Toyoda's cousin Toyoda Eiji was President of the company from 1967 to 1982, as it was becoming a major player in the world market.

Yamagishi Akira (1929–): Chairman of the labor federation Rengo from 1989 to 1995. The original Rengo, founded in 1987, absorbed most public sector unions in the year that he became its head, after some years as a leading figure in *Zendentsu*, the union of NTT, one of the country's strongest unions. He had displayed great political acumen in the 1980s when the public sector faced restructuring, including partial privatization, at the hands of conservative politicians, but Rengo's accomplishments during his years in the chair were generally regarded as disappointing.

Yoshida Shigeru (1878–1967): Prime Minister from 1946 to 1947, and again from 1948 to 1954; one of the architects of postwar Japan. A diplomat before World War II, Yoshida became a leading conservative politician after the defeat, initially as Minister of Foreign Affairs (1945–46). He became leader of the Liberal Party and Prime Minister in 1946, replacing Hatoyama Ichiro, who was "purged" by SCAP before he could take office. Yoshida's government gave way to a Socialist/center-right coalition in 1947, but after its collapse he entered upon six years as Prime Minister (1948–54), cooperating with the Occupation authorities in the "reverse course". This saw the return of many purged politicians and officials to public life; the application of the "Dodge Line" to curb inflation; the use of US bases during the Korean War, which started an economic boom in Japan; the restoration of sovereignty in 1952; and the creation of the Self-Defense Forces in 1954. Yoshida's policy of adhering to the United States and opposing Communism, at home and abroad, has been dubbed the "Yoshida Doctrine", and has influenced Japanese foreign policy ever since he formulated it. Nicknamed "One-Man Yoshida", because of his tendency to make decisions without consultation, he was increasingly resented by his colleagues, who welcomed Hatoyama's successful campaign to replace him as Prime Minister in 1954. Yoshida and his protégé Sato Eisaku refused to join Hatoyama's LDP until 1957. In retirement, Yoshida continued to advise Sato. After his death, he was honored in Japan's first postwar state funeral.

Appendix 4

Political and Economic Institutions

The Political System

The Emperor (*Tenno*): literally, "heavenly king", calling attention to what was probably the *Tenno*'s oldest role, as high priest of the sun-goddess, and imperial ancestor, Amaterasu O-mikami. The Japanese monarchy has survived for at least 1,500 years perhaps largely because *Tenno*, quite unlike western emperors, generally had little political power, and therefore rarely aroused the hostility of rivals. Since Japan's first constitution was promulgated in 1889, modern *Tenno* have been, politically, no more than constitutional monarchs, in the same sense as in the United Kingdom. However, the religious significance of the *Tenno* was exploited by the political elite, and the enforced veneration of portraits of the emperors, in schools and other institutions, went beyond even the British cult of royalty. The era of "Emperor worship", which culminated in 14 years of brutal and disastrous war in the name of Hirohito – even though much of what happened was kept secret from him – ended in January 1946, when he denounced the cult. The definition of the *Tenno* in the 1947 Constitution, as the "symbol of the state and of the unity of the people", would have been appropriate at almost any time in Japanese history, especially since "state" is an inadequate translation of *Nihonkoku*, which means the "country of Japan" in much more than just the political sense.

Parliament (*Kokkai*): The national legislature consists of two houses: the House of Representatives (*Shugi'in*), chaired by its Speaker (*Shugi'in-gicho*), for which there must be a general election at least once every four years; and the House of Councillors (*Sangi'in*), chaired by its President (*Sangi'in-gicho*), of which half the members are elected every three years. The members of these houses are known as *Shugi'in-gi'in* and *Sangi'in-gi'in*, respectively.

National government (*Seifu*): The executive branch is controlled by the Cabinet (*Naikaku*), comprising the Prime Minister (*Sori Daijin* or *Shusho*), with his own Prime Minister's Office (*Sorifu*), and other Cabinet members (*Kakuryo*), who are either Ministers (*Daijin*), or Directors General (*Chokan*) of certain agencies.

There are 12 Ministries, of Foreign Affairs (*Gaimusho* or MoFA), Finance (*Okurasho* or MoF), Education (*Mombusho*), Health and Welfare (*Koseisho* or MHW), Agriculture, Forestry and Fisheries (*Norinsuisansho* or MAFF), International Trade and Industry (*Tsusansho* or MITI), Transport (*Unyusho* or MoT), Posts and Telecommunications (*Yuseisho* or MPT), Labor (*Rodosho*), Construction (*Kensetsusho*), and Home Affairs (*Jichisho* – although this literally means "Ministry of [Local] Autonomy" – or MHA).

The Cabinet generally also includes the heads of the Defense Agency (*Boei-cho*), which supervises the three Self-Defense Forces (*Jieitai*) – Ground, Maritime, and Air; the Economic Planning Agency (*Keizai Kikaku-cho*); the Science and Technology Agency (*Kagaku Gijutsu-cho*); the Environment Agency (*Kankyocho*); two Development Agencies (*Kaihatsucho*), for Hokkaido and Okinawa; and the National Land Agency (*Kokudocho*).

Other leading government bodies include the National Defense Council (*Kokubo Kaigi*); the National Public Safety Commission (*Kokka Koan I'inkai*), which supervises the National Police Agency (*Keisatsucho*); the Fair Trade Commission (*Kosei Torihiki I'inkai*); and the Imperial Household Agency (*Kunaicho*). There are also numerous "special legal entities" attached to ministries or agencies.

Courts (*Saibansho*): Japan has a single national judicial system. The Supreme Court (*Saiko Saibansho*), headed by the Chief Justice (*Saiko Saichokan*), was created by the 1947 Constitution and closely modeled on the US Supreme Court. It has the power of judicial review on the constitutionality of laws and regulations, and hears appeals from inferior courts. Unlike its US model, it also has the power, independent of the government or Parliament, to train, appoint and discipline all the judges in the High Courts (*Koto Saibansho*), District Courts (*Chiho Saibansho*), Family Courts (*Katei Saibansho*) and Summary Courts (*Kan'i Saibansho*).

Local government (*chiho kosei*): Japan has been divided into 47 prefectures since the 1870s: the English word for them reflects the fact that up to 1945 they had "prefects" or governors appointed by the national government (on what was then the French model). In Japanese, however, there are four different terms. The prefectures thus comprise one *to* (capital city) – Tokyo-to; two *fu* (metropolitan centers) – Kyoto-fu and Osaka-fu; one *do* ("circuit") – Hokkaido; and 43 *ken* (ordinary prefectures). Each has an elected governor (*chiji*) and a single-chamber legislative assembly (*gikai*), its own police service and school system, and broadly similar powers in relation to health and welfare, physical planning, and other local services.

Every prefecture includes several *shi* (cities); most also include several *gun* (counties). The larger part of Tokyo-to, and 11 large "designated cities" (*shitei toshi*) – Fukuoka, Hiroshima, Kawasaki, Kitakyushu, Kobe, Kyoto-shi, Nagoya, Osaka-shi, Sapporo, Sendai, and Yokohama – are further divided into *ku* (usually translated "wards", but more like the boroughs of New York or London). The terms *cho* or *machi* (district or neighborhood), or, where appropriate, *son* or *mura* (village or hamlet), are used to indicate the smallest unit of local government within cities and counties. Each has its own "self-government association" (*jichikai*).

Political parties (*seito*): As in most western countries (other than the United States), it has been customary in Japan to view politics as a contest between "left" and "right". Accordingly, parties have been perceived as belonging either to the "conservative camp" (*hoshu jin'ei*), broadly favoring the United States, the free market and the Self-Defense Forces; or to the "progressive camp" (*kakushin jin'ei*), broadly favoring neutralism, greater state intervention in the economy, and a strict interpretation of Article 9 of the Constitution, which renounces war. However, there has been considerable collaboration and compromise, often behind the scenes, between the two "camps"; the "progressive" parties have never in fact been in full agreement on all three of the policies cited; and the division may well have been swept away by the political upheavals of 1993–94, which resulted in collaboration between the largest "conservative" party and the remnant of the largest "progressive" party.

After World War II, the center-right half of the political spectrum was dominated by the Liberal Party (*Jiyuto*) and the Democratic Party (*Minshuto*). Both were created in November 1945, largely by politicians from pre-war parties. (The Democrats were known as Progressives until March 1947, and absorbed a rival group, the People's Cooperative Party, in April 1950.) The Liberal Democratic Party (*Jiyu Minshuto*, *Jiminto*, or LDP), created from a merger of the Liberals and Democrats in November 1955, has survived crises and defections ever since, perhaps largely because of the sharing of posts and resources among the party's various factions, and the maintenance of close ties with business organizations and with leading officials (many of whom have become LDP politicians).

In 1976, the New Liberal Club (*Shin Jiyu Kurabu*), a small dissident group, broke away, only to form a coalition government with the LDP in 1983 and merge back into it in 1986. The Japan New Party (*Nihon Shinto*), formed by Hosokawa Morihiro and others in 1992, was the next group to break away from the LDP, but it had little impact until it joined forces in government (from August 1993 onwards) with older anti-LDP parties (see below), and with two other ex-LDP groups, the New Party Harbinger (*Shinto Sakigake*), and the New Reform Party (*Shinseito*). After the LDP returned to office in June 1994, in coalition with the Social Democrats and the Harbinger Party, the nine remaining opposition parties merged as the New Frontier Party (*Shinshinto*) in December 1994.

While the LDP governed, split, and then returned to government, the center-left in Japanese politics remained disunited and marginalized. For many years the largest opposition party was the Japan Socialist Party (*Nihon Shakaito* or JSP). It was founded in October 1945, claiming continuity from pre-war groups, split into left and right factions in October 1951, but was reunited in October 1955. It lost a large minority of its supporters when the more moderate Democratic Socialist Party (*Minshato* or DSP) was founded in January 1960; lost more members in 1978 to the United Social Democratic Party (*Shaminren* or USDP); and lost still more members, and most of its remaining traditional voters, when it entered government with its former enemy, the LDP, in June 1994. Confusingly, while it changed its name in English to the Social Democratic Party of Japan (SDPJ) in 1991, it did not adopt a Japanese equivalent (*Nihon Shakai Minshuto*) until January 1996. Finally, between the party conference that month and the general election to the House of Representatives in October, about half of the SDPJ's parliamentarians joined in forming yet another new party, the Democratic Party (*Minshuto*). This broadly "progressive" grouping has no connection with its namesake of the late 1940s and early 1950s.

Two other parties should also be mentioned. The Clean Government Party (*Komeito*) was founded in November 1964 by Soka Gakkai, an organization of laypeople supporting the Nichiren sect of Buddhism. The formal link with Soka Gakkai was broken in 1970. Komeito's position on the conventional left/right spectrum has long been unclear and controversial, with each "camp" tending to view it as an ally of the other camp. Finally, there is the Japan Communist Party (*Nihon Kyosanto* or JCP), one of the few parties to retain that name in the developed world. Founded in 1922, and relaunched in October 1945, it is Japan's oldest and most cohesive party. It has never allied or merged with any other group, but factions have broken away from it over the decades, as it has moved from supporting the Soviet Union, through a period of admiring China, to a broadly "Eurocommunist" position adopted in July 1976.

Following the most recent general election, in October 1996, the 500 seats in the House of Representatives, filled by a new electoral system (see Chapter 2), were distributed as follows: LDP 239, New Frontier Party 156, Democratic Party 52, JCP 26, SDPJ 15, Harbinger Party 2, Democratic Reform Party 1, Independents 9. Many individuals, mainly from the three ex-LDP parties, then crossed over to the LDP, restoring its majority by September 1997. Three months later, the New Frontier Party disintegrated back into its constituent elements (the Japan New Party, the New Reform Party, the DSP, the USDP, Komeito, and others). Japanese politics has thus returned, for now, to its normal postwar pattern: a united, pragmatic but broadly center-right LDP in government, facing several mutually mistrustful opposition groups.

Economic Institutions

Central bank: The Bank of Japan (*Nippon Ginko* or BoJ) was founded in 1882 on the model of the Belgian central bank. Its autonomy has been enhanced, first under the Occupation, and again, as part of the "Big Bang" reforms, in 1998.

Stock exchanges (*Shoken torihikijo*): Since securities trading was resumed, under the Securities and Exchange Law 1949, there have been eight self-governing stock exchanges: the Tokyo Stock Exchange, founded in 1878, which accounts for about 75% of all trading in stocks; the Osaka Securities Exchange, which is dominant in derivatives trading; and six much smaller exchanges, in Nagoya, Kyoto, Hiroshima, Fukuoka, Niigata and Sapporo.

Employers' organizations: Three of the four main organizations were established under the Occupation, and have largely maintained the division of labor imposed on them by SCAP. The Committee for Economic Development (*Keizai Doyukai*), founded in April 1946, has successfully advocated the introduction of new production methods, in many cases from the United States, and stresses "harmony" in labor relations. The Federation of Economic Organizations (*Keidanren*), founded in August 1946, comprises a number of industry-specific organizations, lobbying on their behalf and acquiring a reputation as an ally of the LDP. The Japan Federation of Employers' Associations (*Nikkeiren*), founded in August 1948, represents the interests of the large corporations in negotiations with organized labor. In addition, the Japan Chamber of Commerce and Industry (*Nissho*), which traces its origins to 1878, was revived during the 1960s as a frequently effective lobby group for small and medium-sized enterprises.

Forms of business organization: As in western countries, the bulk of economic activity is carried on by joint stock companies (*kabushiki gaisha*, abbreviated in English as KK), but there are also thousands of limited liability companies (*yugen gaisha*), limited partnerships (*goshi gaisha*), and unincorporated private businesses (*kojin kigyo*). Alongside the private sector, there are still several state-owned public corporations (*kodan*). In addition, there are numerous agricultural cooperatives and fisheries cooperatives (see Chapter 5), while the consumers' cooperative movement, originating in Kobe in 1921 and relaunched in 1948, now has more members and higher total turnover than any other such movement in the world.

Labor unions: Most labor unions are organized company by company (rather than industry by industry, as is usual in the West). Sohyo, Domei and other confederations of unions which dominated organized labor during much of the postwar period (see Chapter 13) have now been superseded by three coordinating bodies: the Japanese Trade Union Confederation, as *Rengo,* founded in 1987, has been known in English since its enlargement in 1989; the National Confederation of Trade Unions (*Zenroren*); and the National Trade Union Council (*Zenrokyo*). Among these Rengo, which represents 62% of unionized workers and 14% of the total workforce, is by far the largest and most wideranging. The International Metalworkers Federation-Japan Council (IMF-JC), founded in 1964, has also played a leading role, notably in wage determination.

International Organizations

Postwar Japan began its integration into the system of supranational agencies in 1951, when it was readmitted to the International Labor Organization (ILO). Since then, it has joined many other organizations engaged in worldwide activities, notably the IMF and the World Bank, in 1952; GATT, in 1955; the UN and its specialized agencies, in 1956; the Bank for International Settlements (BIS), in 1970; and the World Trade Organization (WTO), in 1995. In addition, it became the first non-western member of the OECD in 1964, and has been represented at summits of the leading developed nations (initially the Group of Five, now the Group of Seven) since 1975.

In the Asia-Pacific region, Japan has been a member of the Colombo Plan organization, which aids and promotes economic and social development, since 1954. It was a founding member of the Asian Development Bank (ADB), and the Asian and Pacific Council (ASPAC), both in 1966; the Asia-Pacific Economic Cooperation (APEC) conferences, from 1993 onwards; the ASEAN Regional Forum (ARF), set up by the members of the Association of Southeast Asian Nations and 11 other states in 1994; and the Asia-Europe Meetings (ASEM) between East Asian states and EU members, which started in 1996. Japan is also involved in European affairs through its participation in the Group of 24, aiding central and Eastern Europe since 1989; its membership of the European Bank for Reconstruction and Development (EBRD), which began operations in 1990; and its observer status, granted in 1992, at the Conference on (now Organization for) Security and Cooperation in Europe (CSCE/OSCE).

Appendix 5

Bibliography

This bibliography is intended to draw readers' attention to some of the most useful and stimulating English-language books on Japan available in the mid-1990s. It therefore excludes periodicals, websites, and Japanese-language publications. It should be seen as a supplement to the suggestions for further reading at the end of each chapter. The reliability and readability of publications about Japan, scholarly or otherwise, vary enormously, and their writers' claims to expertise or insight should be considered carefully, even skeptically, before any time or money is spent on them. In particular, those books which offer monocausal explanations of everything Japanese – often relying on myths of "national character", and just as often based on travelers' anecdotes, or plagiarism from earlier peddlers of the same myths – should be seen as, at best, mildly entertaining displays of invincible ignorance. Accordingly, this bibliography does not include any books either by such western simplifiers as Ruth Benedict, or by such Japanese practitioners of *Nihonjinron* (discussions of "Japaneseness") as Nakane Chie or Doi Takeo.

Reference

Richard Bowring and Peter Kornicki, editors, *The Cambridge Encyclopedia of Japan*, Cambridge and New York: Cambridge University Press, 1993

A well-presented single-volume reference book, full of useful facts and figures, and reflecting the ever broadening range of specialisms and methodologies among scholars in the field of Japanese studies.

Bureau of Statistics, *Japan Statistical Yearbook*, Tokyo: Office of the Prime Minister, Government of Japan, annual publication

An English-language version of the standard source for most of the statistics that influence government policy and help to shape the image of Japan.

Collcutt, Martin, Marius B. Jansen and Isao Kumakura, *Cultural Atlas of Japan*, Oxford: Phaidon, 1988

Much more than its title may suggest, this beautifully illustrated and very informative book presents the richness of Japanese culture, defined in the broadest terms, from prehistory up to the present day. The economic and political context of cultural activities is not neglected, and the book also includes special features on the Ainu, the No drama, gardens, castles, and many other topics.

The Kodansha Encyclopedia of Japan, Tokyo, New York and London: Kodansha International, 1983

This multi-volume, multi-author work covers almost every conceivable topic related to Japanese society, culture and history, and is copiously illustrated with maps, pictures and tables. Written mainly by American and US-educated Japanese scholars, it is in many ways a distillation of the mainstream "modernization" consensus prevailing when it was compiled.

Perren, Richard, *Japanese Studies from Prehistory to 1990: A Bibliographical Guide*, Manchester: Manchester University Press, 1992

An indispensable aid to further discoveries about Japan, whether for beginners or experts.

General

Beasley, W. G., *The Modern History of Japan*, third edition, London: Weidenfeld and Nicolson, and New York: St Martin's Press, 1981

In this undergraduate textbook, a distinguished British historian of Japan tells the story of Japan's development from the early 19th century to the present day, focusing mainly but by no means exclusively on politics. His presentation of Japan's international relations is particularly lucid.

Booth, Alan, *The Roads to Sata: A 2,000-Mile Walk Through Japan*, New York and Tokyo: Weatherhill, 1985; and *Japan*, revised edition, Hong Kong: The Guidebook Company, 1991

Booth's premature death in 1993 deprived his many admirers of further books like these two. The first is a charming and thoughtful narrative of a journey through the "back country" of Japan, full of insights into a society which Booth refused to reduce to easy generalizations. The second, written and published as a travel guide, is more accurate, and more entertaining, than most such publications. Its sections on major cities and historic sites together form a history of the country, and a basis for Booth's wise observations on its society and culture.

Chamberlain, Basil Hall, *Japanese Things: Being Notes on Various Subjects Connected with Japan*, Rutland, VT, and Tokyo: Charles E. Tuttle, 1971

This is a partially updated paperback edition of *Things Japanese*, first published in 1890 and frequently revised up to 1905. Chamberlain presents an enormous amount of information about Japan as he knew it. Parts of the book are obviously outdated, but most of it remains remarkably relevant and vivid – notably his emphasis on forming one's own opinions, his rejection of "Orientalism" (decades before Edward W. Saïd's important book on the subject), and his shrewd observations on westerners' ambivalence towards Japan.

Livingston, Jon, Joe Moore and Felicia Oldfather, editors, *The Japan Reader*, Volume 1, *Imperial Japan, 1800–1945*, and Volume 2, *Postwar Japan, 1945 to the Present*, New York: Random House, 1973

The editors have selected excerpts from both eyewitness accounts of events and scholarly texts on Japanese history to create an outstanding introductory reader. Their obvious sympathy for the left in Japanese politics makes their book even more valuable, since they present information and opinions that are generally ignored by mainstream scholars and commentators.

Reischauer, Edwin O., *The Japanese Today*, Cambridge, MA: Belknap Press, 1988; and *Japan: The Story of a Nation*, fourth edition, New York: McGraw-Hill, 1990

Among the many books produced by Reischauer, a founding father of Japanese studies at Harvard University, and US Ambassador to Japan in the 1960s, these two volumes are perhaps the classic presentations of his influential view of Japanese history as a trajectory from tradition to modernity under the influence of westernization. Now that the Cold War is over, some of his assumptions look dubious, but his breadth of vision, his attention to detail, and his openness to debate about Japan put many of his critics, and his disciples, to shame.

Sansom, Sir George, *Japan: A Short Cultural History*, revised edition, London: The Cresset Press, 1946; and *A History of Japan*, Volume 1, *To 1334*, Volume 2, *1334–1615*, Volume 3, *1615–1867*, Stanford, CA: Stanford University Press, 1959–63, and Rutland, VT, and Tokyo: Charles E. Tuttle, 1974

Like Basil Hall Chamberlain (see above), Sansom was a British "gentleman scholar" who did most of his work before Japanese studies became an organized academic discipline. Consequently, while some scholars still admire his pioneering work, others dismiss it as "popular" (which, of course, reveals more about them than about Sansom). Yet no other one-volume history of premodern Japan in English surpasses the first book cited here, in range, depth, or readability, while the three volumes of the second book cited have withstood all scholarly challenges to become standard works.

Storry, Richard, *A History of Modern Japan*, London and New York: Penguin Books, 1960 and subsequent revised reprints

Storry wrote this accessible paperback history with a non-scholarly audience in mind. The first one third of the book is a concise, lively account of the country's history up to the Meiji Restoration of 1868 (the event usually taken as the beginning of "modern" Japan). The main body of the book takes the story up to the early 1980s (Storry died in 1982), usefully placing more emphasis on Japan's relations with China and other neighboring states than is usual in books aimed at western readers.

Tames, Richard, *A Traveller's History of Japan*, Moreton-in-Marsh, Gloucestershire: The Windrush Press, 1993

Tames succeeds admirably in summarizing more than 2,000 years of Japanese history in a concise, entertaining, and accurate text, and provides handy brief appendices on such topics as the Japanese language, Buddhism, and national holidays, as well as a concise historical gazetteer and a comprehensive bibliography.

History since 1945

Allinson, Gary, *Japan's Postwar History*, London: UCL Press, and Ithaca, NY: Cornell University Press, 1997

Allinson's wideranging and accessible account seems likely to become the standard textbook on the period. It is especially strong on the growth, development, and recent problems of the economy.

Brackman, Arnold C., *The Other Nuremberg: The Untold Story of the Tokyo War Crimes Trials*, New York: William Morrow, 1987

Brackman draws on his own coverage of the trials as a young reporter, and on many other sources, to provide a detailed and mostly temperate account.

Cohen, Theodore, *Remaking Japan: The American Occupation as New Deal*, New York: Free Press, 1987

An interesting view of the Occupation's purposes and achievements, written by a former SCAP official who helped to reorganize labor/management relations in Japan.

Dore, Ronald P., *Land Reform in Japan*, London: Athlone Press, 1987

A new edition of the classic English-language text on perhaps the single most important and enduring reform introduced during the Occupation period.

Dower, John, *Empire and Aftermath: Yoshida Shigeru and the Japanese Experience, 1878–1954*, Cambridge, MA: Harvard University Press, 1979

An authoritative study of the conservative diplomat and politician who dominated Japanese politics in the Occupation years and after, and helped to initiate many of the policies that would prevail into the 1990s.

Duus, Peter, *The Cambridge History of Japan*, Volume 6: *The Twentieth Century*, Cambridge and New York: Cambridge University Press, 1988

Four of the 14 chapters in this book directly address postwar Japan: Duus's own thoughtful Introduction, Haruhiko Fukui's informative survey of politics between 1945 and 1973, Yutaka Kosai's discussion of the economy during the same period, and Koji Taira's account of industrial relations between 1905 and 1955. Readers interested in developments up to 1945 should find much of interest in the other chapters, as well as in the other five volumes of this series.

Havens, Thomas R. H., *Fire Across the Sea: The Vietnam War and Japan, 1965–1975*, Princeton, NJ: Princeton University Press, 1987

This unusually readable academic text is a comprehensive and balanced survey of the topic, covering the responses of governments, the activities of protest groups, and the bemusement of much of the public, thus tacitly rebutting any lingering notion that the Japanese are politically homogeneous.

Moore, Joe, *Japanese Workers and the Struggle for Power, 1945–1947*, Madison: University of Wisconsin Press, 1983

Moore describes the rise and fall of the "production control" movement, and other radical initiatives by labor unions and protest groups under the Occupation. His study focuses attention on groups and events that have been ignored in most histories of modern Japan, but perhaps also exaggerates their importance.

Packard, George R., *Protest in Tokyo: The Security Treaty Crisis of 1960*, Princeton, NJ: Princeton University Press, 1966

This is a comprehensive and thoughtful study of the revision of the Security Treaty, its stormy passage through the Japanese Parliament, and the huge but unsuccessful movement of opposition to it.

Ward, Robert E., and Sakamoto Yoshikazu, editors, *Democratizing Japan: The Allied Occupation*, Honolulu: University of Hawaii Press, 1987

A collection of competent scholarly essays reviewing the main reforms introduced or encouraged by SCAP, including reflections on their longer-term effects on the economy and society.

Politics and Institutions

Abe Hitoshi, Shindo Muneyuki and Kawato Sadafumi, *The Government and Politics of Japan*, Tokyo: University of Tokyo Press, 1994

Students of Japanese politics will find the similarities and differences between the approach taken by these Japanese authors, and the treatment of the same topics in books by non-Japanese writers, very interesting.

Apter, David E., and Nagayo Sawa, *Against the State: Politics and Social Protest in Japan*, Cambridge, MA: Harvard University Press, 1984

Apter and Sawa deliver less than their ambitious title promises, since their book is in fact a thorough sociological study of just one social protest, the movement against Narita International Airport.

Curtis, Gerald, *The Japanese Way of Politics*, New York: Columbia University Press, 1988

Curtis's distinctive account of Japanese politics as characterized by instability remains useful and informative. An updated edition would be very welcome.

Itoh, Hiroshi, *The Japanese Supreme Court: Constitutional Policies*, New York: Markus Wiener Publishing, 1989

A careful, comprehensive but somewhat dry study, which begins with a brief history and description of the modern judicial system, and then analyzes the postwar Supreme Court's various functions, notably review of the constitutionality of laws, administration of the judicial system, and training of judges. The official English translation of the 1947 Constitution is included as an appendix.

Japan Institute of International Affairs, *White Papers of Japan*, Tokyo: The Japan Institute of International Affairs, annual publication

This is a convenient collection of the Japanese government's policy statements and overviews of its activities. It shoud be noted that all these White Papers, along with a wealth of other government data, are now available on line (see Chapters 5 and 12).

Johnson, Chalmers, *MITI and the Japanese Miracle: The Growth of Industrial Policy 1925–1975*, Stanford, CA: Stanford University Press, 1982; Rutland, VT, and Tokyo: Charles E. Tuttle, 1986

It seems astonishing that no other writer in English has tried to approach the political and economic structures of modern Japan through a detailed study of one of the influential government ministries. Johnson's study is now a classic in its field, covering a wide range of topics in detail, and raising a number of important issues about the distinctiveness of Japanese policy.

Kohno Masaru, *Japan's Postwar Party Politics*, Princeton, NJ: Princeton University Press, 1997

Kohno presents a series of case studies in the creation and development of Japan's political parties, in relation to the 1947 Constitution and the institutions that have been created under it, or have continued in existence from earlier periods. Some readers may find that his interest and expertise in the practical details of political life can make the book seem lacking in theoretical depth.

Kyogoku Jun'ichi, *The Political Dynamics of Japan*, Tokyo: University of Tokyo Press, 1987

Since the original Japanese-language version of this book was published in 1983, some of Kyogoku's material is now outdated. Nevertheless, this is an impressive attempt to describe and explain Japanese politics in its social and economic context, aimed at Japanese undergraduates.

McCormack, Gavan, *The Emptiness of Japanese Affluence*, Armonk, NY: M. E. Sharpe, 1996

McCormack offers a highly critical account of the achievements and the failures of Japan's "construction state", which usefully interrogates the more triumphalist versions of Japan's economic and social development.

Pempel, T. J., editor, *Uncommon Democracies: The One-Party Dominant Regimes*, Ithaca, NY: Cornell University Press, 1990

Pempel is a leading expert on Japanese politics, and the coverage of Japan in this textbook of comparative politics is extensive and stimulating, even though events since the book was written – in Sweden, Italy and Israel, as much as in Japan – have made the concept of "one-party dominant regimes" even more problematic than it was to begin with.

Rosenbluth, Frances McCall, *Financial Politics in Contemporary Japan*, Ithaca, NY: Cornell University Press, 1989

A groundbreaking description and explanation of Japan's financial system as it developed up to the "bubble economy" of the late 1980s, incorporating a mass of interesting data.

Schwartz, Frank J., *Advice and Consent: The Politics of Consultation in Japan*, Cambridge, MA: Harvard University Press, 1997

This is the most up-to-date and the most comprehensive account in English of the Japanese system of advisory councils (*shingikai*) attached to government ministries and agencies. Schwartz insightfully explains their importance as sites for negotiations among interest groups.

Steiner, Kurt, Ellis S. Krauss and Scott Flanagan, editors, *Political Opposition and Local Politics in Japan*, Princeton, NJ: Princeton University Press, 1980

An interesting collection of essays examining the record of political parties, especially the Socialists and Communists, at the sub-national level, and highlighting the opportunities available to local governments, as well as the constraints upon their activities.

Stockwin, J. A. A., Alan Rix, Aurelia George, Daiichi Ito and Martin Collick, editors, *Dynamic and Immobilist Politics in Japan*, London: Macmillan, and Honolulu: University of Hawaii Press, 1988

The editors and other distinguished scholars describe and explore the tensions between groups and institutions promoting or welcoming political and economic change, and those resisting it. Some are less convincing on their chosen topics than others, but their collective concern with the complexities and contradictions of the political system is refreshing.

Stockwin, J. A. A., *Governing Japan: Divided Politics in a Major Economy*, Oxford: Blackwell, 1998 (forthcoming)

Earlier versions of this textbook, under the title *Japan: Divided Politics in a Growth Economy*, became essential reading for students of modern Japanese political economy.

Tsuneoka, Setsuko Norimoto, and others, *Nihonkoku Kempo o Yomu / The Constitution of Japan*, Tokyo: Kashiwashobo, 1993

This dual-language publication (with Japanese and English texts in parallel on facing pages) contains the 1947 Constitution and the "MacArthur Draft" on which it was partly based, as well as three scholarly articles on how the document was compiled, and its impact on contemporary Japan.

Van Wolferen, Karel, *The Enigma of Japanese Power: People and Politics in a Stateless Nation*, London: Macmillan, 1988, and New York: Knopf, 1989

Written by a distinguished and experienced Dutch journalist, this enjoyably polemical book mixes extensive research, personal anecdotes, and some perhaps over-ambitious attempts at political philosophy, to present a view of Japan as a country where "the buck keeps circulating", because the leading political and economic institutions are not answerable to any genuine state, to public opinion, or even to one another.

The Economy

ABARE (Australian Bureau of Agricultural and Resource Economics), *Japanese Agricultural Policies: A Time of Change*, Policy Monograph No. 3, Canberra: Australian Government Publishing Service, 1988

A somewhat dated but still usable overview of the subject.

Abegglen, James, and George Stalk, *Kaisha: The Japanese Corporation*, New York: Basic Books, and Rutland, VT, and Tokyo: Charles E. Tuttle, 1985

Abegglen and Stalk, members of the Boston Consulting Group, present a detailed study of management processes and production systems in large corporations. Their approach is generally favorable, and the book contains a great deal of interesting information, along with some questionable generalizations and such influential jargon as "accelerated learning".

Anchordoguy, Marie, *Computers Inc.: Japan's Challenge to IBM*, Cambridge, MA: Harvard University Press, 1989

The two main strengths of this important scholarly study are its painstaking presentation of the relationships that underlie Japanese industrial policy, and its vivid account of how Japan's computer industry was created and expanded up to the late 1980s.

Aoki Masahiko, *Information, Incentives, and Bargaining in the Japanese Economy*, Cambridge and New York: Cambridge University Press, 1988

Aoki's goal is clearly to present relations between labor and management in Japan in as favorable a light as possible. The result is a one-sided and over-generalized study which, nevertheless, contains much of interest.

Argy, Victor, and Leslie Stein, *The Japanese Economy*, New York: New York University Press, 1997

An up-to-date, wideranging and reliable description of the contemporary Japanese economy, its strengths and its weaknesses, with some sophisticated use of comparative materials.

Cusumano, Michael, *The Japanese Automobile Industry: Technology and Management at Nissan and Toyota*, Cambridge, MA: Harvard University Press, 1985

This impressive historical study extends from the 1930s to the early 1980s, and includes some information on all the major companies, but its main focus is on developments at the two firms mentioned in its subtitle in the 1950s and 1960s, including the introduction of new production techniques, the elimination of labor militancy, and the beginning of serious competition with western rivals.

Emmott, Bill, *The Sun Also Sets: Why Japan Will Not be Number One*, London: Simon and Schuster, 1989

This was among the first of the wave of books questioning the Japanese "economic miracle", and remains one of the best. Emmott's discussion of the main issues is intelligent and well-informed, and he does not indulge in any triumphalism (unlike the general run of such books). He also usefully explodes some persistent myths, for example, about Japan's savings rates, dependence on exports, housing stock, and suicide rates.

Emmott, Bill, *Japan's Global Reach*, London: Century Business, 1992

This is a disappointment after Emmott's previous book, since it often loses sight of the main topic – overseas investments by Japanese companies – in discussions of globalization and other very large questions. However, it contains a great deal of detailed information and some fascinating case studies.

Field, Graham, *Japan's Financial System: Restoration and Reform*, London: Euromoney Publications, 1997

A very informative and judiciously critical discussion of Japan's financial services, focusing on developments in the 1980s and 1990s.

Fingleton, Eamonn, *Blindside: Why Japan is Still On Track to Overtake the United States by the Year 2000*, London: Simon and Schuster, and Boston: Houghton Mifflin, 1995

This lively polemic argues that the Japanese economy is not close to collapse, that the lifetime employment system continues to flourish in the large corporations, and that Japanese industry is still capable of innovation, for example, in developing the magnetic levitation ("maglev") train, flat-screen televisions, car navigation systems, and solar cells. Fingleton occasionally seems over-optimistic, but provides plenty of food for thought.

Hirschmeier, J., and T. Yui, *The Development of Japanese Business, 1600–1973*, London: Allen and Unwin, 1979

A thorough academic text which helps to place discussions of contemporary Japanese business practices in their proper historic context, making comparisons and contrasts with practices in other countries on the basis of detailed research, not *a priori* assertions.

Kamata Satoshi, *Japan in the Passing Lane*, London: Allen and Unwin, 1984

A very critical description of daily life on the Toyota production line, by a reporter who worked on it under cover, this book remains at the center of controversy over whether it is wholly reliable, and whether Kamata's experience is representative.

McMillan, Charles J., *The Japanese Industrial System*, second edition, Berlin and New York: Walter de Gruyter, 1996

The latest edition of an academic study of Japan's production system which has become a standard textbook.

Morita Akio, with Edwin M. Reingold and Shimomura Mitsuko, *Made in Japan*, New York: E. P. Dutton, 1986, London: Collins, 1987, and Tokyo and New York: Weatherhill, 1987

The co-founder of Sony cannot be expected to write an objective history of his company. Instead, he offers an engaging and often revealing celebration of himself, his colleagues and their achievements, as well as some interesting general reflections on Japanese society and economics.

Nakamura Takafusa, *The Postwar Japanese Economy: Its Development and Structure, 1937–1994*, second edition, Tokyo: University of Tokyo Press, 1995

A welcome update of Nakamura's study, retaining its comprehensive account of the era of high growth, and adding copious data on developments since the early 1980s.

Ohno Taichi [Ono Tai'ichi], *Toyota Production System: Beyond Large-scale Production*, Cambridge, MA: Productivity Press, 1988

This is an English translation of a Japanese book, published in 1978, by the Toyota executive most closely associated with the development of the "just in time" (*kamban*) system. Ono gives an absorbing (although naturally one-sided) account of the details of the system, and its benefits for management and workers.

Toyoda Eiji, *Toyota – Fifty Years in Motion*, Tokyo, New York and London: Kodansha International, 1987

An obviously uncritical but readable presentation of Toyota's rise to become one of the largest car-makers in the world.

Vogel, Ezra F., *Japan as Number One: Lessons for America*, Cambridge, MA: Harvard University Press, 1979

Vogel's once popular book was the most influential of the huge number of books that sought to draw lessons for the West from Japan's economic achievements. Much of the information that forms the basis of his enthusiastic approach to Japanese institutions is now outdated, but his discussion is not as one-sided as later critics have suggested, and the book remains readable.

Whitaker, D. H., *Small Firms in the Japanese Economy*, Cambridge and New York: Cambridge University Press, 1997

Whitaker usefully draws attention to the companies that account for the bulk of Japan's output, in a refreshing corrective to the fixation of all too many western commentators on the larger companies that occupy the more visible section of the "dual economy". The book includes some considered comparisons and contrasts with the small business sector in the United Kingdom.

Yamamura Kozo and Yasuba Yasukichi, editors, *The Political Economy of Japan*, Volume 1, *The Domestic Transformation*, Stanford, CA: Stanford University Press, 1987

Like the other two volumes in the series (Kumon and Rosovsky on society and culture, Inoguchi and Okamoto on international relations), this was intended to be a collection of authoritative papers by leading scholars, both Japanese and western. The papers inevitably vary in readability and relevance, but the volume as a whole is worth consulting, as a summary of one very influential approach to the Japanese economy.

Society and Culture

Bayley, David H., *Forces of Order: Policing Modern Japan*, Berkeley, Los Angeles: University of California Press, 1991

A revised and updated version of a sociological text first published in 1976, this examines the police from the top down and the bottom up, and includes careful comparisons and contrasts with policing in the United States.

Brown, Clair, Yoshifumi Nakata, Michael Reich, and Lloyd Ulman, *Work and Pay in the United States and Japan*, New York and Oxford: Oxford University Press, 1997

A thorough and up-to-date comparative study, covering relations between labor and management, wage-setting procedures and structures, and many other aspects of working life in the world's two largest economies.

Buruma, Ian, *A Japanese Mirror: Heroes and Villains of Japanese Culture,* London: Jonathan Cape, 1984

One of several books by Buruma on Japan and other Asian societies, this is an entertaining and wideranging survey of Japanese myths and popular culture, even though Buruma occasionally exaggerates the differences between Japanese people and other human beings, and pays only fitful attention to historical changes (as opposed to continuities).

Campbell, John C., *How Policies Change: The Japanese Government and the Aging Society*. Princeton, NJ: Princeton University Press, 1992

Campbell, one of the few experts on the Japanese welfare system writing in English, describes and explains political responses to the social impact of aging populations, mainly in Japan, but also in the United States. The topic is also likely, of course, to become increasingly important in the politics of other advanced industrial nations.

Clark, Rodney, *The Japanese Company*, New Haven, CT: Yale University Press, and Rutland, VT, and Tokyo: Charles E. Tuttle, 1979

The emphasis of this study is on the Japanese employment system, and the role of companies seen in the wider social context. Clark's conclusions have been very influential, and the book remains readable, although some would question whether his generalizations about corporate culture can adequately capture the variety and complexity of company cultures in a dual economy which is rapidly changing.

Cole, Robert E., *Japanese Blue Collar: The Changing Tradition*, Berkeley, Los Angeles: University of California Press, 1971

This classic study of Japanese manufacturing workers, based on Cole's observations as a temporary participant at a company making automobile parts, still has a great deal to offer, even in the late 1990s, as a counterbalance to the numerous books presenting management-oriented and, often, highly abstract versions of Japanese labor relations.

Cole, Robert E., *Work, Mobility and Participation: A Comparative Study of American and Japanese Industry*, Berkeley, Los Angeles and London: University of California Press, 1979

A thorough, rigorous and still thoughtprovoking survey of similarities and differences between working conditions in the world's two largest economies, grounded in case studies.

Crump, Thomas, *The Death of an Emperor: Japan at the Crossroads*, reissued with a new Afterword, Oxford and New York: Oxford University Press, 1991

Originally published in 1989, this is a very rewarding study of the significance of the Japanese monarchy, the life, death, and funeral of Hirohito, the relations between the state and Shinto, and a number of other related subjects. Its value is enhanced by the fact that Crump, a cultural anthropologist who lives and works in Amsterdam, is in a position to extend his discussion of western attitudes to Japan beyond those prevailing in English-speaking countries.

Cutts, Robert L., *An Empire of Schools: Japan's Universities and the Molding of a National Power Elite*, Armonk, NY: M. E. Sharpe, 1997

This looks likely to become the standard text in English on the role of the leading universities in training Japan's decision-makers and opinion-formers.

Dale, Peter, *The Myth of Japanese Uniqueness*, London: Croom Helm, and New York: St Martin's Press, 1986

A sometimes startling exploration of the more extreme examples of the *Nihonjinron* ("Japaneseness") literature that warps perceptions both inside Japan and overseas, this would be even more valuable if Dale had made concessions to readers unfamiliar with his sometimes arcane scholarly vocabulary.

Denoon, Donald, Mark Hudson, Gavan McCormack and Tessa Morris-Suzuki, editors, *Multicultural Japan*, Cambridge and New York: Cambridge University Press, 1996

A collection of lively and enlightening papers on various aspects of contemporary Japan, emphasizing the diversity of ethnicities and cultures within a society that is often presented as simply unitary.

DeVos, George, and Wagatsumi Hiroshi, *Japan's Invisible Race*, Berkeley, Los Angeles: University of California Press, 1972

This remains a remarkable account of the *burakumin*, the hereditary outcasts of Japanese society whose ill-treatment by some sections of the mainstream population continues today.

Dore, Ronald, *British Factory – Japanese Factory: The Origins of National Diversity in Industrial Relations*, London: Allen and Unwin, and Berkeley, Los Angeles: University of California Press, 1973; reissued with a new Afterword, 1990

As in his other books, Dore here moves – perhaps too quickly and confidently – from the micro level (a richly realistic discussion of factory life) to the macro level (broad statements of theory, presented with a characteristic air of world-weariness). Nevertheless, Dore is an outstandingly knowledgable observer of Japanese society, and his ideas deserve to be taken seriously.

Fukutake Tadashi, *The Japanese Social Structure: Its Evolution in the Modern Century*, translated and with a foreword by Ronald Dore, second edition, Tokyo: University of Tokyo Press, 1989

Fukutake, a prominent sociologist, discusses the transformation of Japanese society lucidly and evenhandedly, although (as Dore points out in his enjoyably argumentative foreword) he does not hesitate to suggest where and how things have gone wrong.

Gordon, Andrew, *The Evolution of Labor Relations in Japan: Heavy Industry, 1853–1955*, Cambridge, MA: Harvard University Press, 1985

In this book, the standard text in English on the topic, Gordon traces the main changes and continuities in relations between labor and management in the leading manufacturing industries, from the earliest stages of industrialization up to the introduction of the institutions and practices that were to remain in place for most of the postwar period.

Jungk, Robert, *Children of the Ashes: The People of Hiroshima After the Bomb*, updated edition, London: Granada Publishing, 1985

A sober and convincing study of the effects of the first atomic bomb, by a prominent historian of science, focusing on three individuals, and including reflections on a visit to Hiroshima in 1980.

Kaplan, David E., and Andrew Marshall, *The Cult at the End of the World: The Incredible Story of Aum*, London: Hutchinson, and New York: Crown, 1996

The authors' background in journalism, and their sensationalist subtitle, may mislead some potential readers. In fact, their book is a very readable and far-reaching account of the religious group which released poison gas on the Tokyo subway in 1995, and committed other serious crimes under the direction of its leader, Asahara Shoko. As the authors point out, incompetent police, religious fanatics and alienated scientists are by no means unique to Japan, and Aum, despite its presentation in most western media, is neither laughable nor inexplicable.

Kitagawa, Joseph M., *Religion in Japanese History*, New York: Columbia University Press, 1966; reissued with a new preface and updated chronology, 1990

This is still the standard text in English on its subject, which Kitagawa explores comprehensively, and with something close to genuine objectivity (it is impossible to guess what his own religious beliefs are). The final chapter details new developments after World War II, including the formal separation of Shinto from the state, the explosion of new religions, and the varied fortunes of the numerous sects of Buddhism.

Koike Kazuo, *Understanding Industrial Relations in Japan*, translated by Mary Saso, London: Macmillan, and New York: St Martin's Press, 1988

A fascinating exploration of a number of important issues in relations between management and labor in Japan, by a prominent and expert scholar.

Kosaka Kenji, editor, *Social Stratification in Contemporary Japan*, London and New York: Routledge, 1994

This collection of sociological studies presents a wealth of information and ideas on a topic that has received little attention from scholars and other observers of Japan in the English-speaking world.

Krauss, Ellis S., Thomas P. Rohlen and Patricia G. Steinhoff, editors, *Conflict in Japan*, Honolulu: University of Hawaii Press, 1984

This very wideranging but somewhat inchoate collection of scholarly essays includes Rohlen's valuable analysis of the politics of education, and Steinhoff's history of student movements, as well as useful chapters on other types of conflict in an allegedly harmonious and conformist society.

Kumazawa Makoto, *Portraits of the Japanese Workplace: Labor Movements, and Managers*, edited by Andrew Gordon; translated by Andrew Gordon and Mikiso Hane, Boulder, CO: Westview Press, 1996

Gordon and Hane, who have made their own important contributions to the study of Japanese labor history, here make available in English some of the work of Kumazawa, a prominent critic of the state of industrial relations in his country. Even those who disagree with his approach may well find his observations relevant and challenging.

Kumon Shumpei and Henry Rosovsky, editors, *The Political Economy of Japan*, Volume 3, *Cultural and Social Dynamics*, Stanford, CA: Stanford University Press, 1992

Perhaps surprisingly, the scholars, both Japanese and western, who contributed papers to this ambitious volume (in the same series as Yamamura and Yasuba on the economy, and Kumon and Rosovsky on international relations) mostly fail to say much that is interesting or new about what should have been a stimulating range of topics. The book is saved from being a major disappointment by the sheer bulk of data that it contains, and by its usefulness as a gateway to the many other books on Japanese society and culture.

Mouer, Ross, and Sugimoto Yoshio, *Images of Japanese Society: A Study in the Structure of Social Reality*, London, New York, Sydney and Henley: KPI, 1986

A groundbreaking scholarly text which surveys competing "consensus" and "conflict" images of Japan, relates them to conservative and radical approaches to sociology and politics, analyzes and demolishes some leading examples of mainstream Japanology, and then offers a more complex methodology, along with observations on the future of Japanese studies. The book is aimed at the authors' colleagues in the academy, rather than at general readers, but is more accessible, and often more entertaining, than most such texts.

Richie, Donald, *Geisha, Gangster, Neighbor, Nun: Scenes from Japanese Lives*, Tokyo, London and New York: Kodansha International, 1991

In a series of short sketches of 23 famous individuals, and 25 "ordinary" ones, Richie, an American writer who has lived in Japan since the Occupation years, reveals and celebrates the sheer variety of their experiences and interests.

Rohlen, Thomas P., *Japan's High Schools*, Berkeley, Los Angeles: University of California Press, 1983

Like so many other western observers of Japan, Rohlen is sometimes too eager to fit what he has seen into preconceptions about Japanese culture, and those who have spent more time in such schools than he was able to will find some of his conclusions unconvincing. However, as the only widely available book in English on the topic, this has become a standard text. Its strengths lie in Rohlen's direct observations of five schools, and his sympathetic and well-informed discussion of their distinctive features.

Seidensticker, Edward, *Tokyo Rising*, New York: Knopf, 1990

This is a sequel to the author's *Low City, High City* (New York: Knopf, 1983). It continues the story of Japan's capital city from the Kanto earthquake of 1923 onwards, concentrating, as in the previous book, on its rapidly changing topography, popular culture, and local politics. Seidensticker, a distinguished translator of Japanese literature, is an enjoyably opinionated and impassioned guide to one of the world's great cities.

Smith, Robert J., *Kurusu: The Price of Progress in a Japanese Village, 1951–1975*, Stanford, CA: Stanford University Press, and Folkestone, Kent: Wm Dawson, 1978

This humane and memorable book brings the transformation of postwar Japan to life by describing one small village in great detail, from farms, schools, factories and hotels, to rice-cookers, coffins, clothing, and even the effect of brightly lit store windows on insects and crops. Smith indulges in nostalgia for dying communal traditions, but also reports the villagers' own views of the changes that they have seen, and deftly sketches in the larger political and economic context.

Ventura, Rey, *Underground in Japan*, London: Jonathan Cape, 1992

On its first appearance, some readers questioned the veracity of this first-person narrative of life as an illegal immigrant from the Philippines. However, it is now generally accepted as that rarity in English, a view of everyday life in urban Japan from a perspective that few Japanese, and even fewer westerners, would otherwise be able to share.

International Relations

Curtis, Gerald L., editor, *Japan's Foreign Policy after the Cold War: Coping with Change*, New York: M. E. Sharpe, 1993

A number of distinguished scholars address the main issues raised by the changes in international relations, and in Japan's position, in recent years. The effect is occasionally as if the reader is eavesdropping on obscure academic debates, but the book as a whole is more accessible, and more stimulating, than many in this field.

Darby, James, editor, *Japan and the European Periphery*, London: Macmillan, and New York: St Martin's Press, 1996

Darby and his contributors examine relations between Japan and Europe, with an emphasis on the "other Europe", those parts of the continent that are not within the EU. The book is mainly concerned to describe and explain the development of economic relations.

Dore, Ronald, *Japan, Internationalism and the UN*, London and New York: Routledge, 1997

Dore, who has written prolifically across an impressively wide range of topics related to Japan, here turns his skeptical eye on Japan's international relations. As always, his conclusions may sometimes be questionable, but his erudition and insight are admirable.

Inoguchi Takashi and Daniel I. Okimoto, editors, *The Political Economy of Japan*, Volume 2, *The Changing International Context*, Stanford, CA: Stanford University Press, 1988

This is a comprehensive and very informative collection of papers by prominent US and Japanese scholars, focusing on the postwar period. As with the other two volumes in the series (Yamamura and Yasuba on the economy, and Kumon and Rosovsky on society and culture), most of the papers are characterized by loyalty to the mainstream academic consensus, and a certain fondness for ambitious theorizing, but there is still a great deal of worthwhile material here.

Inoguchi Takashi, *Japan's Foreign Policy in an Era of Global Change*, London and New York: Pinter, 1993

Inoguchi, a leading figure in the study of Japan's international relations, here surveys the opportunities and challenges facing the country's policy-makers in the 1990s, including the impact of globalization and of increasing activism at the UN and other forums.

Lincoln, Edward J., *Japan's New Global Role*, Washington, DC: The Brookings Institution, 1993

Lincoln provides a competent analysis of the changes in Japan's international relations following the economic boom of the 1980s, which made it the world's largest creditor nation and the largest donor of overseas development assistance.

Nester, William, *European Power and the Japanese Challenge*, London: Macmillan, and New York: New York University Press, 1993

Nester's main concern is with the structural problems of European political systems and economies (and especially those in the EU), but his book also contains some thoughtful observations on Japan.

Reischauer, Edwin O., *The United States and Japan*, 3rd edition, New York: Knopf, 1981

Like Reischauer's other books, this was written on the basis of assumptions – about the Cold War, and about a linear process of "modernization" inexorably transforming Japan – which can seem dated or irrelevant. Yet Reischauer's enormous knowledge of his subject, and his evident concern for the future good of Japan, and of the western country which is both its closest ally and most serious rival, still make this book worth reading.

Rothacher, Albrecht, *Economic Diplomacy between the European Community and Japan*, Aldershot: Gower, 1983

Very much a "top down" history of the period, this has become a standard textbook on how relations between Japan and the EC developed as the Japanese economy grew and the EC itself expanded in membership and functions.

Wilkinson, Endymion, *Japan versus the West*, London and New York: Penguin Books, 1991

This is the latest version of a readable study by a leading British expert on EU affairs. Wilkinson's main emphasis remains on Europe and on postwar developments, but he also provides some deft summaries of earlier periods, and brings together information and ideas from an eclectic range of sources.

Index

Index

3/18/04
7/5/05